Student Solutions Manual and S

Numerical Analysis

NINTH EDITION

Richard L. Burden
Youngstown State University

J. Douglas Faires
Youngstown State University

Prepared by

Richard L. Burden
Youngstown State University

J. Douglas Faires
Youngstown State University

BROOKS/COLE
CENGAGE Learning

Australia • Brazil • Japan • Korea • Mexico • Singapore • Spain • United Kingdom • United States

For product information and technology assistance, contact us at
**Cengage Learning Customer & Sales Support,
1-800-354-9706**

For permission to use material from this text or product, submit all requests online at **www.cengage.com/permissions**
Further permissions questions can be emailed to
permissionrequest@cengage.com

ISBN-13: 978-0-538-73563-6
ISBN-10: 0-538-73563-5

Brooks/Cole
20 Channel Center Street
Boston, MA 02210
USA

Cengage Learning is a leading provider of customized learning solutions with office locations around the globe, including Singapore, the United Kingdom, Australia, Mexico, Brazil, and Japan. Locate your local office at:
www.cengage.com/global

Cengage Learning products are represented in Canada by Nelson Education, Ltd.

To learn more about Brooks/Cole, visit
www.cengage.com/brookscole

Purchase any of our products at your local college store or at our preferred online store
www.cengagebrain.com

Printed in the United States of America
1 2 3 4 5 6 7 14 13 12 11 10

Contents

Preface

This Student Solutions Manual and Study Guide for *Numerical Analysis*, Ninth Edition, by Burden and Faires contains representative exercises that have been worked out in detail for all the techniques discussed in the book. Particular attention was paid to ensure that the exercises solved in the Guide are those requiring insight into the theory and methods discussed in the book. Although the answers to the odd exercises are also in the back of the book, the results listed in this Study Guide generally go well beyond those in the book.

For this edition we have added a number of exercises to the text that involve the use of a computer algebra system (CAS). We chose Maple as our standard CAS, because their *NumericalAnalysis* package parallels the algorithms in this book. However, any of the common computer algebra systems, such Mathematica, MATLAB, and the public domain system, Sage, can be used with satisfaction. In our recent teaching of the course we have found that students understood the concepts better when they worked through the algorithms step-by-step, but let a computer algebra system do the tedious computation.

It has been our practice to include structured algorithms in our Numerical Analysis book for all the techniques discussed in the text. The algorithms are given in a form that can be coded in any appropriate programming language, by students with even a minimal amount of programming expertise.

At the website for the book,

http://www.math.ysu.edu/~faires/Numerical-Analysis/

you will find code for all the algorithms written in the programming languages FORTRAN, Pascal, C, Java. You will also find code in the form of worksheets for the computer algebra systems, Maple, MATLAB, and Mathematica. For this edition we have rewritten all the Maple programs to reflect the *NumericalAnalysis* package and the numerous changes that have been made to this system.

The website contains additional information about the book, and will be updated regularly to reflect any modifications that might be made. For example, we will place there any errata we are aware of, as well as responses to questions from users of the book concerning interpretations of the exercises and appropriate applications of the techniques.

We hope our Guide helps you with your study of Numerical Analysis. If you have any suggestions for improvements that can be incorporated into future editions of the book or the supplements, we would be most grateful to receive your comments. We can be most easily contacted by electronic mail at the addresses listed below.

Youngstown State University

Richard L. Burden
burden@math.ysu.edu

August 14, 2010

J. Douglas Faires
faires@math.ysu.edu

Mathematical Preliminaries

Exercise Set 1.1, page 14

1. d. Show that the equation $x - (\ln x)^x = 0$ has at least one solution in the interval $[4, 5]$.

SOLUTION: It is not possible to algebraically solve for the solution x, but this is not required in the problem, we must show only that a solution exists. Let

$$f(x) = x - (\ln x)^x = x - \exp(x(\ln(\ln x))).$$

Since f is continuous on $[4, 5]$ with $f(4) \approx 0.3066$ and $f(5) \approx -5.799$, the Intermediate Value Theorem 1.11 implies that a number x must exist in $(4, 5)$ with $0 = f(x) = x - (\ln x)^x$.

2. c. Find intervals that contain a solution to the equation $x^3 - 2x^2 - 4x + 3 = 0$.

SOLUTION: Let $f(x) = x^3 - 2x^2 - 4x + 3$. The critical points of f occur when

$$0 = f'(x) = 3x^2 - 4x - 4 = (3x + 2)(x - 2);$$

that is, when $x = -\frac{2}{3}$ and $x = 2$. Relative maximum and minimum values of f can occur only at these values. There are at most three solutions to $f(x) = 0$, because $f(x)$ is a polynomial of degree three. Since $f(-2) = -5$ and $f\left(-\frac{2}{3}\right) \approx 4.48$; $f(0) = 3$ and $f(1) = -2$; and $f(2) = -5$ and $f(4) = 19$; solutions lie in the intervals $[-2, -2/3]$, $[0, 1]$, and $[2, 4]$.

4. a. Find $\max\limits_{0 \le x \le 1} |f(x)|$ when $f(x) = (2 - e^x + 2x)/3$.

SOLUTION: First note that $f'(x) = (-e^x + 2)/3$, so the only critical point of f occurs at $x = \ln 2$, which lies in the interval $[0, 1]$. The maximum for $|f(x)|$ must consequently be

$$\max\{|f(0)|, |f(\ln 2)|, |f(1)|\} = \max\{1/3, (2 \ln 2)/3, (4 - e)/3\} = (2 \ln 2)/3.$$

5. Use the Intermediate Value Theorem 1.11 and Rolle's Theorem 1.7 to show that the graph of $f(x) = x^3 + 2x + k$ crosses the x-axis exactly once, regardless of the value of the constant k.

SOLUTION: For $x < 0$, we have $f(x) < 2x + k < 0$, provided that $x < -\frac{1}{2}k$. Similarly, for $x > 0$, we have $f(x) > 2x + k > 0$, provided that $x > -\frac{1}{2}k$. By Theorem 1.11, there exists a number c with $f(c) = 0$.

If $f(c) = 0$ and $f(c') = 0$ for some $c' \ne c$, then by Theorem 1.7, there exists a number p between c and c' with $f'(p) = 0$. However, $f'(x) = 3x^2 + 2 > 0$ for all x. This gives a contradiction to the statement that $f(c) = 0$ and $f(c') = 0$ for some $c' \ne c$. Hence there is exactly one number c with $f(c) = 0$.

1

9. Find the second Taylor polynomial for $f(x) = e^x \cos x$ about $x_0 = 0$, and:

a. Use $P_2(0.5)$ to approximate $f(0.5)$, find an upper bound for $|f(0.5) - P_2(0.5)|$, and compare this to the actual error.

b. Find a bound for the error $|f(x) - P_2(x)|$, for x in $[0, 1]$.

c. Approximate $\int_0^1 f(x) \, dx$ using $\int_0^1 P_2(x) \, dx$.

d. Find an upper bound for the error in part (c).

SOLUTION: Since

$$f'(x) = e^x(\cos x - \sin x), \quad f''(x) = -2e^x(\sin x), \quad \text{and} \quad f'''(x) = -2e^x(\sin x + \cos x),$$

we have $f(0) = 1$, $f'(0) = 1$, and $f''(0) = 0$. So

$$P_2(x) = 1 + x \quad \text{and} \quad R_2(x) = \frac{-2e^\xi(\sin \xi + \cos \xi)}{3!} x^3.$$

a. We have $P_2(0.5) = 1 + 0.5 = 1.5$ and

$$|f(0.5) - P_2(0.5)| \leq \max_{\xi \in [0.0.5]} \left| \frac{-2e^\xi(\sin \xi + \cos \xi)}{3!} (0.5)^2 \right| \leq \frac{1}{3}(0.5)^2 \max_{\xi \in [0,0.5]} |e^\xi(\sin \xi + \cos \xi)|.$$

To maximize this quantity on $[0, 0.5]$, first note that $D_x e^x(\sin x + \cos x) = 2e^x \cos x > 0$, for all x in $[0, 0.5]$. This implies that the maximum and minimum values of $e^x(\sin x + \cos x)$ on $[0, 0.5]$ occur at the endpoints of the interval, and

$$e^0(\sin 0 + \cos 0) = 1 < e^{0.5}(\sin 0.5 + \cos 0.5) \approx 2.24.$$

Hence

$$|f(0.5) - P_2(0.5)| \leq \frac{1}{3}(0.5)^3(2.24) \approx 0.0932.$$

b. A similar analysis to that in part (a) gives, for all $x \in [0, 1]$,

$$|f(x) - P_2(x)| \leq \frac{1}{3}(1.0)^3 e^1(\sin 1 + \cos 1) \approx 1.252.$$

c.

$$\int_0^1 f(x) \, dx \approx \int_0^1 1 + x \, dx = \left[x + \frac{x^2}{2} \right]_0^1 = \frac{3}{2}.$$

d. From part (b),

$$\int_0^1 |R_2(x)| \, dx \leq \int_0^1 \frac{1}{3} e^1(\cos 1 + \sin 1)x^3 \, dx = \int_0^1 1.252 x^3 \, dx = 0.313.$$

Since

$$\int_0^1 e^x \cos x \, dx = \left[\frac{e^x}{2}(\cos x + \sin x) \right]_0^1 = \frac{e}{2}(\cos 1 + \sin 1) - \frac{1}{2}(1 + 0) \approx 1.378,$$

the actual error is $|1.378 - 1.5| \approx 0.12$.

14. Use the error term of a Taylor polynomial to estimate the error involved in using $\sin x \approx x$ to approximate $\sin 1°$.

SOLUTION: First we need to convert the degree measure for the sine function to radians. We have $180° = \pi$ radians, so $1° = \frac{\pi}{180}$ radians. Since $f(x) = \sin x$, $f'(x) = \cos x$, $f''(x) = -\sin x$, and $f'''(x) = -\cos x$, we have $f(0) = 0$, $f'(0) = 1$, and $f''(0) = 0$. The approximation $\sin x \approx x$ is given by $f(x) \approx P_2(x)$ and $R_2(x) = -\frac{\cos \xi}{3!} x^3$. If we use the bound $|\cos \xi| \le 1$, then

$$\left| \sin \left(\frac{\pi}{180} \right) - \frac{\pi}{180} \right| = \left| R_2 \left(\frac{\pi}{180} \right) \right| = \left| \frac{-\cos \xi}{3!} \left(\frac{\pi}{180} \right)^3 \right| \le 8.86 \times 10^{-7}.$$

16. Let $f(x) = e^{x/2} \sin \frac{x}{3}$.

a. Use Maple to determine the third Maclaurin polynomial $P_3(x)$.

b. Find $f^{(4)}(x)$ and bound the error $|f(x) - P_3(x)|$ on $[0, 1]$.

SOLUTION: **a.** Define $f(x)$ by

$f := exp \left(\frac{x}{2} \right) \cdot sin \left(\frac{x}{3} \right)$

$$f := e^{(1/2)x} \sin \left(\frac{1}{3} x \right)$$

Then find the first three terms of the Taylor series with

$g := taylor(f, x = 0, 4)$

$$g := \frac{1}{3} x + \frac{1}{6} x^2 + \frac{23}{648} x^3 + O\left(x^4 \right)$$

Extract the third Maclaurin polynomial with

$p3 := convert(g, polynom)$

$$p3 := \frac{1}{3} x + \frac{1}{6} x^2 + \frac{23}{648} x^3$$

b. Determine the fourth derivative.

$f4 := diff(f, x, x, x, x)$

$$f4 := -\frac{119}{1296} e^{(1/2x)} \sin \left(\frac{1}{3} x \right) + \frac{5}{54} e^{(1/2x)} \cos \left(\frac{1}{3} x \right)$$

Find the fifth derivative.

$f5 := diff(f4, x)$

$$f5 := -\frac{199}{2592} e^{(1/2x)} \sin \left(\frac{1}{3} x \right) + \frac{61}{3888} e^{(1/2x)} \cos \left(\frac{1}{3} x \right)$$

See if the fourth derivative has any critical points in $[0, 1]$.

$p := fsolve(f5 = 0, x, 0..1)$

$$p := .6047389076$$

The extreme values of the fourth derivative will occur at $x = 0, 1$, or p.

$c1 := evalf(subs(x = p, f4))$

$$c1 := .09787176213$$

$c2 := evalf(subs(x = 0, f4))$

$$c2 := .09259259259$$

$c3 := evalf(subs(x = 1, f4))$

$$c3 := .09472344463$$

The maximum absolute value of $f^{(4)}(x)$ is c_1 and the error is given by

$error := c1/24$

$$error := .004077990089$$

24. In Example 3 it is stated that x we have $|\sin x| \leq |x|$. Use the following to verify this statement.

    a. Show that for all $x \geq 0$ the function $f(x) = x - \sin x$ is non-decreasing, which implies that $\sin x \leq x$ with equality only when $x = 0$.

    b. Use the fact that the sine function is odd to reach the conclusion.

SOLUTION: First observe that for $f(x) = x - \sin x$ we have $f'(x) = 1 - \cos x \geq 0$, because $-1 \leq \cos x \leq 1$ for all values of x. Also, the statement clearly holds when $|x| \geq \pi$, because $|\sin x| \leq 1$.

a. The observation implies that $f(x)$ is non-decreasing for all values of x, and in particular that $f(x) > f(0) = 0$ when $x > 0$. Hence for $x \geq 0$, we have $x \geq \sin x$, and when $0 \leq x \leq \pi$, we have $|\sin x| = \sin x \leq x = |x|$.

b. When $-\pi < x < 0$, we have $\pi \geq -x > 0$. Since $\sin x$ is an odd function, the fact (from part (a)) that $\sin(-x) \leq (-x)$ implies that $|\sin x| = -\sin x \leq -x = |x|$.

As a consequence, for all real numbers x we have $|\sin x| \leq |x|$.

28. Suppose $f \in C[a,b]$, and that x_1 and x_2 are in $[a,b]$.

    a. Show that a number ξ exists between x_1 and x_2 with

$$f(\xi) = \frac{f(x_1) + f(x_2)}{2} = \frac{1}{2}f(x_1) + \frac{1}{2}f(x_2).$$

    b. Suppose that c_1 and c_2 are positive constants. Show that a number ξ exists between x_1 and x_2 with

$$f(\xi) = \frac{c_1 f(x_1) + c_2 f(x_2)}{c_1 + c_2}.$$

    c. Give an example to show that the result in part (b) does not necessarily hold when c_1 and c_2 have opposite signs with $c_1 \neq -c_2$.

SOLUTION:

a. The number

$$\frac{1}{2}(f(x_1) + f(x_2))$$

is the average of $f(x_1)$ and $f(x_2)$, so it lies between these two values of f. By the Intermediate Value Theorem 1.11 there exist a number ξ between x_1 and x_2 with

$$f(\xi) = \frac{1}{2}(f(x_1) + f(x_2)) = \frac{1}{2}f(x_1) + \frac{1}{2}f(x_2).$$

b. Let $m = \min\{f(x_1), f(x_2)\}$ and $M = \max\{f(x_1), f(x_2)\}$. Then $m \le f(x_1) \le M$ and $m \le f(x_2) \le M$, so

$$c_1 m \le c_1 f(x_1) \le c_1 M \quad \text{and} \quad c_2 m \le c_2 f(x_2) \le c_2 M.$$

Thus

$$(c_1 + c_2)m \le c_1 f(x_1) + c_2 f(x_2) \le (c_1 + c_2)M$$

and

$$m \le \frac{c_1 f(x_1) + c_2 f(x_2)}{c_1 + c_2} \le M.$$

By the Intermediate Value Theorem 1.11 applied to the interval with endpoints x_1 and x_2, there exists a number ξ between x_1 and x_2 for which

$$f(\xi) = \frac{c_1 f(x_1) + c_2 f(x_2)}{c_1 + c_2}.$$

c. Let $f(x) = x^2 + 1$, $x_1 = 0$, $x_2 = 1$, $c_1 = 2$, and $c_2 = -1$. Then $f(x) > 0$ for all values of x, but

$$\frac{c_1 f(x_1) + c_2 f(x_2)}{c_1 + c_2} = \frac{2(1) - 1(2)}{2 - 1} = 0.$$

Exercise Set 1.2, page 28

2. c. Find the largest interval in which p^* must lie to approximate $\sqrt{2}$ with relative error at most 10^{-4}.

SOLUTION: We need

$$\frac{|p^* - \sqrt{2}|}{|\sqrt{2}|} \le 10^{-4}, \quad \text{so} \quad |p^* - \sqrt{2}| \le \sqrt{2} \times 10^{-4};$$

that is,

$$-\sqrt{2} \times 10^{-4} \le p^* - \sqrt{2} \le \sqrt{2} \times 10^{-4}.$$

This implies that p^* must be in the interval $\left(\sqrt{2}(0.9999), \sqrt{2}(1.0001)\right)$.

5. e. Use three-digit rounding arithmetic to compute

$$\frac{\frac{13}{14} - \frac{6}{7}}{2e - 5.4},$$

and determine the absolute and relative errors.

SOLUTION: Using three-digit rounding arithmetic gives $\frac{13}{14} = 0.929$, $\frac{6}{7} = 0.857$, and $e = 2.72$. So

$$\frac{13}{14} - \frac{6}{7} = 0.0720 \quad \text{and} \quad 2e - 5.4 = 5.44 - 5.40 = 0.0400.$$

Hence

$$\frac{\frac{13}{14} - \frac{6}{7}}{2e - 5.4} = \frac{0.0720}{0.0400} = 1.80.$$

The correct value is approximately 1.954, so the absolute and relative errors to three digits are

$$|1.80 - 1.954| = 0.154 \quad \text{and} \quad \frac{|1.80 - 1.954|}{1.954} = 0.0788,$$

respectively.

7. e. Repeat Exercise 5(e) using three-digit chopping arithmetic.

SOLUTION: Using three-digit chopping arithmetic gives $\frac{13}{14} = 0.928$, $\frac{6}{7} = 0.857$, and $e = 2.71$. So

$$\frac{13}{14} - \frac{6}{7} = 0.0710 \quad \text{and} \quad 2e - 5.4 = 5.42 - 5.40 = 0.0200.$$

Hence

$$\frac{\frac{13}{14} - \frac{6}{7}}{2e - 5.4} = \frac{0.0710}{0.0200} = 3.55.$$

The correct value is approximately 1.954, so the absolute and relative errors to three digits are

$$|3.55 - 1.954| = 1.60, \quad \text{and} \quad \frac{|3.55 - 1.954|}{1.954} = 0.817,$$

respectively. The results in Exercise 5(e) were considerably better.

9. a. Use the first three terms of the Maclaurin series for the arctangent function to approximate $\pi = 4 \left[\arctan \frac{1}{2} + \arctan \frac{1}{3} \right]$, and determine the absolute and relative errors.

SOLUTION: Let $P(x) = x - \frac{1}{3}x^3 + \frac{1}{5}x^5$. Then $P\left(\frac{1}{2}\right) = 0.464583$ and $P\left(\frac{1}{3}\right) = 0.3218107$, so

$$\pi = 4 \left[\arctan \frac{1}{2} + \arctan \frac{1}{3} \right] \approx 3.145576.$$

The absolute and relative errors are, respectively,

$$|\pi - 3.145576| \approx 3.983 \times 10^{-3} \quad \text{and} \quad \frac{|\pi - 3.145576|}{|\pi|} \approx 1.268 \times 10^{-3}.$$

12. Let

$$f(x) = \frac{e^x - e^{-x}}{x}.$$

a. Find $\lim_{x \to 0} f(x)$.

b. Use three-digit rounding arithmetic to evaluate $f(0.1)$.

c. Replace each exponential function with its third Maclaurin polynomial and repeat part (b).

SOLUTION: **a.** Since $\lim_{x \to 0} e^x - e^{-x} = 1 - 1 = 0$ and $\lim_{x \to 0} x = 0$, we can use L'Hospitals Rule to give

$$\lim_{x \to 0} \frac{e^x - e^{-x}}{x} = \lim_{x \to 0} \frac{e^x + e^{-x}}{1} = \frac{1 + 1}{1} = 2.$$

b. With three-digit rounding arithmetic we have $e^{0.100} = 1.11$ and $e^{-0.100} = 0.905$, so

$$f(0.100) = \frac{1.11 - 0.905}{0.100} = \frac{0.205}{0.100} = 2.05.$$

c. The third Maclaurin polynomials give

$$e^x \approx 1 + x + \frac{1}{2}x^2 + \frac{1}{6}x^3 \quad \text{and} \quad e^{-x} \approx 1 - x + \frac{1}{2}x^2 - \frac{1}{6}x^3,$$

so

$$f(x) \approx \frac{\left(1 + x + \frac{1}{2}x^2 + \frac{1}{6}x^3\right) - \left(1 - x + \frac{1}{2}x^2 - \frac{1}{6}x^3\right)}{x} = \frac{2x + \frac{1}{3}x^3}{x} = 2 + \frac{1}{3}x^2.$$

Thus, with three-digit rounding, we have

$$f(0.100) \approx 2 + \frac{1}{3}(0.100)^2 = 2 + (0.333)(0.001) = 2.00 + 0.000333 = 2.00.$$

15. c. Find the decimal equivalent of the floating-point machine number

0   01111111111   0101001100000000000000000000000000000000000000000000.

SOLUTION: This binary machine number is the decimal number

$$+ 2^{1023-1023}\left(1 + \left(\frac{1}{2}\right)^2 + \left(\frac{1}{2}\right)^4 + \left(\frac{1}{2}\right)^7 + \left(\frac{1}{2}\right)^8\right)$$

$$= 2^0\left(1 + \frac{1}{4} + \frac{1}{16} + \frac{1}{128} + \frac{1}{256}\right) = 1 + \frac{83}{256} = 1.32421875.$$

16. c. Find the decimal equivalents of the next largest and next smallest floating-point machine number to

0   01111111111   0101001100000000000000000000000000000000000000000000.

SOLUTION: The next smallest machine number is

0   01111111111   0101001011111111111111111111111111111111111111111111
$$= 1.32421875 - 2^{1023-1023}\left(2^{-52}\right)$$
$$= 1.3242187499999997779553950749686919152736663818359375,$$

and next largest machine number is

0   01111111111   0101001100000000000000000000000000000000000000000001
$$= 1.32421875 + 2^{1023-1023}\left(2^{-52}\right)$$
$$= 1.3242187500000002220446049250313080847263336181640625.$$

21. a. Show that the polynomial nesting technique can be used to evaluate

$$f(x) = 1.01e^{4x} - 4.62e^{3x} - 3.11e^{2x} + 12.2e^x - 1.99.$$

b. Use three-digit rounding arithmetic and the formula given in the statement of part (a) to evaluate $f(1.53)$.

c. Redo the calculations in part (b) using the nesting form of $f(x)$ that was found in part (a).

d. Compare the approximations in parts (b) and (c).

SOLUTION: **a.** Since $e^{nx} = (e^x)^n$, we can write

$$f(x) = ((((1.01)e^x - 4.62)\, e^x - 3.11)\, e^x + 12.2)\, e^x - 1.99.$$

b. Using $e^{1.53} = 4.62$ and three-digit rounding gives $e^{2(1.53)} = (4.62)^2 = 21.3$, $e^{3(1.53)} = (4.62)^2(4.62) = (21.3)(4.62) = 98.4$, and $e^{4(1.53)} = (98.4)(4.62) = 455$. So

$$f(1.53) = 1.01(455) - 4.62(98.4) - 3.11(21.3) + 12.2(4.62) - 1.99$$
$$= 460 - 455 - 66.2 + 56.4 - 1.99$$
$$= 5.00 - 66.2 + 56.4 - 1.99$$
$$= -61.2 + 56.4 - 1.99 = -4.80 - 1.99 = -6.79.$$

c. We have

$$f(1.53) = (((1.01)4.62 - 4.62)4.62 - 3.11)4.62 + 12.2)4.62 - 1.99$$
$$= (((4.67 - 4.62)4.62 - 3.11)4.62 + 12.2)4.62 - 1.99$$
$$= ((0.231 - 3.11)4.62 + 12.2)4.62 - 1.99$$
$$= (-13.3 + 12.2)4.62 - 1.99 = -7.07.$$

d. The exact result is 7.61, so the absolute errors in parts (b) and (c) are, respectively, $|-6.79 + 7.61| = 0.82$ and $|-7.07 + 7.61| = 0.54$. The relative errors are, respectively, 0.108 and 0.0710.

24. Suppose that $fl(y)$ is a k-digit rounding approximation to y. Show that

$$\left| \frac{y - fl(y)}{y} \right| \le 0.5 \times 10^{-k+1}.$$

SOLUTION: We will consider the solution in two cases, first when $d_{k+1} \le 5$, and then when $d_{k+1} > 5$.

When $d_{k+1} \le 5$, we have

$$\left| \frac{y - fl(y)}{y} \right| = \frac{0.d_{k+1} \ldots \times 10^{n-k}}{0.d_1 \ldots \times 10^n} \le \frac{0.5 \times 10^{-k}}{0.1} = 0.5 \times 10^{-k+1}.$$

When $d_{k+1} > 5$, we have

$$\left| \frac{y - fl(y)}{y} \right| = \frac{(1 - 0.d_{k+1} \ldots) \times 10^{n-k}}{0.d_1 \ldots \times 10^n} < \frac{(1 - 0.5) \times 10^{-k}}{0.1} = 0.5 \times 10^{-k+1}.$$

Hence the inequality holds in all situations.

28. Show that both sets of data given in the opening application for this chapter can give values of T that are consistent with the ideal gas law.

SOLUTION: For the initial data, we have

$$0.995 \leq P \leq 1.005, \quad 0.0995 \leq V \leq 0.1005,$$

$$0.082055 \leq R \leq 0.082065, \quad \text{and} \quad 0.004195 \leq N \leq 0.004205.$$

This implies that
$$287.61 \leq T \leq 293.42.$$

Since $15°$ Celsius = 288.16 kelvin, we are within the bound. When P is doubled and V is halved,

$$1.99 \leq P \leq 2.01 \quad \text{and} \quad 0.0497 \leq V \leq 0.0503,$$

so
$$286.61 \leq T \leq 293.72.$$

Since $19°$ Celsius = 292.16 kelvin, we are again within the bound. In either case it is possible that the actual temperature is 290.15 kelvin = $17°$ Celsius.

Exercise Set 1.3, page 39

3. a. Determine the number n of terms of the series

$$\arctan x = \lim_{n \to \infty} P_n(x) = \sum_{i=1}^{\infty} (-1)^{i+1} \frac{x^{2i-1}}{(2i-1)}$$

that are required to ensure that $|4P_n(1) - \pi| < 10^{-3}$.

b. How many terms are required to ensure the 10^{-10} accuracy needed for an approximation to π?

SOLUTION: **a.** Since the terms of the series

$$\pi = 4 \arctan 1 = 4 \sum_{i=1}^{\infty} (-1)^{i+1} \frac{1}{2i-1}$$

alternate in sign, the error produced by truncating the series at any term is less than the magnitude of the next term. To ensure significant accuracy, we need to choose n so that

$$\frac{4}{2(n+1)-1} < 10^{-3} \quad \text{or} \quad 4000 < 2n+1.$$

So $n \geq 2000$.

b. In this case, we need

$$\frac{4}{2(n+1)-1} < 10^{-10} \quad \text{or} \quad n > 20{,}000{,}000{,}000.$$

Clearly, a more rapidly convergent method is needed for this approximation.

5. Another formula for computing π can be deduced from the identity

$$\frac{\pi}{4} = 4 \arctan \frac{1}{5} - \arctan \frac{1}{239}.$$

Determine the number of terms that must be summed to ensure an approximation to π to within 10^{-3}.

SOLUTION: The identity implies that

$$\pi = 4 \sum_{i=1}^{\infty} (-1)^{i+1} \frac{1}{5^{2i-1}(2i-1)} - \sum_{i=1}^{\infty} (-1)^{i+1} \frac{1}{239^{2i-1}(2i-1)}$$

The second sum is much smaller than the first sum. So we need to determine the minimal value of i so that the $i+1$st term of the first sum is less than 10^{-3}. We have

$$i := 1 : \frac{4}{5^1(1)} = \frac{4}{5}, \quad i = 2 : \frac{4}{5^3(3)} = \frac{4}{375} \quad \text{and} \quad i = 3 : \frac{4}{5^5(5)} = \frac{4}{15625} = 2.56 \times 10^{-4}.$$

So 3 terms are sufficient.

8. a. How many calculations are needed to determine a sum of the form

$$\sum_{i=1}^{n} \sum_{j=1}^{i} a_i b_j ?$$

b. Re-express the series in a way that will reduce the number of calculations needed to determine this sum.

SOLUTION: **a.** For each i, the inner sum $\sum_{j=1}^{i} a_i b_j$ requires i multiplications and $i-1$ additions, for a total of

$$\sum_{i=1}^{n} i = \frac{n(n+1)}{2} \quad \text{multiplications} \quad \text{and} \quad \sum_{i=1}^{n} i - 1 = \frac{n(n+1)}{2} - n \quad \text{additions.}$$

Once the n inner sums are computed, $n-1$ additions are required for the final sum.

The final total is:

$$\frac{n(n+1)}{2} \quad \text{multiplications} \quad \text{and} \quad \frac{(n+2)(n-1)}{2} \quad \text{additions.}$$

b. By rewriting the sum as

$$\sum_{i=1}^{n} \sum_{j=1}^{i} a_i b_j = \sum_{i=1}^{n} a_i \sum_{j=1}^{i} b_j,$$

we can significantly reduce the amount of calculation. For each i, we now need $i-1$ additions to sum b_j's for a total of

$$\sum_{i=1}^{n} i - 1 = \frac{n(n+1)}{2} - n \quad \text{additions.}$$

Once the b_j's are summed, we need n multiplications by the a_i's, followed by $n-1$ additions of the products.

The total additions by this method is still $\frac{1}{2}(n+2)(n-1)$, but the number of multiplications has been reduced from $\frac{1}{2}n(n+1)$ to n.

10. Devise an algorithm to compute the real roots of a quadratic equation in the most efficient manner.

SOLUTION: The following algorithm uses the most effective formula for computing the roots of a quadratic equation.

INPUT A, B, C.
OUTPUT x_1, x_2.

Step 1 If $A = 0$ then

         if $B = 0$ then   OUTPUT ('NO SOLUTIONS');
                       STOP.
             else set $x_1 = -C/B$;
                       OUTPUT ('ONE SOLUTION',x_1);
                       STOP.

Step 2 Set $D = B^2 - 4AC$.

Step 3 If $D = 0$ then set $x_1 = -B/(2A)$;
                       OUTPUT ('MULTIPLE ROOTS', x_1);
                       STOP.

Step 4 If $D < 0$ then set
$$b = \sqrt{-D}/(2A);$$
$$a = -B/(2A);$$
OUTPUT ('COMPLEX CONJUGATE ROOTS');
$$x_1 = a + bi;$$
$$x_2 = a - bi;$$
OUTPUT (x_1, x_2);
STOP.

Step 5 If $B \geq 0$ then set
$$d = B + \sqrt{D};$$
$$x_1 = -2C/d;$$
$$x_2 = -d/(2A)$$
     else set
$$d = -B + \sqrt{D};$$
$$x_1 = d/(2A);$$
$$x_2 = 2C/d.$$

Step 6 OUTPUT (x_1, x_2);
       STOP.

15. Suppose that as x approaches zero,

$$F_1(x) = L_1 + O\left(x^\alpha\right) \quad \text{and} \quad F_2(x) = L_2 + O\left(x^\beta\right).$$

Let c_1 and c_2 be nonzero constants, and define

$$F(x) = c_1 F_1(x) + c_2 F_2(x) \quad \text{and} \quad G(x) = F_1(c_1 x) + F_2(c_2 x).$$

Show that if $\gamma = \text{minimum}\ \{\alpha, \beta\}$, then as x approaches zero,

a. $F(x) = c_1 L_1 + c_2 L_2 + O\left(x^\gamma\right)$

b. $G(x) = L_1 + L_2 + O\left(x^\gamma\right)$

SOLUTION: Suppose for sufficiently small $|x|$ we have positive constants k_1 and k_2 independent of x, for which

$$|F_1(x) - L_1| \leq K_1|x|^\alpha \quad \text{and} \quad |F_2(x) - L_2| \leq K_2|x|^\beta.$$

Let $c = \max\left(|c_1|, |c_2|, 1\right)$, $K = \max\left(K_1, K_2\right)$, and $\delta = \max\left(\alpha, \beta\right)$.

a. We have

$$\begin{aligned}
|F(x) - c_1 L_1 - c_2 L_2| &= |c_1(F_1(x) - L_1) + c_2(F_2(x) - L_2)| \\
&\leq |c_1||K_1||x|^\alpha + |c_2||K_2||x|^\beta \\
&\leq cK\left(|x|^\alpha + |x|^\beta\right) \\
&\leq cK|x|^\gamma\left(1 + |x|^{\delta-\gamma}\right) \leq K|x|^\gamma,
\end{aligned}$$

for sufficiently small $|x|$. Thus, $F(x) = c_1 L_1 + c_2 L_2 + O\left(x^\gamma\right)$.

b. We have

$$\begin{aligned}
|G(x) - L_1 - L_2| &= |F_1(c_1 x) + F_2(c_2 x) - L_1 - L_2| \\
&\leq K_1|c_1 x|^\alpha + K_2|c_2 x|^\beta \\
&\leq Kc^\delta\left(|x|^\alpha + |x|^\beta\right) \\
&\leq Kc^\delta|x|^\gamma\left(1 + |x|^{\delta-\gamma}\right) \leq K''|x|^\gamma,
\end{aligned}$$

for sufficiently small $|x|$. Thus, $G(x) = L_1 + L_2 + O\left(x^\gamma\right)$.

16. Consider the Fibonacci sequence defined by $F_0 = 1$, $F_1 = 1$, and $F_{n+2} = F_{n+1} + F_n$, if $n \geq 0$. Define $x_n = F_{n+1}/F_n$. Assuming that $\lim_{n\to\infty} x_n = x$ converges, show that the limit is the golden ratio: $x = \left(1 + \sqrt{5}\right)/2$.

SOLUTION: Since

$$\lim_{n\to\infty} x_n = \lim_{n\to\infty} x_{n+1} = x \quad \text{and} \quad x_{n+1} = 1 + \frac{1}{x_n},$$

we have

$$x = 1 + \frac{1}{x}, \quad \text{which implies that} \quad x^2 - x - 1 = 0.$$

The only positive solution to this quadratic equation is $x = \left(1 + \sqrt{5}\right)/2$.

17. The Fibonacci sequence also satisfies the equation

$$F_n \equiv \tilde{F}_n = \frac{1}{\sqrt{5}}\left[\left(\frac{1 + \sqrt{5}}{2}\right)^n - \left(\frac{1 - \sqrt{5}}{2}\right)^n\right].$$

a. Write a Maple procedure to calculate F_{100}.

b. Use Maple with the default value of *Digits* followed by *evalf* to calculate $\tilde{F}_{100}$.

c. Why is the result from part (a) more accurate than the result from part (b)?

d. Why is the result from part (b) obtained more rapidly than the result from part (a)?

e. What results when you use the command *simplify* instead of *evalf* to compute $\tilde{F}_{100}$?

SOLUTION: **a.** To save space we will show the Maple output for each step in one line. Maple would produce this output on separate lines. The procedure for calculating the terms of the sequence are:

$n := 98; f := 1; s := 1$

$$n := 98 \quad f := 1 \quad s := 1$$

for i from 1 to n do

$\quad l := f + s; f := s; s := l; od :$

$$l := 2 \quad f := 1 \quad s := 2$$
$$l := 3 \quad f := 2 \quad s := 3$$
$$\vdots$$
$$l := 218922995834555169026 \quad f := 135301852344706746049 \quad s := 218922995834555169026$$
$$l := 354224848179261915075$$

b. We have

$$F100 := \frac{1}{sqrt(5)} \left(\left(\frac{(1 + sqrt(5))}{2} \right)^{100} - \left(\frac{1 - sqrt(5)}{2} \right)^{100} \right)$$

$$F100 := \frac{1}{\sqrt{5}} \left(\left(\frac{1}{2} + \frac{1}{2}\sqrt{5} \right)^{100} - \left(\frac{1}{2} - \frac{1}{2}\sqrt{5} \right)^{100} \right)$$

evalf(F100)

$$0.3542248538 \times 10^{21}$$

c. The result in part (a) is computed using exact integer arithmetic, and the result in part (b) is computed using ten-digit rounding arithmetic.

d. The result in part (a) required traversing a loop 98 times.

e. The result is the same as the result in part (a).

Solutions of Equations of One Variable

Exercise Set 2.1, page 54

1. Use the Bisection method to find p_3 for $f(x) = \sqrt{x} - \cos x$ on $[0, 1]$.

   SOLUTION: Using the Bisection method gives $a_1 = 0$ and $b_1 = 1$, so $f(a_1) = -1$ and $f(b_1) = 0.45970$. We have

   $$p_1 = \frac{1}{2}(a_1 + b_1) = \frac{1}{2} \quad \text{and} \quad f(p_1) = -0.17048 < 0.$$

   Since $f(a_1) < 0$ and $f(p_1) < 0$, we assign $a_2 = p_1 = 0.5$ and $b_2 = b_1 = 1$. Thus

   $$f(a_2) = -0.17048 < 0, \quad f(b_2) = 0.45970 > 0, \quad \text{and} \quad p_2 = \frac{1}{2}(a_2 + b_2) = 0.75.$$

   Since $f(p_2) = 0.13434 > 0$, we have $a_3 = 0.5$; $b_3 = p_3 = 0.75$ so that

   $$p_3 = \frac{1}{2}(a_3 + b_3) = 0.625.$$

2. **a.** Let $f(x) = 3(x+1)\left(x - \frac{1}{2}\right)(x - 1)$. Use the Bisection method on the interval $[-2, 1.5]$ to find p_3.

   SOLUTION: Since

   $$f(x) = 3(x + 1)\left(x - \frac{1}{2}\right)(x - 1),$$

   we have the following sign graph for $f(x)$:

   Thus, $a_1 = -2$, with $f(a_1) < 0$, and $b_1 = 1.5$, with $f(b_1) > 0$. Since $p_1 = -\frac{1}{4}$, we have $f(p_1) > 0$. We assign $a_2 = -2$, with $f(a_2) < 0$, and $b_2 = -\frac{1}{4}$, with $f(b_2) > 0$. Thus, $p_2 = -1.125$ and $f(p_2) < 0$. Hence, we assign $a_3 = p_2 = -1.125$ and $b_3 = -0.25$. Then $p_3 = -0.6875$.

15

8. a. Sketch the graphs of $y = x$ and $y = \tan x$.

b. Use the Bisection method to find an approximation to within 10^{-5} to the first positive value of x with $x = \tan x$.

SOLUTION:

a. The graphs of $y = x$ and $y = \tan x$ are shown in the figure. From the graph it appears that the graphs cross near $x = 4.5$.

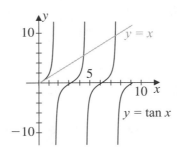

b. Because $g(x) = x - \tan x$ has

$$g(4.4) \approx 1.303 > 0 \quad \text{and} \quad g(4.6) \approx -4.260 < 0,$$

the fact that g is continuous on $[4.4, 4.6]$ gives us a reasonable interval to start the bisection process. Using Algorithm 2.1 gives $p_{16} = 4.4934143$, which is accurate to within 10^{-5}.

11. Let $f(x) = (x + 2)(x + 1)x(x - 1)^3(x - 2)$. To which zero of f does the Bisection method converge for the following intervals?

a. $[-3, 2.5]$

c. $[-1.75, 1.5]$

SOLUTION: Since

$$f(x) = (x + 2)(x + 1)x(x - 1)^3(x - 2),$$

we have the following sign graph for $f(x)$.

$x + 2$	$- - - - 0\ +$
$x + 1$	$- - - - - - - - 0\ + + + + + + + + + + + + + + + + + +$
x	$- - - - - - - - - - - - 0\ + + + + + + + + + + + + +$
$(x - 1)^3$	$- - - - - - - - - - - - - - - - - 0\ + + + + + + + + +$
$x - 2$	$- 0\ + + + + + +$
$f(x)$	$- - - - 0\ + + + 0\ - - - 0\ + + + 0\ - - - 0\ + + + + + +$

a. The interval $[-3, 2.5]$ contains all 5 zeros of f. For $a_1 = -3$, with $f(a_1) < 0$, and $b_1 = 2.5$, with $f(b_1) > 0$, we have $p_1 = (-3 + 2.5)/2 = -0.25$, so $f(p_1) < 0$. Thus we assign $a_2 = p_1 = -0.25$, with $f(a_2) < 0$, and $b_2 = b_1 = 2.5$, with $f(b_1) > 0$.

Hence $p_2 = (-0.25 + 2.5)/2 = 1.125$ and $f(p_2) < 0$. Then we assign $a_3 = 1.125$, with $f(a_3) < 0$, and $b_3 = 2.5$, with $f(b_3) > 0$. Since $[1.125, 2.5]$ contains only the zero 2, the method converges to 2.

c. The interval $[-1.75, 1.5]$ contains the zeros $-1, 0, 1$. For $a_1 = -1.75$, with $f(a_1) > 0$, and $b_1 = 1.5$, with $f(b_1) < 0$, we have $p_1 = (-1.75 + 1.5)/2 = -0.125$ and $f(p_1) < 0$. Then we assign $a_2 = a_1 = -1.75$, with $f(a_1) > 0$, and $b_2 = p_1 = -0.125$, with $f(b_2) < 0$. Since $[-1.75, -0.125]$ contains only the zero -1, the method converges to -1.

12. Use the Bisection Algorithm to find an approximation to $\sqrt{3}$ that is accurate to within 10^{-4}.

SOLUTION: The function defined by $f(x) = x^2 - 3$ has $\sqrt{3}$ as its only positive zero. Applying the Bisection method to this function on the interval $[1, 2]$ gives $\sqrt{3} \approx p_{14} = 1.7320$. Using a smaller starting interval would decrease the number of iterations that are required.

14. Use Theorem 2.1 to find a bound for the number of iterations needed to approximate a solution to the equation $x^3 + x - 4 = 0$ on the interval $[1, 4]$ to an accuracy of 10^{-3}.

SOLUTION: First note that the particular equation plays no part in finding the bound; all that is needed is the interval and the accuracy requirement. To find an approximation that is accurate to within 10^{-3}, we need to determine the number of iterations n so that

$$|p - p_n| < \frac{b - a}{2^n} = \frac{4 - 1}{2^n} < 0.001; \quad \text{that is,} \quad 3 \times 10^3 < 2^n.$$

As a consequence, a bound for the number of iterations is $n \geq 12$. Applying the Bisection Algorithm gives $p_{12} = 1.3787$.

17. Define the sequence $\{p_n\}$ by $p_n = \sum_{k=1}^{n} \frac{1}{k}$. Show that $\lim_{n\to\infty} (p_n - p_{n-1}) = 0$, even though the sequence $\{p_n\}$ diverges.

SOLUTION: Since $p_n - p_{n-1} = 1/n$, we have $\lim_{n\to\infty}(p_n - p_{n-1}) = 0$. However, p_n is the nth partial sum of the divergent *harmonic* series. The harmonic series is the classic example of a series whose terms go to zero, but not rapidly enough to produce a convergent series. There are many proofs of the divergence of this series, any calculus text should give at least two. One proof will simply analyze the partial sums of the series and another is based on the Integral Test.

The point of the problem is not the fact that this particular sequence diverges, it is that a test for an approximate solution to a root based on the condition that $|p_n - p_{n-1}|$ is small should always be suspect. Consecutive terms of a sequence might be close to each other, but not sufficiently close to the actual solution you are seeking.

19. A trough of water of length $L = 10$ feet has a cross section in the shape of a semicircle with radius $r = 1$ foot. When filled with water to within a distance h of the top, the volume $V = 12.4$ ft^3 of the water is given by the formula

$$12.4 = 10 \left[0.5\pi - \arcsin h - h \left(1 - h^2\right)^{1/2} \right]$$

Determine the depth of the water to within 0.01 feet.

SOLUTION: Applying the Bisection Algorithm on the interval $[0, 1]$ to the function

$$f(h) = 12.4 - 10 \left[0.5\pi - \arcsin h - h \left(1 - h^2 \right)^{1/2} \right]$$

gives $h \approx p_{13} = 0.1617$, so the depth is $r - h \approx 1 - 0.1617 = 0.8383$ feet.

Exercise Set 2.2, page 64

3. The following methods are proposed to compute $21^{1/3}$. Rank them in order, based on their apparent speed of convergence, assuming $p_0 = 1$.

a. $p_n = \dfrac{20p_{n-1} + 21/p_{n-1}^2}{21}$

b. $p_n = p_{n-1} - \dfrac{p_{n-1}^3 - 21}{3p_{n-1}^2}$

c. $p_n = p_{n-1} - \dfrac{p_{n-1}^4 - 21p_{n-1}}{p_{n-1}^2 - 21}$

d. $p_n = \left(\dfrac{21}{p_{n-1}} \right)^{1/2}$

SOLUTION: **a.** Since

$$p_n = \frac{20p_{n-1} + 21/p_{n-1}^2}{21}, \quad \text{we have} \quad g(x) = \frac{20x + 21/x^2}{21} = \frac{20}{21}x + \frac{1}{x^2},$$

and $g'(x) = \dfrac{20}{21} - \dfrac{2}{x^3}$. Thus, $g'\left(21^{1/3}\right) = \dfrac{20}{21} - \dfrac{2}{21} = \dfrac{6}{7} \approx 0.857$.

b. Since

$$p_n = p_{n-1} - \frac{p_{n-1}^3 - 21}{3p_{n-1}^2}, \quad \text{we have} \quad g(x) = x - \frac{x^3 - 21}{3x^2} = x - \frac{1}{3}x + \frac{7}{x^2} = \frac{2}{3}x + \frac{7}{x^2}$$

and $g'(x) = \dfrac{2}{3} - \dfrac{7}{x^3}$. Thus, $g'\left(21^{1/3}\right) = \dfrac{2}{3} - \dfrac{1}{3} = \dfrac{1}{3} = 0.33\overline{3}$.

c. Since

$$p_n = p_{n-1} - \frac{p_{n-1}^4 - 21p_{n-1}}{p_{n-1}^2 - 21},$$

we have

$$g(x) = x - \frac{x^4 - 21x}{x^2 - 21} = \frac{x^3 - 21x - x^4 + 21x}{x^2 - 21} = \frac{x^3 - x^4}{x^2 - 21}$$

and

$$g'(x) = \frac{\left(x^2 - 21\right)\left(3x^2 - 4x^3\right) - \left(x^3 - x^4\right)2x}{\left(x^2 - 21\right)^2} = \frac{3x^4 - 63x^2 - 4x^5 + 84x^3 - 2x^4 + 2x^5}{\left(x^2 - 21\right)^2}$$

$$= \frac{-2x^5 + x^4 + 84x^3 - 63x^2}{\left(x^2 - 21\right)^2}.$$

Thus $g'\left(21^{1/3}\right) \approx 5.706 > 1$.

d. Since

$$p_n = \left(\frac{21}{p_{n-1}}\right)^{1/2}, \quad \text{we have} \quad g(x) = \left(\frac{21}{x}\right)^{1/2} = \frac{\sqrt{21}}{x^{1/2}},$$

and $g'(x) = \frac{-\sqrt{21}}{2x^{3/2}}$. Thus, $g'\left(21^{1/3}\right) = -\frac{1}{2}$.

The order of convergence would likely be (b), (d), (a). Choice (c) will not likely converge.

9. Use a fixed-point iteration method to determine an approximation to $\sqrt{3}$ that is accurate to within 10^{-4}.

SOLUTION: As always with fixed-point iteration, the trick is to choose the fixed-point problem that will produce rapid convergence.

Recalling the solution to Exercise 12 in Section 2.1, we need to convert the root-finding problem $f(x) = x^2 - 3$ into a fixed-point problem. One successful solution is to write

$$0 = x^2 - 3 \quad \text{as} \quad x = \frac{3}{x},$$

then add x to both sides of the latter equation and divide by 2. This gives $g(x) = 0.5\left(x + \frac{3}{x}\right)$, and for $p_0 = 1.0$, we have $\sqrt{3} \approx p_4 = 1.73205$.

12. c. Determine a fixed-point function g and an appropriate interval that produces an approximation to a positive solution of $3x^2 - e^x = 0$ that is accurate to within 10^{-5}.

SOLUTION: There are numerous possibilities:

For $g(x) = \sqrt{\frac{1}{3}e^x}$ on $[0, 1]$ with $p_0 = 1$, we have $p_{12} = 0.910015$.

For $g(x) = \ln 3x^2$ on $[3, 4]$ with $p_0 = 4$, we have $p_{16} = 3.733090$.

14. Use a fixed-point iteration method to determine a solution accurate to within 10^{-4} for $x = \tan x$, for x in $[4, 5]$.

SOLUTION: Using $g(x) = \tan x$ and $p_0 = 4$ gives $p_1 = g(p_0) \approx 1.158$, which is not in the interval $[4, 5]$. So we need a different fixed-point function. If we note that $x = \tan x$ implies that

$$\frac{1}{x} = \frac{1}{\tan x} \quad \text{and define} \quad g(x) = x + \frac{1}{\tan x} - \frac{1}{x}$$

we obtain, again with $p_0 = 4$:

$$p_1 \approx 4.61369, \quad p_2 = 4.49596, \quad p_3 = 4.49341 \quad \text{and} \quad p_4 = 4.49341.$$

Because p_3 and p_4 agree to five decimal places it is reasonable to assume that these values are sufficiently accurate.

18. a. Show that Theorem 2.3 is true if $|g'(x)| \le k$ is replaced by the statement "$g'(x) \le k < 1$, for all $x \in [a, b]$".

b. Show that Theorem 2.4 may not hold when $|g'(x)| \le k$ is replaced by the statement "$g'(x) \le k < 1$, for all $x \in [a, b]$".

SOLUTION: **a.** The proof of existence is unchanged. For uniqueness, suppose p and q are fixed points in $[a, b]$ with $p \neq q$. By the Mean Value Theorem, a number ξ in (a, b) exists with

$$p - q = g(p) - g(q) = g'(\xi)(p - q) \leq k(p - q) < p - q,$$

giving the same contradiction as in Theorem 2.3.

b. For Theorem 2.4, consider $g(x) = 1 - x^2$ on $[0, 1]$. The function g has the unique fixed point $p = \frac{1}{2}\left(-1 + \sqrt{5}\right)$. With $p_0 = 0.7$, the sequence eventually alternates between numbers close to 0 and to 1, so there is no convergence.

19. **a.** Use Theorem 2.4 to show that the sequence

$$x_n = \frac{1}{2}x_{n-1} + \frac{1}{x_{n-1}}$$

converges for any $x_0 > 0$.

b. Show that if $0 < x_0 < \sqrt{2}$, then $x_1 > \sqrt{2}$.

c. Show that the sequence in (a) converges for every $x_0 > 0$.

SOLUTION: **a.** First let $g(x) = x/2 + 1/x$. For $x \neq 0$, we have $g'(x) = 1/2 - 1/x^2$. If $x > \sqrt{2}$, then $1/x^2 < 1/2$, so $g'(x) > 0$. Also, $g\left(\sqrt{2}\right) = \sqrt{2}$.

Suppose, as is the assumption given in part (a), that $x_0 > \sqrt{2}$. Then

$$x_1 - \sqrt{2} = g(x_0) - g\left(\sqrt{2}\right) = g'(\xi)\left(x_0 - \sqrt{2}\right),$$

where $\sqrt{2} < \xi < x_0$. Thus, $x_1 - \sqrt{2} > 0$ and $x_1 > \sqrt{2}$. Further,

$$x_1 = \frac{x_0}{2} + \frac{1}{x_0} < \frac{x_0}{2} + \frac{1}{\sqrt{2}} = \frac{x_0 + \sqrt{2}}{2},$$

and $\sqrt{2} < x_1 < x_0$. By an inductive argument, we have

$$\sqrt{2} < x_{m+1} < x_m < \ldots < x_0.$$

Thus, $\{x_m\}$ is a decreasing sequence that has a lower bound and must therefore converge. Suppose $p = \lim_{m \to \infty} x_m$. Then

$$p = \lim_{m \to \infty}\left(\frac{x_{m-1}}{2} + \frac{1}{x_{m-1}}\right) = \frac{p}{2} + \frac{1}{p}.$$

Thus

$$p = \frac{p}{2} + \frac{1}{p}, \quad \text{which implies that} \quad p^2 = 2,$$

so $p = \pm\sqrt{2}$. Since $x_m > \sqrt{2}$ for all m, $\lim_{m \to \infty} x_m = \sqrt{2}$.

b. Consider the situation when $0 < x_0 < \sqrt{2}$, which is the situation in part (b). Then we have

$$0 < \left(x_0 - \sqrt{2}\right)^2 = x_0^2 - 2x_0\sqrt{2} + 2,$$

so

$$2x_0\sqrt{2} < x_0^2 + 2 \quad \text{and} \quad \sqrt{2} < \frac{x_0}{2} + \frac{1}{x_0} = x_1.$$

c. To complete the problem, we consider the three possibilities for $x_0 > 0$.

Case 1: $x_0 > \sqrt{2}$, which by part (a) implies that $\lim_{m \to \infty} x_m = \sqrt{2}$.

Case 2: $x_0 = \sqrt{2}$, which implies that $x_m = \sqrt{2}$ for all m and that $\lim_{m \to \infty} x_m = \sqrt{2}$.

Case 3: $0 < x_0 < \sqrt{2}$, which implies that $\sqrt{2} < x_1$ by part (b). Thus

$$0 < x_0 < \sqrt{2} < x_{m+1} < x_m < \ldots < x_1 \quad \text{and} \quad \lim_{m \to \infty} x_m = \sqrt{2}.$$

In any situation, the sequence converges to $\sqrt{2}$, and rapidly, as we will discover in the Section 2.3.

24. Suppose that the function g has a fixed-point at p, that $g \in C[a, b]$, and that g' exists in (a, b). Show that if $|g'(p)| > 1$, then the fixed-point sequence will fail to converge for any initial choice of p_0, except if $p_n = p$ for some value of n.

SOLUTION: Since g' is continuous at p and $|g'(p)| > 1$, by letting $\epsilon = |g'(p)| - 1$ there exists a number $\delta > 0$ such that

$$|g'(x) - g'(p)| < \varepsilon = |g'(p)| - 1,$$

whenever $0 < |x - p| < \delta$. Since

$$|g'(x) - g'(p)| \geq |g'(p)| - |g'(x)|,$$

for any x satisfying $0 < |x - p| < \delta$, we have

$$|g'(x)| \geq |g'(p)| - |g'(x) - g'(p)| > |g'(p)| - (|g'(p)| - 1) = 1.$$

If p_0 is chosen so that $0 < |p - p_0| < \delta$, we have by the Mean Value Theorem that

$$|p_1 - p| = |g(p_0) - g(p)| = |g'(\xi)||p_0 - p|,$$

for some ξ between p_0 and p. Thus, $0 < |p - \xi| < \delta$ and

$$|p_1 - p| = |g'(\xi)||p_0 - p| > |p_0 - p|.$$

This means that when an approximation gets close to p, but is not equal to p, the succeeding terms of the sequence move away from p. So the sequence cannot converge to p.

Exercise Set 2.3, page 75

1. Let $f(x) = x^2 - 6$ and $p_0 = 1$. Use Newton's method to find p_2.

SOLUTION: Let $f(x) = x^2 - 6$. Then $f'(x) = 2x$, and Newton's method becomes

$$p_n = p_{n-1} - \frac{f(p_{n-1})}{f'(p_{n-1})} = p_{n-1} - \frac{p_{n-1}^2 - 6}{2p_{n-1}}.$$

With $p_0 = 1$, we have

$$p_1 = p_0 - \frac{p_0^2 - 6}{2p_0} = 1 - \frac{1 - 6}{2} = 1 + 2.5 = 3.5$$

and

$$p_2 = p_1 - \frac{p_1^2 - 6}{2p_1} = 3.5 - \frac{3.5^2 - 6}{2(3.5)} = 2.60714.$$

3. Let $f(x) = x^2 - 6$. With $p_0 = 3$ and $p_1 = 2$, find p_3 for **(a)** the Secant method and **(b)** the method of False Position.

    c. Which method gives better results?

    SOLUTION: The formula for both the Secant method and the method of False Position is

    $$p_n = p_{n-1} - \frac{f(p_{n-1})(p_{n-1} - p_{n-2})}{f(p_{n-1}) - f(p_{n-2})}.$$

    a. The Secant method:

    With $p_0 = 3$ and $p_1 = 2$, we have $f(p_0) = 9 - 6 = 3$ and $f(p_1) = 4 - 6 = -2$. The Secant method gives

    $$p_2 = p_1 - \frac{f(p_1)(p_1 - p_0)}{f(p_1) - f(p_0)} = 2 - \frac{(-2)(2-3)}{-2-3} = 2 - \frac{2}{-5} = 2.4$$

    and $f(p_2) = 2.4^2 - 6 = -0.24$. Then we have

    $$p_3 = p_2 - \frac{f(p_2)(p_2 - p_1)}{f(p_2) - f(p_1)} = 2.4 - \frac{(-0.24)(2.4-2)}{(-0.24-(-2))} = 2.4 - \frac{-0.096}{1.76} = 2.45454.$$

    b. The method of False Position:

    With $p_0 = 3$ and $p_1 = 2$, we have $f(p_0) = 3$ and $f(p_1) = -2$. As in the Secant method (part (a)), $p_2 = 2.4$ and $f(p_2) = -0.24$. Since $f(p_1) < 0$ and $f(p_2) < 0$, the method of False Position requires a reassignment of p_1. Then p_1 is changed to p_0 so that $p_1 = 3$, with $f(p_1) = 3$, and $p_2 = 2.4$, with $f(p_2) = -0.24$. We calculate p_3 by

    $$p_3 = p_2 - \frac{f(p_2)(p_2 - p_1)}{f(p_2) - f(p_1)} = 2.4 - \frac{(-0.24)(2.4-3)}{-0.24-3} = 2.4 - \frac{0.144}{-3.24} = 2.44444.$$

    c. Since $\sqrt{6} \approx 2.44949$, the accuracy of the approximations is the same. Continuing to more approximations would show that the Secant method is better.

5. **c.** Apply Newton's method to find a solution to $x - \cos x = 0$ in the interval $[0, \pi/2]$ that is accurate to within 10^{-4}.

    SOLUTION: With $f(x) = x - \cos x$, we have $f'(x) = 1 + \sin x$, and the sequence generated by Newton's method is

    $$p_n = p_{n-1} - \frac{p_{n-1} - \cos p_{n-1}}{1 + \sin p_{n-1}}.$$

    For $p_0 = 0$, we have $p_1 = 1$, $p_2 = 0.75036$, $p_3 = 0.73911$, and $p_4 = 0.73909$.

7. **c.** Apply the Secant method to find a solution to $x - \cos x = 0$ in the interval $[0, \pi/2]$ that is accurate to within 10^{-4}.

    SOLUTION: The Secant method approximations are generated by the sequence

    $$p_n = p_{n-1} - \frac{(p_{n-1} - \cos p_{n-1})(p_{n-1} - p_{n-2})}{(p_{n-1} - \cos p_{n-1}) - (p_{n-2} - \cos p_{n-2})}.$$

7)

Using the endpoints of the intervals as p_0 and p_1, we have the entries in the following table.

n	p_n
0	0
1	1.5707963
2	0.6110155
3	0.7232695
4	0.7395671
5	0.7390834
6	0.7390851

9. c. Apply the method of False Position to find a solution to $x - \cos x = 0$ in the interval $[0, \pi/2]$ that is accurate to within 10^{-4}.

SOLUTION: The method of False Position approximations are generated using this same formula as in Exercise 7, but incorporates the additional bracketing test. Using the endpoints of the intervals as p_0 and p_1, we have the entries in the following table.

n	p_n
0	0
1	1.5707963
2	0.6110155
3	0.7232695
4	0.7372659
5	0.7388778
6	0.7390615
7	0.7390825

13. Apply Newton's method to find a solution, accurate to within 10^{-4}, to the value of x that produces the closest point on the graph of $y = x^2$ to the point $(1, 0)$.

SOLUTION: The distance between an arbitrary point (x, x^2) on the graph of $y = x^2$ and the point $(1, 0)$ is

$$d(x) = \sqrt{(x - 1)^2 + (x^2 - 0)^2} = \sqrt{x^4 + x^2 - 2x + 1}.$$

Because a derivative is needed to find the critical points of d, it is easier to work with the square of this function,

$$f(x) = [d(x)]^2 = x^4 + x^2 - 2x + 1,$$

whose minimum will occur at the same value of x as the minimum of $d(x)$. To minimize $f(x)$ we need x so that $0 = f'(x) = 4x^3 + 2x - 2$.

Applying Newton's method to find the root of this equation with $p_0 = 1$ gives $p_5 = 0.589755$. The point on the graph of $y = x^2$ that is closest to $(1, 0)$ has the approximate coordinates $(0.589755, 0.347811)$.

16. Use Newton's method to solve for roots of

$$0 = \frac{1}{2} + \frac{1}{4}x^2 - x\sin x - \frac{1}{2}\cos 2x.$$

SOLUTION: Newton's method with $p_0 = \frac{\pi}{2}$ gives $p_{15} = 1.895488$ and with $p_0 = 5\pi$ gives $p_{19} = 1.895489$. With $p_0 = 10\pi$, the sequence does not converge in 200 iterations.

The results do not indicate the fast convergence usually associated with Newton's method because the function and its derivative have the same roots. As we approach a root, we are dividing by numbers with small magnitude, which increases the round-off error.

19. Explain why the iteration equation for the Secant method should not be used in the algebraically equivalent form

$$p_n = \frac{f(p_{n-1})p_{n-2} - f(p_{n-2})p_{n-1}}{f(p_{n-1}) - f(p_{n-2})}.$$

SOLUTION: This formula incorporates the subtraction of nearly equal numbers in both the numerator and denominator when p_{n-1} and p_{n-2} are nearly equal. The form given in the Secant Algorithm subtracts a correction from a result that should dominate the calculations. This is always the preferred approach.

22. Use Maple to determine how many iterations of Newton's method with $p_0 = \pi/4$ are needed to find a root of $f(x) = \cos x - x$ to within 10^{-100}.

SOLUTION: We first define $f(x)$ and $f'(x)$ with

$f := x- > cos(x) - x$

$$f := x \to \cos(x) - x$$

and

$fp := x- > (D)(f)(x)$

$$fp := x \to -\sin(x) - 1$$

We wish to use 100-digit rounding arithmetic so we set

$Digits := 100; p0 := Pi/4$

$$Digits := 100$$

$$p0 := \frac{1}{4}\pi$$

for n from 1 *to* 7 *do*

$p1 := evalf(p0 - f(p0)/fp(p0))$

$err := abs(p1 - p0)$

$p0 := p1$

od

This gives

$$p_7 = .7390851332151606416553120876738734040134117589007574649656806357732846548835475945993761069317665319,$$

which is accurate to 10^{-100}.

23. The function defined by $f(x) = \ln\left(x^2 + 1\right) - e^{0.4x} \cos \pi x$ has an infinite number of zeros.

    a. Approximate the only negative zero to within 10^{-6}.

    b. Approximate the four smallest positive zeros to within 10^{-6}.

    c. Find an initial approximation for the nth smallest positive zero.

    d. Approximate the 25th smallest positive zero to within 10^{-6}.

    SOLUTION: The key to this problem is recognizing the behavior of $e^{0.4x}$. When x is negative, this term goes to zero, so $f(x)$ is dominated by $\ln\left(x^2 + 1\right)$. However, when x is positive, $e^{0.4x}$ dominates the calculations, and $f(x)$ will be zero approximately when this term makes no contribution; that is, when $\cos \pi x = 0$. This occurs when $x = n/2$ for a positive integer n. Using this information to determine initial approximations produces the following results:

    a. We can use $p_0 = -0.5$ to find the sufficiently accurate $p_3 = -0.4341431$.

    b. We can use: $p_0 = 0.5$ to give $p_3 = 0.4506567$; $p_0 = 1.5$ to give $p_3 = 1.7447381$; $p_0 = 2.5$ to give $p_5 = 2.2383198$; and $p_0 = 3.5$ to give $p_4 = 3.7090412$.

    c. In general, a reasonable initial approximation for the nth positive root is $n - 0.5$.

    d. Let $p_0 = 24.5$. A sufficiently accurate approximation to the 25th smallest positive zero is $p_2 = 24.4998870$.

    Graphs for various parts of the region are shown below.

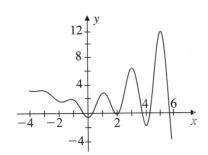 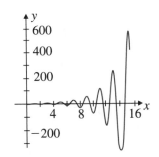 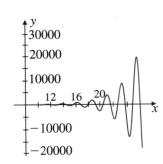

26. Determine the minimal annual interest rate i at which an amount $P = \$1500$ per month can be invested to accumulate an amount $A = \$750,000$ at the end of 20 years based on the annuity due equation

$$A = \frac{P}{i}\left[(1 + i)^n - 1\right].$$

SOLUTION: This is simply a root-finding problem where the function is given by

$$f(i) = A - \frac{P}{i}[(1 + i)^n - 1] = 750000 - \frac{1500}{(i/12)}\left[(1 + i/12)^{(12)(20)} - 1\right].$$

Notice that n and i have been adjusted because the payments are made monthly rather than yearly. The approximate solution to this equation can be found by any method in this section. Newton's method is a bit cumbersome for this problem, since the derivative of f is complicated. The Secant method would be a likely choice. The minimal annual interest is approximately 6.67%.

28. A drug administered to a patient produces a concentration in the blood stream given by $c(t) = Ate^{-t/3}$ mg/mL, t hours after A units have been administered. The maximum safe concentration is 1 mg/mL.

a. What amount should be injected to reach this safe level, and when does this occur?

b. When should an additional amount be administered, if it is administered when the level drops to 0.25 mg/mL?

c. Assuming 75% of the original amount is administered in the second injection, when should a third injection be given?

SOLUTION: **a.** The maximum concentration occurs when

$$0 = c'(t) = A\left(1 - \frac{t}{3}\right)e^{-t/3}.$$

This happens when $t = 3$ hours, and since the concentration at this time will be $c(3) = 3Ae^{-1}$, we need to administer $A = \frac{1}{3}e$ units.

b. We need to determine t so that

$$0.25 = c(t) = \left(\frac{1}{3}e\right)te^{-t/3}.$$

This occurs when t is 11 hours and 5 minutes; that is, when $t = 11.08\overline{3}$ hours.

c. We need to find t so that

$$0.25 = c(t) = \left(\frac{1}{3}e\right)te^{-t/3} + 0.75\left(\frac{1}{3}e\right)(t - 11.08\overline{3})e^{-(t-11.08\overline{3})/3}.$$

This occurs after 21 hours and 14 minutes.

29. Let $f(x) = 3^{3x+1} - 7 \cdot 5^{2x}$.

a. Use the Maple commands *solve* and *fsolve* to try to find all roots of f.

b. Plot $f(x)$ to find initial approximations to roots of f.

c. Use Newton's method to find the zeros of f to within 10^{-16}.

d. Find the exact solutions of $f(x) = 0$ algebraically.

SOLUTION: **a.** First define the function by

$f := x-> 3^{3x+1} - 7 \cdot 5^{2x}$

$$f := x \to 3^{(3x+1)} - 7\,5^{2x}$$

$solve(f(x) = 0, x)$

$$-\frac{\ln(3/7)}{\ln(27/25)}$$

$fsolve(f(x) = 0, x)$

$$\text{fsolve}(3^{(3x+1)} - 7\,5^{(2x)} = 0, x)$$

The procedure *solve* gives the exact solution, and *fsolve* fails because the negative x-axis is an asymptote for the graph of $f(x)$.

b. Using the Maple command $plot(\{f(x)\}, x = 9.5..11.5)$ produces the following graph.

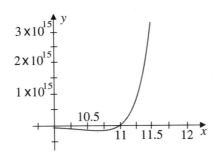

c. Define $f'(x)$ using

$fp := x- > (D)(f)(x)$

$$fp := x \to 3\,3^{(3x+1)} \ln(3) - 14\,5^{(2x)} \ln(5)$$

$Digits := 18; p0 := 11$

$$Digits := 18$$
$$p0 := 11$$

for i *from* 1 *to* 5 *do*

$p1 := evalf(p0 - f(p0)/fp(p0))$

$err := abs(p1 - p0)$

$p0 := p1$

od

The results are given in the following table.

i	p_i	$|p_i - p_{i-1}|$
1	11.0097380401552503	0.0097380401552503
2	11.0094389359662827	0.0002991041889676
3	11.0094386442684488	$0.2916978339\ 10^{-6}$
4	11.0094386442681716	$0.2772\ 10^{-12}$
5	11.0094386442681716	0

d. We have $3^{3x+1} = 7 \cdot 5^{2x}$. Taking the natural logarithm of both sides gives

$$(3x + 1)\ln 3 = \ln 7 + 2x \ln 5.$$

Thus

$$3x \ln 3 - 2x \ln 5 = \ln 7 - \ln 3, \quad x(3\ln 3 - 2\ln 5) = \ln\frac{7}{3},$$

and

$$x = \frac{\ln 7/3}{\ln 27 - \ln 25} = \frac{\ln 7/3}{\ln 27/25} = -\frac{\ln 3/7}{\ln 27/25}.$$

This agrees with part (a).

Exercise Set 2.4, page 85

1. a. Use Newton's method to find a solution accurate to within 10^{-5} for $x^2 - 2xe^{-x} + e^{-2x} = 0$, where $0 \le x \le 1$.

SOLUTION: Since

$$f(x) = x^2 - 2xe^{-x} + e^{-2x} \quad \text{and} \quad f'(x) = 2x - 2e^{-x} + 2xe^{-x} - 2e^{-2x},$$

the iteration formula is

$$p_n = p_{n-1} - \frac{f(p_{n-1})}{f'(p_{n-1})} = p_{n-1} - \frac{p_{n-1}^2 - 2p_{n-1}e^{-p_{n-1}} + e^{-2p_{n-1}}}{2p_{n-1} - 2e^{-p_{n-1}} + 2p_{n-1}e^{-p_{n-1}} - 2e^{-2p_{n-1}}}.$$

With $p_0 = 0.5$, we have

$$p_1 = 0.5 - (0.01134878)/(-0.3422895) = 0.5331555.$$

Continuing in this manner, $p_{13} = 0.567135$ is accurate to within 10^{-5}.

3. a. Repeat Exercise 1(a) using the modified Newton-Raphson method described in Eq. (2.13). Is there an improvement in speed or accuracy over Exercise 1?

SOLUTION: Since

$$f(x) = x^2 - 2xe^{-x} + e^{-2x},$$
$$f'(x) = 2x - 2e^{-x} + 2xe^{-x} - 2e^{-2x},$$

and

$$f''(x) = 2 + 4e^{-x} - 2xe^{-x} + 4e^{-2x},$$

the iteration formula is

$$p_n = p_{n-1} - \frac{f(p_{n-1})f'(p_{n-1})}{[f'(p_{n-1})]^2 - f(p_{n-1})f''(p_{n-1})}.$$

With $p_0 = 0.5$, we have $f(p_0) = 0.011348781$, $f'(p_0) = -0.342289542$, $f''(p_0) = 5.291109744$ and

$$p_1 = 0.5 - \frac{(0.01134878)(-0.342289542)}{(-0.342289542)^2 - (0.011348781)(5.291109744)} = 0.5680137.$$

Continuing in this manner, $p_3 = 0.567143$ is accurate to within 10^{-5}, which is considerably better than in Exercise 1.

6. a. Show that the sequence $p_n = 1/n$ converges linearly to $p = 0$, and determine the number of terms required to have $|p_n - p| < 5 \times 10^{-2}$.

SOLUTION: First note that $\displaystyle \lim_{n \to \infty} \frac{1}{n} = 0$. Since

$$\lim_{n \to \infty} \frac{|p_{n+1} - p|}{|p_n - p|} = \lim_{n \to \infty} \frac{1/(n+1)}{1/n} = \lim_{n \to \infty} \frac{n}{n+1} = 1,$$

the convergence is linear. To have $|p_n - p| < 5 \times 10^{-2}$, we need $1/n < 0.05$, which implies that $n > 20$.

8. Show that:

a. The sequence $p_n = 10^{-2^n}$ converges quadratically to zero;

b. The sequence $p_n = 10^{-n^k}$ does not converge to zero quadratically, regardless of the size of $k > 1$.

SOLUTION:

a. Since

$$\lim_{n \to \infty} \frac{|p_{n+1} - 0|}{|p_n - 0|^2} = \lim_{n \to \infty} \frac{10^{-2^{n+1}}}{(10^{-2^n})^2} = \lim_{n \to \infty} \frac{10^{-2^{n+1}}}{10^{-2 \cdot 2^n}} = \lim_{n \to \infty} \frac{10^{-2^{n+1}}}{10^{-2^{n+1}}} = 1,$$

the sequence is quadratically convergent.

b. For any $k > 1$,

$$\lim_{n \to \infty} \frac{|p_{n+1} - 0|}{|p_n - 0|^2} = \lim_{n \to \infty} \frac{10^{-(n+1)^k}}{\left(10^{-n^k}\right)^2} = \lim_{n \to \infty} \frac{10^{-(n+1)^k}}{10^{-2n^k}} = \lim_{n \to \infty} 10^{2n^k - (n+1)^k}$$

diverges. So the sequence $p_n = 10^{-n^k}$ does not converge quadratically for any $k > 1$.

10. Show that the fixed-point method

$$g(x) = x - \frac{mf(x)}{f'(x)}$$

has $g'(p) = 0$, if p is a zero of f of multiplicity m.

SOLUTION: If f has a zero of multiplicity m at p, then a function q exists with

$$f(x) = (x - p)^m q(x), \quad \text{where} \quad \lim_{x \to p} q(x) \neq 0.$$

Since

$$f'(x) = m(x - p)^{m-1} q(x) + (x - p)^m q'(x),$$

we have

$$g(x) = x - \frac{mf(x)}{f'(x)} = x - \frac{m(x - p)^m q(x)}{m(x - p)^{m-1} q(x) + (x - p)^m q'(x)},$$

which reduces to

$$g(x) = x - \frac{m(x - p)q(x)}{mq(x) + (x - p)q'(x)}.$$

Differentiating this expression and evaluating at $x = p$ gives

$$g'(p) = 1 - \frac{mq(p)[mq(p)]}{[mq(p)]^2} = 0.$$

If f''' is continuous, Theorem 2.9 implies that this sequence produces quadratic convergence once we are close enough to the solution p.

12. Suppose that f has m continuous derivatives. Show that f has a zero of multiplicity m at p if and only if

$$0 = f(p) = f'(p) = \cdots = f^{(m-1)}(p), \quad \text{but} \quad f^{(m)}(p) \neq 0.$$

SOLUTION: If f has a zero of multiplicity m at p, then f can be written as

$$f(x) = (x - p)^m q(x), \quad \text{for } x \neq p, \text{ where } \lim_{x \to p} q(x) \neq 0.$$

Thus

$$f'(x) = m(x - p)^{m-1} q(x) + (x - p)^m q'(x)$$

and $f'(p) = 0$. Also

$$f''(x) = m(m - 1)(x - p)^{m-2} q(x) + 2m(x - p)^{m-1} q'(x) + (x - p)^m q''(x)$$

and $f''(p) = 0$.

In general, for $k \leq m$,

$$f^{(k)}(x) = \sum_{j=0}^{k} \binom{k}{j} \frac{d^j (x - p)^m}{dx^j} q^{(k-j)}(x)$$

$$= \sum_{j=0}^{k} \binom{k}{j} m(m - 1) \cdots (m - j + 1)(x - p)^{m-j} q^{(k-j)}(x).$$

Thus, for $0 \leq k \leq m - 1$, we have $f^{(k)}(p) = 0$, but

$$f^{(m)}(p) = m! \lim_{x \to p} q(x) \neq 0.$$

Conversely, suppose that $f(p) = f'(p) = \ldots = f^{(m-1)}(p) = 0$ and $f^{(m)}(p) \neq 0$. Consider the $(m - 1)$th Taylor polynomial of f expanded about p :

$$f(x) = f(p) + f'(p)(x - p) + \ldots + \frac{f^{(m-1)}(p)(x - p)^{m-1}}{(m - 1)!} + \frac{f^{(m)}(\xi(x))(x - p)^m}{m!}$$

$$= (x - p)^m \frac{f^{(m)}(\xi(x))}{m!},$$

where $\xi(x)$ is between x and p. Since $f^{(m)}$ is continuous, let

$$q(x) = \frac{f^{(m)}(\xi(x))}{m!}.$$

Then $f(x) = (x - p)^m q(x)$ and

$$\lim_{x \to p} q(x) = \frac{f^{(m)}(p)}{m!} \neq 0.$$

So p is a zero of multiplicity m.

14. Show that the Secant method converges of order α, where $\alpha = \left(1 + \sqrt{5}\right)/2$, the golden ratio.

SOLUTION: Let $e_n = p_n - p$. If

$$\lim_{n \to \infty} \frac{|e_{n+1}|}{|e_n|^\alpha} = \lambda > 0,$$

then for sufficiently large values of n, $|e_{n+1}| \approx \lambda |e_n|^\alpha$. Thus

$$|e_n| \approx \lambda |e_{n-1}|^\alpha \quad \text{and} \quad |e_{n-1}| \approx \lambda^{-1/\alpha} |e_n|^{1/\alpha}.$$

The hypothesis that for some constant C and sufficiently large n we have $|p_{n+1} - p| \approx C|p_n - p|\,|p_{n-1} - p|$, gives

$$\lambda|e_n|^\alpha \approx C|e_n|\lambda^{-1/\alpha}|e_n|^{1/\alpha}, \quad \text{so} \quad |e_n|^\alpha \approx C\lambda^{-1/\alpha-1}|e_n|^{1+1/\alpha}.$$

Since the powers of $|e_n|$ must agree,

$$\alpha = 1 + 1/\alpha \quad \text{and} \quad \alpha = \frac{1 + \sqrt{5}}{2}.$$

This number, the *Golden Ratio*, appears in numerous situations in mathematics and in art.

Exercise Set 2.5, page 90

2. Apply Newton's method to approximate a root of

$$f(x) = e^{6x} + 3(\ln 2)^2 e^{2x} - \ln 8 e^{4x} - (\ln 2)^3 = 0.$$

Generate terms until $|p_{n+1} - p_n| < 0.0002$, and construct the Aitken's Δ^2 sequence $\{\hat{p}_n\}$.

SOLUTION: Applying Newton's method with $p_0 = 0$ requires finding $p_{16} = -0.182888$. For the Aitken's Δ^2 sequence, we have sufficient accuracy with $\hat{p}_6 = -0.183387$. Newton's method fails to converge quadratically because there is a multiple root.

3. Let $g(x) = \cos(x - 1)$ and $p_0^{(0)} = 2$. Use Steffensen's method to find $p_0^{(1)}$.

SOLUTION: With $g(x) = \cos(x - 1)$ and $p_0^{(0)} = 2$, we have

$$p_1^{(0)} = g\left(p_0^{(0)}\right) = \cos(2 - 1) = \cos 1 = 0.5403023$$

and

$$p_2^{(0)} = g\left(p_1^{(0)}\right) = \cos(0.5403023 - 1) = 0.8961867.$$

Thus

$$p_0^{(1)} = p_0^{(0)} - \frac{\left(p_1^{(0)} - p_0^{(0)}\right)^2}{p_2^{(0)} - 2p_1^{(0)} - 2p_1^{(0)} + p_0^{(0)}} \qquad \text{Aitken's } \Delta^2 \text{ method}$$

$$= 2 - \frac{(0.5403023 - 2)^2}{0.8961867 - 2(0.5403023) + 2} = 2 - 1.173573 = 0.826427.$$

5. Steffensen's method is applied to a function $g(x)$ using $p_0^{(0)} = 1$ and $p_2^{(0)} = 3$ to obtain $p_0^{(1)} = 0.75$. What could $p_1^{(0)}$ be?

SOLUTION: Steffensen's method uses the formula

$$p_1^{(0)} = p_0^{(0)} - \frac{\left(p_1^{(0)} - p_0^{(0)}\right)^2}{p_2^{(0)} - 2p_1^{(0)} + p_0^{(0)}}.$$

Substituting for $p_0^{(0)}$, $p_2^{(0)}$, and $p_0^{(1)}$ gives

$$0.75 = 1 - \frac{\left(p_1^{(0)} - 1\right)^2}{3 - 2p_1^{(0)} + 1}, \quad \text{that is,} \quad 0.25 = \frac{\left(p_1^{(0)} - 1\right)^2}{4 - 2p_1^{(0)}}.$$

Thus

$$1 - \frac{1}{2}p_1^{(0)} = \left(p_1^{(0)}\right)^2 - 2p_1^{(0)} + 1, \quad \text{so} \quad 0 = \left(p_1^{(0)}\right)^2 - 1.5p_1^{(0)},$$

and $p_1^{(0)} = 1.5$ or $p_1^{(0)} = 0$.

11. b. Use Steffensen's method to approximate the solution to within 10^{-5} of $x = 0.5(\sin x + \cos x)$, where $g(x) = 0.5(\sin x + \cos x)$.

SOLUTION: With $g(x) = 0.5(\sin x + \cos x)$, we have

$$p_0^{(0)} = 0, \ p_1^{(0)} = g(0) = 0.5,$$

$$p_2^{(0)} = g(0.5) = 0.5(\sin 0.5 + \cos 0.5) = 0.678504051,$$

$$p_0^{(1)} = p_0^{(0)} - \frac{\left(p_1^{(0)} - p_0^{(0)}\right)^2}{p_2^{(0)} - 2p_1^{(0)} + p_0^{(0)}} = 0.777614774,$$

$$p_1^{(1)} = g\left(p_0^{(1)}\right) = 0.707085363,$$

$$p_2^{(1)} = g\left(p_1^{(1)}\right) = 0.704939584,$$

$$p_0^{(2)} = p_0^{(1)} - \frac{\left(p_1^{(1)} - p_0^{(1)}\right)^2}{p_2^{(1)} - 2p_1^{(1)} + p_0^{(1)}} = 0.704872252,$$

$$p_1^{(2)} = g\left(p_0^{(2)}\right) = 0.704815431,$$

$$p_2^{(2)} = g\left(p_1^{(2)}\right) = 0.704812197,$$

$$p_0^{(3)} = p_0^{(2)} - \frac{\left(p_1^{(2)} - p_0^{(2)}\right)^2}{p_2^{(2)} - 2p_1^{(2)} + p_0^{(2)}} = 0.704812002,$$

$$p_1^{(3)} = g\left(p_0^{(3)}\right) = 0.704812002,$$

and

$$p_2^{(3)} = g\left(p_1^{(3)}\right) = 0.704812197.$$

Since $p_2^{(3)}$, $p_1^{(3)}$, and $p_0^{(3)}$ all agree to within 10^{-5}, we accept $p_2^{(3)} = 0.704812197$ as an answer that is accurate to within 10^{-5}.

14. a. Show that a sequence $\{p_n\}$ that converges to p with order $\alpha > 1$ converges superlinearly to p.

b. Show that the sequence $p_n = \dfrac{1}{n^n}$ converges superlinearly to 0, but does not converge of order α for any $\alpha > 1$.

SOLUTION: Since $\{p_n\}$ converges to p with order $\alpha > 1$, a positive constant λ exists with

$$\lambda = \lim_{n \to \infty} \frac{|p_{n+1} - p|}{|p_n - p|^\alpha}.$$

Hence

$$\lim_{n\to\infty}\left|\frac{p_{n+1}-p}{p_n-p}\right| = \lim_{n\to\infty}\frac{|p_{n+1}-p|}{|p_n-p|^\alpha}\cdot|p_n-p|^{\alpha-1} = \lambda\cdot 0 = 0 \quad\text{and}\quad \lim_{n\to\infty}\frac{p_{n+1}-p}{p_n-p} = 0.$$

This implies that $\{p_n\}$ that converges superlinearly to p.

b. The sequence converges $p_n = \dfrac{1}{n^n}$ superlinearly to zero because

$$\lim_{n\to\infty}\frac{1/(n+1)^{(n+1)}}{1/n^n} = \lim_{n\to\infty}\frac{n^n}{(n+1)^{(n+1)}}$$

$$= \lim_{n\to\infty}\left(\frac{n}{n+1}\right)^n\frac{1}{n+1}$$

$$= \lim_{n\to\infty}\left(\frac{1}{(1+1/n)^n}\right)\frac{1}{n+1} = \frac{1}{e}\cdot 0 = 0.$$

However, for $\alpha > 1$, we have

$$\lim_{n\to\infty}\frac{1/(n+1)^{(n+1)}}{(1/n^n)^\alpha} = \lim_{n\to\infty}\frac{n^{\alpha n}}{(n+1)^{(n+1)}}$$

$$= \lim_{n\to\infty}\left(\frac{n}{n+1}\right)^n\frac{n^{(\alpha-1)n}}{n+1}$$

$$= \lim_{n\to\infty}\left(\frac{1}{(1+1/n)^n}\right)\lim_{n\to\infty}\frac{n^{(\alpha-1)n}}{n+1} = \frac{1}{e}\cdot\infty = \infty.$$

So the sequence does not converge of order α for any $\alpha > 1$.

17. Let $P_n(x)$ be the nth Taylor polynomial for $f(x) = e^x$ expanded about $x_0 = 0$.

a. For fixed x, show that $p_n = P_n(x)$ satisfies the hypotheses of Theorem 2.14.

b. Let $x = 1$, and use Aitken's Δ^2 method to generate the sequence $\hat{p}_0, \hat{p}_1, \ldots, \hat{p}_8$.

c. Does Aitken's Δ^2 method accelerate the convergence in this situation?

SOLUTION: **a.** Since

$$p_n = P_n(x) = \sum_{k=0}^{n}\frac{1}{k!}x^k,$$

we have

$$p_n - p = P_n(x) - e^x = \frac{-e^\xi}{(n+1)!}x^{n+1},$$

where ξ is between 0 and x. Thus, $p_n - p \neq 0$, for all $n \geq 0$. Further,

$$\frac{p_{n+1}-p}{p_n-p} = \frac{\frac{-e^{\xi_1}}{(n+2)!}x^{n+2}}{\frac{-e^\xi}{(n+1)!}x^{n+1}} = \frac{e^{(\xi_1-\xi)}x}{n+2},$$

where ξ_1 is between 0 and 1. Thus

$$\lambda = \lim_{n\to\infty}\frac{e^{(\xi_1-\xi)}x}{n+2} = 0 < 1.$$

b. The sequence has the terms shown in the following tables.

n	0	1	2	3	4	5	6
p_n	1	2	2.5	$2.\overline{6}$	$2.708\overline{3}$	$2.71\overline{6}$	$2.71805\overline{5}$
$\hat{p}_n$	3	2.75	$2.7\overline{2}$	2.71875	$2.718\overline{3}$	2.7182870	2.7182823

n	7	8	9	10
p_n	2.7182539	2.7182787	2.7182815	2.7182818
$\hat{p}_n$	2.7182818	2.7182818		

c. Aitken's Δ^2 method gives quite an improvement for this problem. For example, $\hat{p}_6$ is accurate to within 5×10^{-7}. We need p_{10} to have this accuracy.

Exercise Set 2.6, page 100

2. b. Use Newton's method to approximate, to within 10^{-5}, the real zeros of

$$P(x) = x^4 - 2x^3 - 12x^2 + 16x - 40.$$

Then reduce the polynomial to lower degree, and determine any complex zeros.

SOLUTION: Applying Newton's method with $p_0 = 1$ gives the sufficiently accurate approximation $p_7 = -3.548233$. When $p_0 = 4$, we find another zero to be $p_5 = 4.381113$. If we divide $P(x)$ by

$$(x + 3.548233)(x - 4.381113) = x^2 - 0.832880x - 15.54521,$$

we find that

$$P(x) \approx \left(x^2 - 0.832880x - 15.54521\right)\left(x^2 - 1.16712x + 2.57315\right).$$

The complex roots of the quadratic on the right can be found by the quadratic formula and are approximately $0.58356 \pm 1.49419i$.

4. b. Use Müller's method to find the real and complex zeros of

$$P(x) = x^4 - 2x^3 - 12x^2 + 16x - 40.$$

SOLUTION: The following table lists the initial approximation and the roots. The first initial approximation was used because $f(0) = -40$, $f(1) = -37$, and $f(2) = -56$ implies that there is a minimum in $[0, 2]$. This is confirmed by the complex roots that are generated.

The second initial approximations are used to find the real root that is known to lie between 4 and 5, due to the fact that $f(4) = -40$ and $f(5) = 115$.

The third initial approximations are used to find the real root that is known to lie between -3 and -4, since $f(-3) = -61$ and $f(-4) = 88$.

p_0	p_1	p_2	Approximated Roots	Complex Conjugate Root
0	1	2	$p_7 = 0.583560 - 1.494188i$	$0.583560 + 1.494188i$
2	3	4	$p_6 = 4.381113$	
-2	-3	-4	$p_5 = -3.548233$	

5. b. Find the zeros and critical points of

$$f(x) = x^4 - 2x^3 - 5x^2 + 12x - 5,$$

and use this information to sketch the graph of f.

SOLUTION: There are at most four real zeros of f and $f(0) < 0$, $f(1) > 0$, and $f(2) < 0$. This, together with the fact that $\lim_{x \to \infty} f(x) = \infty$ and $\lim_{x \to -\infty} f(x) = \infty$, implies that these zeros lie in the intervals $(-\infty, 0)$, $(0, 1)$, $(1, 2)$, and $(2, \infty)$. Applying Newton's method for various initial approximations in these intervals gives the approximate zeros: 0.5798, 1.521, 2.332, and -2.432. To find the critical points, we need the zeros of

$$f'(x) = 4x^3 - 6x^2 - 10x + 12.$$

Since $x = 1$ is quite easily seen to be a zero of $f'(x)$, the cubic equation can be reduced to a quadratic to find the other two zeros: 2 and -1.5.

Since the quadratic formula applied to

$$0 = f''(x) = 12x^2 - 12x - 10$$

gives $x = 0.5 \pm \left(\sqrt{39}/6 \right)$, we also have the points of inflection.

A sketch of the graph of f is given below.

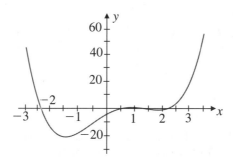

9. Find a solution, accurate to within 10^{-4}, to the problem

$$600x^4 - 550x^3 + 200x^2 - 20x - 1 = 0, \quad \text{for } 0.1 \le x \le 1$$

by using the various methods in this chapter.

SOLUTION:

a. Bisection method: For $p_0 = 0.1$ and $p_1 = 1$, we have $p_{14} = 0.23233$.

b. Newton's method: For $p_0 = 0.55$, we have $p_6 = 0.23235$.

c. Secant method: For $p_0 = 0.1$ and $p_1 = 1$, we have $p_8 = 0.23235$.

d. Method of False Position: For $p_0 = 0.1$ and $p_1 = 1$, we have $p_{88} = 0.23025$.

e. Müller's method: For $p_0 = 0$, $p_1 = 0.25$, and $p_2 = 1$, we have $p_6 = 0.23235$.

Notice that the method of False Position for this problem was considerably less effective than both the Secant method and the Bisection method.

11. A can in the shape of a right circular cylinder must have a volume of 1000 cm^3. To form seals, the top and bottom must have a radius 0.25 cm more than the radius and the material for the side must be 0.25 cm longer than the circumference of the can. Minimize the amount of material that is required.

SOLUTION: Since the volume is given by

$$V = 1000 = \pi r^2 h,$$

we have $h = 1000/ \left(\pi r^2\right)$. The amount of material required for the top of the can is $\pi(r + 0.25)^2$, and a similar amount is needed for the bottom. To construct the side of the can, the material needed is $(2\pi r + 0.25)h$. The total amount of material $M(r)$ is given by

$$M(r) = 2\pi(r + 0.25)^2 + (2\pi r + 0.25)h = 2\pi(r + 0.25)^2 + 2000/r + 250/\pi r^2.$$

Thus

$$M'(r) = 4\pi(r + 0.25) - 2000/r^2 - 500/(\pi r^3).$$

Solving $M'(r) = 0$ for r gives $r \approx 5.363858$. Evaluating $M(r)$ at this value of r gives the minimal material needed to construct the can:

$$M(5.363858) \approx 573.649 \text{ cm}^2.$$

12. Leonardo of Pisa (Fibonacci) found the base 60 approximation

$$1 + 22 \left(\frac{1}{60}\right) + 7 \left(\frac{1}{60}\right)^2 + 42 \left(\frac{1}{60}\right)^3 + 33 \left(\frac{1}{60}\right)^4 + 4 \left(\frac{1}{60}\right)^5 + 40 \left(\frac{1}{60}\right)^6$$

as a root of the equation

$$x^3 + 2x^2 + 10x = 20.$$

How accurate was his approximation?

SOLUTION: The decimal equivalent of Fibonacci's base 60 approximation is 1.3688081078532, and Newton's Method gives 1.36880810782137 with a tolerance of 10^{-16}. So Fibonacci's answer was correct to within 3.2×10^{-11}. This is the most accurate approximation to an irrational root of a cubic polynomial that is known to exist, at least in Europe, before the sixteenth century. Fibonacci probably learned the technique for approximating this root from the writings of the great Persian poet and mathematician Omar Khayyám.

Interpolation and Polynomial Approximation

Exercise Set 3.1, page 114

2. a. For the given functions $f(x) = \sin \pi x$, let $x_0 = 1$, $x_1 = 1.25$, and $x_2 = 1.6$. Construct interpolation polynomials of degree at most one and at most two to approximate $f(1.4)$, and find the absolute error.

SOLUTION: We have

$$P_1(x) = \frac{(x - 1.6)}{(1.25 - 1.6)} f(1.25) + \frac{(x - 1.25)}{(1.6 - 1.25)} f(1.6) = -0.6969992408x + 0.1641422691,$$

with $P_1(1.4) = -0.811656668$ and $|f(1.4) - P_1(1.45)| = 0.139399849$;

$$P_2(x) = \frac{(x - x_1)(x - x_2)}{(x_0 - x_1)(x_0 - x_2)} f(x_0) + \frac{(x - x_0)(x - x_2)}{(x_1 - x_0)(x_1 - x_2)} f(x_1) + \frac{(x - x_0)(x - x_1)}{(x_2 - x_0)(x_2 - x_1)} f(x_2)$$
$$= 3.552379809x^2 - 10.82128170x + 7.2689018882$$

with $P_2(1.4) = -0.9182280623$ and $|f(1.4) - P_2(1.4)| = 0.328284543 \times 10^{-1}$.

4. a. Use Theorem 3.3 to find an error bound for the approximations in Exercise 2(a).

SOLUTION: For the linear polynomial we use $x_1 = 1.25$ and $x_2 = 1.6$.

$$|f(1.4) - P_1(1.4)| = \left| \frac{f''(\xi)}{2!} \right| |x - x_1||x - x_2| = 0.5| - \pi^2 \sin(\pi\xi)||1.4 - 1.25||1.4 - 1.6|$$
$$\leq 0.5\pi^2(0.15)(0.2) \leq 0.14804407.$$

For the second degree polynomial we use $x_1 = 1.25$, $x_2 = 1.6$ and $x_0 = 1.0$.

$$|f(1.4) - P_2(1.4)| = \left| \frac{f^{(3)}(\xi)}{3!} \right| |x - x_1||x - x_2||x_0 - x|$$
$$= \frac{1}{6}| - \pi^3 \cos(\pi\xi)||1.4 - 1.25||1.4 - 1.6||1.4 - 1.0| \leq 0.062012553.$$

5. a. Use appropriate Lagrange interpolating polynomials of degrees one, two, and three to approximate $f(8.4)$ if $f(8.1) = 16.94410$, $f(8.3) = 17.56492$, $f(8.6) = 18.50515$, and $f(8.7) = 18.82091$

SOLUTION: With $x_0 = 8.1$, $y_0 = 16.94410$, $x_1 = 8.3$, $y_1 = 17.56492$, $x_2 = 8.6$, $y_2 = 18.50515$, $x_3 = 8.73$ and $y_3 = 18.82091$ we have

$$P_1(x) = 3.1341x - 8.44811, \quad P_2(x) = 0.06x^2 + 2.1201x - 4.16531$$

and

$$P_3(x) = -0.002083333334x^3 + 0.1120833334x^2 + 1.686204167x - 2.96077250$$

These interpolation polynomials give the following results.

n	$x_0, x_1, \ldots, x_n$	$P_n(8.4)$
1	8.3, 8.6	17.87833
2	8.3, 8.6, 8.7	17.87716
3	8.3, 8.6, 8.7, 8.1	17.87714

7. a. The data for Exercise 5(a) were generated using $f(x) = x \ln x$. Use the error formula to find a bound for the error, and compare the bound to the actual error for the cases $n = 1$ and $n = 2$.

SOLUTION: For the linear polynomial we use $x_1 = 8.3$ and $x_2 = 8.6$.

$$|f(8.4) - P_1(8.4)| = \left| \frac{f''(\xi)}{2!} \right| |x - x_1||x - x_2| = 0.5 \left| \frac{1}{\xi} \right| |8.4 - 8.3||8.4 - 8.6| \leq 0.5 \frac{1}{8.3}(0.1)(0.2) \leq 0.001204819$$

For the second degree polynomial we use $x_1 = 8.3$, $x_2 = 8.6$ and $x_0 = 8.1$.

$$|f(8.4) - P_2(8.4)| = \left| \frac{f^{(3)}(\xi)}{3!} \right| |x - x_1||x - x_2||x_0 - x|$$

$$= \frac{1}{6} \left| -\frac{1}{\xi^2} \right| |8.4 - 8.3||8.4 - 8.6||8.4 - 8.1| \leq 01.5241579 \times 10^{-5}.$$

The actual errors are

$$|f(8.4) - P_1(8.4)| = 0.0011836 \quad \text{and} \quad |f(8.4) - P_2(8.4)| = 0.00001367.$$

9. Let $P_3(x)$ be the interpolating polynomial for the data $(0, 0)$, $(0.5, y)$, $(1, 3)$, and $(2, 2)$. Find y if the coefficient of x^3 in $P_3(x)$ is 6.

SOLUTION: Solving for $P_3(x)$ gives

$$P_3(x) = \frac{(x - 0.5)(x - 1)(x - 2)}{(0 - 0.5)(0 - 1)(0 - 2)} \cdot 0 + \frac{(x - 0)(x - 1)(x - 2)}{(0.5 - 0)(0.5 - 1)(0.5 - 2)} \cdot y + \frac{(x - 0)(x - 0.5)(x - 2)}{(1 - 0)(1 - 0.5)(1 - 2)} \cdot 3$$

$$+ \frac{(x - 0)(x - 0.5)(x - 1)}{(2 - 0)(2 - 0.5)(2 - 1)} \cdot 2$$

$$= \frac{8}{3}x(x - 1)(x - 2)y - 6x(x - 0.5)(x - 2) + \frac{2}{3}x(x - 0.5)(x - 1)$$

$$= \left(\frac{8}{3}y - 6 + \frac{2}{3} \right) x^3 + \cdots.$$

Since the coefficient of x^3 is given to be 6, we have $6 = \frac{8}{3}y - 6 + \frac{2}{3}$, so $\frac{34}{3} = \frac{8}{3}y$ and $y = 4.25$.

11. Use the following values and four-digit rounding arithmetic to construct a third Lagrange polynomial approximation to $f(1.09)$. The function being approximated is $f(x) = \log_{10}(\tan x)$. Use this knowledge to find a bound for the error in the approximation.

$$f(1.00) = 0.1924, \quad f(1.05) = 0.2414, \quad f(1.10) = 0.2933, \quad f(1.15) = 0.3492$$

SOLUTION: The Lagrange coefficient polynomials are

$$L_0(x) = -1333(x - 1.05)(x - 1.10)(x - 1.15),$$
$$L_1(x) = 4000(x - 1.00)(x - 1.10)(x - 1.15),$$
$$L_2(x) = -4000(x - 1.00)(x - 1.05)(x - 1.15),$$
$$L_3(x) = 1333(x - 1.00)(x - 1.05)(x - 1.10),$$

so

$$L_0(1.4) = -0.03199, \quad L_1(1.4) = 0.2160, \quad L_2(1.4) = 0.8640, \quad \text{and} \quad L_3(1.4) = -0.04799.$$

This gives

$$P_3(1.4) = -0.03199(0.1924) + 0.2160(0.2414) + 0.8640(0.2933) - 0.04799(0.3492) = 0.2826.$$

Since $f(x) = \log_{10}(\tan x)$, and we have

$$f^{(4)}(x) = -12\frac{\left(1 + (\tan x)^2\right)^2}{\ln 10} + 8\frac{(\tan x)^2\left(1 + (\tan x)^2\right)}{\ln 10}$$
$$+ 16\frac{\left(1 + (\tan x)^2\right)^3}{(\tan x)^2 \ln 10} - 6\frac{\left(1 + (\tan x)^2\right)^4}{(\tan x)^4 \ln 10}.$$

To find $\max_{1 \leq x \leq 1.15} |f^{(4)}(x)|$, we differentiate $f^{(4)}(x)$ to obtain

$$f^{(5)}(x) = -32\frac{\left(1 + (\tan x)^2\right)^2 \tan x}{\ln 10} + 16\frac{(\tan x)^3\left(1 + (\tan x)^2\right)}{\ln 10} + 96\frac{\left(1 + (\tan x)^2\right)^3}{\tan x \ln 10}$$
$$- 80\frac{\left(1 + (\tan x)^2\right)^4}{(\tan x)^3 \ln 10} + 24\frac{\left(1 + (\tan x)^2\right)^5}{(\tan x)^5 \ln 10}.$$

We find that $f^{(5)}(x) = 0$ does not have a solution in $[1, 1.15]$ and

$$\max_{1 \leq x \leq 1.15} |f^{(4)}(x)| = |f^{(4)}(1.15)| = 81.5082927.$$

Thus we have the error bound

$$|f(x) - P_3(x)| \leq \frac{1}{24}\left[\max_{1 \leq x \leq 1.5} \left|f^{(4)}(x)\right| \left|(1.09 - 1)(1.09 - 1.05)(1.09 - 1.1)(1.09 - 1.15)\right| \right]$$
$$\leq \frac{(81.5082927)(0.09)(0.04)(0.01)(0.06)}{24} = 7.3357 \times 10^{-6}.$$

The approximation is not within the theoretical error bound because of the inaccuracy of the data and the insufficient precision of the arithmetic.

12. Use the Lagrange polynomial of degree three and four-digit chopping arithmetic to approximate $\cos 0.750$, based on the following data:

$$\cos 0.698 = 0.7661, \quad \cos 0.768 = 0.7193, \quad \cos 0.733 = 0.7432, \quad \cos 0.803 = 0.6946.$$

SOLUTION: The four-digit chopping calculations for the Lagrange polynomial of degree three at 0.750 are

$$P_3(0.750) = \frac{(0.750 - 0.733)(0.750 - 0.768)(0.750 - 0.803)}{(0.698 - 0.733)(0.698 - 0.768)(0.698 - 0.803)}(0.7661)$$

$$+ \frac{(0.750 - 0.698)(0.750 - 0.768)(0.750 - 0.803)}{(0.733 - 0.698)(0.733 - 0.768)(0.733 - 0.803)}(0.7432)$$

$$+ \frac{(0.750 - 0.698)(0.750 - 0.733)(0.750 - 0.803)}{(0.768 - 0.698)(0.768 - 0.733)(0.768 - 0.803)}(0.7193)$$

$$+ \frac{(0.750 - 0.698)(0.750 - 0.733)(0.750 - 0.768)}{(0.803 - 0.698)(0.803 - 0.733)(0.803 - 0.768)}(0.6946)$$

$$= \frac{(0.0170)(-0.0180)(-0.0530)}{(-0.0350)(-0.0700)(-0.1050)}(0.7661) + \frac{(0.0520)(-0.0180)(-0.0530)}{(0.0350)(-0.0350)(-0.0700)}(0.7432)$$

$$+ \frac{(0.0520)(0.0170)(-0.0530)}{(0.0700)(0.0350)(-0.0350)}(0.7193) + \frac{(0.0520)(0.0170)(-0.0180)}{(0.1050)(0.0700)(0.0350)}(0.6946)$$

$$= -\frac{0.00001621}{0.0002572}(0.7661) + \frac{0.00004960}{0.00008575}(0.7432) + \frac{0.00004685}{0.00008575}(0.7193) - \frac{0.00001591}{0.0002572}(0.6946)$$

$$= -\frac{0.00001241}{0.0002572} + \frac{0.00003686}{0.00008575} + \frac{0.00003369}{0.00008575} - \frac{0.00001105}{0.0002572}$$

$$= -0.04825 + 0.4298 + 0.3928 - 0.04296 = 0.7313.$$

The actual error is 0.0004, and an error bound is 2.7×10^{-8}. The discrepancy between the error bound and the actual error is due to the fact that since the data is accurate only to four decimal places, significant round-off error occurs in the computation of the approximation .

15. Repeat Exercise 11 using Maple and ten-digit rounding arithmetic.

SOLUTION: The Lagrange coefficient polynomials are now

$$L_0(x) = -1333.333333(x - 1.05)(x - 1.10)(x - 1.15),$$
$$L_1(x) = 4000.000000(x - 1.00)(x - 1.10)(x - 1.15),$$
$$L_2(x) = -4000.000000(x - 1.00)(x - 1.05)(x - 1.15),$$
$$L_3(x) = 1333.333333(x - 1.00)(x - 1.05)(x - 1.10),$$

so

$$L_0(1.4) = -0.03199999999, \quad L_1(1.4) = 0.2160000000, \quad L_2(1.4) = 0.8640000000,$$

and

$$L_3(1.4) = -0.04799999999.$$

This gives

$$P_3(1.4) = -0.03199999999(0.1924) + 0.2160000000(0.2414) + 0.8640000000(0.2933)$$
$$- 0.04799999999(0.3492) = 0.2826352000.$$

The error is now 0.0000038645.

17. Determine the largest possible step size that can be used to construct an eight-decimal-place table of common logarithms on the interval $[1, 10]$ if linear interpolation is to give 10^{-6} accuracy.

SOLUTION: We need to choose the step size h so that we have a linear interpolation error bound less than 10^{-6}. Therefore, we need to have

$$\left| \frac{D_x^2 \log_{10}(\xi(x))}{2!} \frac{h^2}{4} \right| < 0.000001,$$

where h is the step size. Since

$$D_x \log x = \frac{1}{x \ln 10} \quad \text{and} \quad D_x^2 \log x = \frac{-1}{x^2 \ln 10},$$

the maximum absolute value of the second derivative is $1/\ln 10$, which occurs when $x = 1$, and h needs to satisfy

$$\frac{1}{\ln 10} \frac{h^2}{8} < 0.000001.$$

This implies that the largest possible step size is approximately 0.004291932, so 0.004 would be a reasonable choice.

22. Show that $\max_{x_j \le x \le x_{j+1}} |g(x)| = \dfrac{h^2}{4}$, where $g(x) = (x - jh)(x - (j+1)h)$.

SOLUTION: Since $g'(x) = 2x - jh - (j+1)h$, we have $g'(x) = 0$ if and only if $x = \left(j + \frac{1}{2}\right) h$. The maximum of g can occur only at an endpoint or a critical point, so

$$\max |g(x)| = \max \left\{ |g(jh)|, \left| g\left(\left(j + \frac{1}{2}\right)h \right) \right|, |g((j+1)h)| \right\} = \max \left(0, \frac{h^2}{4} \right) = \frac{h^2}{4}.$$

Exercise Set 3.2, page 123

2. b. Use Neville's method to obtain the approximations for Lagrange interpolating polynomials of degrees one, two, and three to approximate $f(0)$ if $f(-0.5) = 1.93750$, $f(-0.25) = 1.33203$, $f(0.25) = 0.800781$, and $f(0.5) = 0.687500$.

SOLUTION: Neville's method gives the table

$x_0 = -0.5$	$P_0 = 1.93750$			
$x_1 = -0.25$	$P_1 = 1.33203$	$P_{0,1} = 0.72656$		
$x_2 = 0.25$	$P_2 = 0.800780$	$P_{1,2} = 1.06641$	$P_{0,1,2} = 0.95312$	
$x_3 = 0.5$	$P_3 = 0.687500$	$P_{2,3} = 0.91406$	$P_{1,2,3} = 1.01562$	$P_{0,1,2,3} = 0.98437$

The interpolation polynomials are

$$P_1(x) = -1.06250x + 1.06641 \quad \text{with} \quad P_1(0.0) = 1.06641,$$
$$P_2(x) = 1.81251x^2 - 1.06250x + 0.95312 \quad \text{with} \quad P_2(0.0) = 0.95312,$$
$$P_3(x) = -1.0000x^3 + 1.31250x^2 - 1.00000x + 0.98437$$

with $P_3(0.0) = 0.98437$, and

$$|f(0.0) - P_1(0.0)| = 0.06641, \quad |f(0.0) - P_2(0.0)| = 0.0468760, \quad \text{and} \quad |f(0.0) - P_3(0.0)| = 0.01563.$$

3. **a.** Use Neville's method with the function $f(x) = 3^x$ and the nodes $x_0 = -2$, $x_1 = -1$, $x_2 = 0$, $x_3 = 1$, and $x_4 = 2$ to approximate $\sqrt{3}$.

b. Use Neville's method with the function $f(x) = \sqrt{x}$ and the nodes $x_0 = 0$, $x_1 = 1$, $x_2 = 2$, $x_3 = 4$, and $x_4 = 5$ to approximate $\sqrt{3}$.

c. Which approximation is more accurate?

SOLUTION: **a.** The expected best approximation for this data is $\sqrt{3} \approx P_4(1/2)$. To construct $P_4(1/2)$, we use the data in the following table:

$x_0 = -2$	$P_0 = 0.1111111$			
$x_1 = -1$	$P_1 = 0.3333333$	$P_{0,1} = 0.6666667$		
$x_2 = 0$	$P_2 = 1$	$P_{1,2} = 1.333333$	$P_{0,1,2} = 1.5$	
$x_3 = 1$	$P_3 = 3$	$P_{2,3} = 2$	$P_{1,2,3} = 1.833333$	$P_{0,1,2,3} = 1.777778$
$x_4 = 2$	$P_4 = 9$	$P_{3,4} = 0$	$P_{2,3,4} = 1.5$	$P_{1,2,3,4} = 1.666667$

This gives

$$\sqrt{3} \approx P_4(1/2) = P_{0,1,2,3,4} = 1.708\overline{3}.$$

b. The expected best approximation for this data is $\sqrt{3} \approx P_4(3)$. To construct $P_4(3)$, we use the data in the following table:

$x_0 = 0$	$P_0 = 0$			
$x_1 = 1$	$P_1 = 1$	$P_{0,1} = 3$		
$x_2 = 2$	$P_2 = 1.414214$	$P_{1,2} = 1.828427$	$P_{0,1,2} = 1.242641$	
$x_3 = 4$	$P_3 = 2$	$P_{2,3} = 1.707107$	$P_{1,2,3} = 1.747547$	$P_{0,1,2,3} = 1.621320$
$x_4 = 5$	$P_4 = 2.236068$	$P_{3,4} = 1.763932$	$P_{2,3,4} = 1.726049$	$P_{1,2,3,4} = 1.736798$

This gives

$$\sqrt{3} \approx P_4(3) = P_{0,1,2,3,4} = 1.690607.$$

c. Since to six decimal places $\sqrt{3} = 1.7332051$, the approximation in part (a) is more accurate than in part (b), but neither is very good.

6. Neville's method is used to approximate $f(0.5)$, giving the following table. Determine $P_2 = f(0.7)$.

$$
\begin{array}{llll}
x_0 = 0 & P_0 = 0 & & \\
x_1 = 0.4 & P_1 = 2.8 & P_{01} = 3.5 & \\
x_2 = 0.7 & P_2 & P_{12} & P_{012} = \frac{27}{7}
\end{array}
$$

SOLUTION: We use the formula

$$
P_{0,1,2} = \frac{(x - x_0)P_{1,2} - (x - x_2)P_{0,1}}{x_2 - x_0}
$$

to obtain $P_{1,2}$. After substitution we have

$$
\frac{27}{7} = \frac{(0.5 - 0)P_{1,2} - (0.5 - 0.7)3.5}{0.7 - 0} \quad \text{or} \quad \frac{27}{7} = \frac{0.5P_{1,2} + 0.7}{0.7},
$$

so $P_{1,2} = 4$. We then use the formula

$$
P_{1,2} = \frac{(x - x_1)P_2 - (x - x_2)P_1}{x_2 - x_1}
$$

to obtain P_2. After substitution we have

$$
4 = \frac{(0.5 - 0.4)P_2 - (0.5 - 0.7)2.8}{0.3} \quad \text{and} \quad 1.2 = 0.1P_2 + 0.56,
$$

so $P_2 = f(0.7) = 6.4$.

10. Neville's Algorithm is used to approximate $f(0)$ given $f(-2)$, $f(-1)$, $f(1)$, and $f(2)$. Suppose $f(-1)$ was overstated by 2 and $f(1)$ was understated by 3. Determine the error in the original calculation of the value of the interpolating polynomial to approximate $f(0)$.

SOLUTION: The errors created in Neville's method are shown in the table. We are given $e_1 = 2$ and $e_2 = -3$ and assume that $e_0 = e_3 = 0$.

$$
\begin{array}{llllll}
x_0 = -2 & e_0 = 0 & & & & \\
x_1 = -1 & e_1 = 2 & e_{0,1} = 4 & & & \\
x_2 = 1 & e_2 = -3 & e_{1,2} = -\frac{1}{2} & e_{0,1,2} = 1 & & \\
x_3 = 2 & e_3 = 0 & e_{2,3} = -6 & e_{1,2,3} = -\frac{7}{3} & e_{0,1,2,3} = -\frac{2}{3}
\end{array}
$$

The original $P_{0,1,2,3}$ is in error by the amount $-\frac{2}{3}$.

13. Construct an algorithm for implementing inverse interpolation.

SOLUTION: Use Algorithm 3.1 with the following changes.

INPUT numbers $y_0, y_1, ..., y_n$; values $x_0, x_1, ..., x_n$ as the first column $Q_{0,0}, Q_{1,0}, ..., Q_{n,0}$ of Q.
OUTPUT the table Q with $Q_{n,n}$ approximating $f^{-1}(0)$.

Step 1 For $i = 1, 2, ..., n$
  for $j = 1, 2, ..., i$
  set

$$Q_{i,j} = \frac{y_i Q_{i-1,j-1} - y_{i-j} Q_{i,j-1}}{y_i - y_{i-j}}.$$

Exercise Set 3.3, page 133

1. **a.** Use Eq. (3.10) or Algorithm 3.2 to construct interpolating polynomials of degree one, two, and three for the following data to approximate $f(8.4)$ if $f(8.1) = 16.94410$, $f(8.3) = 17.56492$, $f(8.6) = 18.50515$, and $f(8.7) = 18.82091$

   SOLUTION: We have

$$f[x_0, x_1] = \frac{f[x_1] - f[x_0]}{(x_1 - x_0)} = 3.10410,$$

$$f[x_1, x_2] = \frac{f[x_2] - f[x_1]}{(x_2 - x_1)} = 3.13410,$$

$$f[x_0, x_1, x_2] = \frac{f[x_1, x_2] - f[x_0, x_1]}{(x_2 - x_0)} = 0.06000,$$

$$f[x_2, x_3] = \frac{f[x_3] - f[x_2]}{(x_3 - x_2)} = 3.15760,$$

$$f[x_1, x_2, x_3] = \frac{f[x_2, x_3] - f[x_1, x_2]}{(x_3 - x_1)} = 0.05875, \quad \text{and}$$

$$f[x_0, x_1, x_2, x_3] = \frac{f[x_1, x_2, x_3] - f[x_0, x_1, x_2]}{(x_3 - x_0)} = -0.002083.$$

   So

$$P_1(x) = 16.94410 + 3.104100000(x - x_0); \quad P_1(8.4) = 17.87533000;$$
$$P_2(x) = P_1(x) + 0.06000000000(x - x_0)(x - x_1); \quad P_2(8.4) = 17.87713000;$$
$$P_3(x) = P_2(x) - 0.002083(x - x_0))(x - x_1)(x - x_2); \quad P_3(8.4) = 17.87714250.$$

4. **a.** Use the Newton forward-difference formula to construct interpolating polynomials of degree one, two, and three for the following data to approximate $f(0.43)$ if $f(0) = 1$, $f(0.25) = 1.64872$, $f(0.5) = 2.71828$, and $f(0.75) = 4.48169$

   SOLUTION: The first divided differences are

$$\delta f(x_0) = f(x_1) - f(x_0) = 0.64872, \quad \delta f(x_1) = 1.06956, \quad \delta f(x_2) = 1.76341,$$

   The second divided differences are

$$\delta^2 f(x_0) = \delta f(x_1) - \delta f(x_0) = 0.42084, \quad \delta^2 f(x_1) = 0.69385,$$

and the third divided difference is

$$\delta^3 f(x_0) = \delta^2 f(x_1) - \delta^2 f(x_1) = 0.27301.$$

In the following equations, we have $s = (1/h)(x - x_0)$.

$$P_1(s) = 1.0 + 0.64872s, \quad P_1(0.43) = 2.11580$$
$$P_2(s) = P_1(s) + 0.21042s(s-1), \quad P_2(0.43) = 2.37638$$
$$P_3(s) = P_2(s) + 0.04550s(s-1)(s-2), \quad P_3(0.43) = 2.36060$$

6. a. Use the Newton back-difference formula to construct interpolating polynomials of degree one, two, and three for the following data to approximate $f(0.43)$ if $f(0) = 1$, $f(0.25) = 1.64872$, $f(0.5) = 2.71828$, and $f(0.75) = 4.48169$.

SOLUTION: We have

$$\nabla f(x_1) = f(x_1) - f(x_0) = 0.64872, \quad \nabla f(x_2) = 1.06956, \quad \nabla f(x_3) = 1.76341.$$

The second divided differences are

$$\nabla^2 f(x_2) = \nabla f(x_2) - \nabla f(x_1) = 0.42084, \quad \nabla^2 f(x_3) = 0.69385,$$

and the third divided difference is

$$\nabla^3 f(x_3) = \nabla^2 f(x_3) - \nabla^2 f(x_2) = 0.27301.$$

In the following equations, we have $x = 0.43$, $h = 0.25$, and $s = (1/h)(x - x_3) = -1.28$.

$$P_1(x) = f_3(x) - (-s)\nabla f(x_3) = 2.22453;$$
$$P_2(x) = P_1(x) + (-1)^2 \frac{(-s(-s-1))}{2} \nabla^2 f(x_3) = 2.34886;$$
$$P_3(x) = P_2(x) + (-1)^3 \frac{(-s(-s-1)(-s-2))}{6} \nabla^3 f(x_3) = 2.36060.$$

7. a. Use Algorithm 3.2 to construct the interpolating polynomial of degree three for the unequally spaced points given in the following table:

x	$f(x)$
-0.1	5.30000
0.0	2.00000
0.2	3.19000
0.3	1.00000

b. Add $f(0.35) = 0.97260$ to the table, and construct the interpolating polynomial of degree four.

SOLUTION:

a. Using Newton Divided-Difference Formula we have

$$x_0 = -0.1, \quad x_1 = 0.0, \quad x_2 = 0.2, \quad x_3 = 0.3;$$

$$f[x_0] = 5.3, \quad f[x_1] = 2.0, \quad f[x_2] = 3.19, \quad f[x_3] = 1.0,$$
$$f[x_0, x_1] = -33.00000, \quad f[x_1, x_2] = 5.95000 \quad f[x_2, x_3] = -21.90000,$$
$$f[x_0, x_1, x_2] = 129.83333, \quad f[x_1, x_2, x_3] = -92.83333, \quad \text{and} \quad f[x_0, x_1, x_2, x_3] = -556.66667.$$

This gives

$$P_3(x) = 5.3 - 33(x + 0.1) + 129.83333(x + 0.1)x - 556.66667(x + 0.1)x(x - 0.2)$$
$$P_4(x) = P_3(x) + 2730.243387(x + 0.1)x(x - 0.2)(x - 0.3)$$

11. a. Show that the divided-difference polynomials

$$P(x) = 3 - 2(x + 1) + 0(x + 1)(x) + (x + 1)(x)(x - 1)$$

and

$$Q(x) = -1 + 4(x + 2) - 3(x + 2)(x + 1) + (x + 2)(x + 1)(x)$$

both interpolate the data given in the following table.

x	-2	-1	0	1	2
$f(x)$	-1	3	1	-1	3

b. Why does part (a) not violate the uniqueness property of interpolating polynomials?

SOLUTION: We have

$$P(-2) = 3 - 2(-2 + 1) + 0 + (-2 + 1)(-2)(-2 - 1) = -1,$$
$$P(-1) = 3 - 2(0) + 0 + 0 = 3,$$
$$P(0) = 3 - 2(0 + 1) + 0 + 0 = 1,$$
$$P(1) = 3 - 2(1 + 1) + 0 + 0 = -1,$$
$$P(2) = 3 - 2(2 + 1) + 0 + (2 + 1)(2)(2 - 1) = 3;$$

and

$$Q(-2) = -1 + 0 + 0 + 0 = -1,$$
$$Q(-1) = -1 + 4(-1 + 2) + 0 + 0 = 3,$$
$$Q(0) = -1 + 4(0 + 2) - 3(0 + 2)(0 + 1) + 0 = 1,$$
$$Q(1) = -1 + 4(1 + 2) - 3(1 + 2)(1 + 1) + (1 + 2)(1 + 1)(1) = -1,$$
$$Q(2) = -1 + 4(2 + 2) - 3(2 + 2)(2 + 1) + (2 + 2)(2 + 1)(2) = 3.$$

Thus, $P(x)$ and $Q(x)$ are both polynomials of degree 3 that interpolate the given data.

b. This does not violate the uniqueness property of interpolating polynomials because $P(x)$ and $Q(x)$ are the same. Simplifying both $P(x)$ and $Q(x)$ to the form $a_0 + a_1 x + a_2 x^2 + a_3 x^3$ shows that they are identical.

12. Compute $\Delta^2 P(10)$ for the fourth-degree polynomial $P(x)$ given that $\Delta^4 P(0) = 24$, $\Delta^3 P(0) = 6$, and $\Delta^2 P(0) = 0$, where $\Delta P(x) = P(x + 1) - P(x)$.

SOLUTION: Since $h = 1$ and $x_0 = 0$ for this problem, $x = x_0 + sh = s$ in Newton's forward difference formula, and since $P(s)$ is a polynomial of degree 4, Newton's forward difference formula gives

$$P(s) = \binom{s}{0}P(0) + \binom{s}{1}\Delta P(0) + \binom{s}{2}\Delta^2 P(0) + \binom{s}{3}\Delta^3 P(0) + \binom{s}{4}\Delta^4 P(0)$$

$$= P(0) + s\Delta P(0) + \frac{1}{2}s(s-1)\Delta^2 P(0) + \frac{1}{6}s(s-1)(s-2)\Delta^3 P(0)$$

$$+ \frac{1}{24}s(s-1)(s-2)(s-3)\Delta^4 P(0)$$

$$= P(0) + s\Delta P(0) + s(s-1)(s-2) + s(s-1)(s-2)(s-3).$$

This gives

$$P(10) = P(0) + 10\Delta P(0) + 10(9)(8) + 10(9)(8)(7) = P(0) + 10\Delta P(0) + 5760,$$
$$P(11) = P(0) + 11\Delta P(0) + 8910,$$

and

$$P(12) = P(0) + 12P(0) + 13200.$$

Hence

$$\Delta P(11) = P(12) - P(11) = \Delta P(0) + 4290,$$
$$\Delta P(10) = P(11) - P(10) = \Delta P(0) + 3150,$$

and

$$\Delta^2 P(10) = \Delta P(11) - \Delta P(10) = 1140.$$

13. Suppose that $P(x)$ is a polynomial of unspecified degree for which $P(0) = 2$, $P(1) = -1$, $P(2) = 4$, and all third-order forward differences are 1. Determine the coefficient of x^2 in $P(x)$.

SOLUTION: Since $x_0 = 0$ and $h = 1$ for this problem, $x = x_0 + sh = s$ in Newton's forward difference formula. We have

$$P(0) = 2, \; \Delta P(0) = P(1) - P(0) = -3,$$
$$\Delta^2 P(0) = (P(2) - P(1)) - (P(1) - P(0)) = 5 - (-3) = 8,$$

and

$$\Delta^3 P(0) = 1, \text{ and } \Delta^4 P(0) = \Delta^5 P(0) = \cdots = 0.$$

Using Newton's forward difference formula, we have

$$P(x) = P(0) + x\Delta P(0) + \frac{1}{2}x(x-1)\Delta^2 P(0) + \frac{1}{6}x(x-1)(x-2)\Delta^3 P(0)$$

$$= 2 - 3x + 4x(x-1) + \frac{1}{6}x(x-1)(x-2)$$

$$= 2 - \frac{20}{3}x + \frac{7}{2}x^2 + \frac{1}{6}x^3.$$

The coefficient of x^2 is 3.5.

15. The Newton forward divided-difference formula is used to approximate $f(0.3)$ given the following data.

x	0.0	0.2	0.4	0.6
$f(x)$	15.0	21.0	30.0	51.0

Suppose it is discovered that $f(0.4)$ was understated by 10 and $f(0.6)$ was overstated by 5. By what amount should the approximation to $f(0.3)$ be changed?

SOLUTION: The following is the incorrect divided-difference table.

$$
\begin{array}{llll}
x_0 = 0.0 & f[x_0] \\
x_1 = 0.2 & f[x_1] \\
x_2 = 0.4 & f[x_2] - 10 & f[x_1, x_2] - 50 & f[x_0, x_1, x_2] - 125 \\
x_3 = 0.6 & f[x_3] + 5 & f[x_2, x_3] + 75 & f[x_1, x_2, x_3] + 312.5 & f[x_0, x_1, x_2, x_3] + 729.1\overline{6}
\end{array}
$$

The incorrect polynomial $Q(x)$ is given by

$$
\begin{aligned}
Q(x) &= f[x_0] + f[x_0, x_1](x - x_0) + (f[x_0, x_1, x_2] - 125)(x - x_0)(x - x_1) \\
&\quad + (f[x_0, x_1, x_2, x_3] + 125)(x - x_0)(x - x_1)(x - x_2) \\
&= P_3(x) - 125(x - x_0)(x - x_1) + 729.1\overline{6}(x - x_0)(x - x_1)(x - x_2) \\
&= P_3(x) - 125x(x - 0.2) + 729.1\overline{6}x(x - 0.2)(x - 0.4).
\end{aligned}
$$

Thus

$$
Q(0.3) = P_3(0.3) - 125(0.3)(0.1) + 729.1\overline{6}(0.3)(0.1)(-0.1) = P_3(0.3) - 5.9375.
$$

To obtain the approximation $P_3(0.3)$ to $f(0.3)$, we need to add 5.9375 to $Q(0.3)$, the number we originally obtained.

17. For a function f, the forward divided-differences are given by the following table.

$$
\begin{array}{llll}
x_0 = 0 & f[x_0] \\
 & & f[x_0, x_1] \\
x_1 = 0.4 & f[x_1] & & f[x_0, x_1, x_2] = \frac{50}{7} \\
 & & f[x_1, x_2] = 10 \\
x_2 = 0.7 & f[x_2] = 6
\end{array}
$$

Determine the missing entries in the table.

SOLUTION: We have the formula

$$
f[x_0, x_1, x_2] = \frac{f[x_1, x_2] - f[x_0, x_1]}{x_2 - x_0},
$$

and substituting gives

$$\frac{50}{7} = \frac{1}{0.7}(10 - f[x_0, x_1]). \quad \text{Thus} \quad f[x_0, x_1] = -0.7\frac{50}{7} + 10 = 5.$$

Using the formula

$$f[x_1, x_2] = \frac{f[x_2] - f[x_1]}{x_2 - x_1}$$

and substituting gives

$$10 = \frac{1}{0.3}(6 - f[x_1]). \quad \text{Thus} \quad f[x_1] = 6 - 3 = 3.$$

Further,

$$f[x_0, x_1] = \frac{f[x_1] - f[x_0]}{x_1 - x_0},$$

so

$$5 = \frac{1}{0.4}(3 - f[x_0]) \quad \text{and} \quad f[x_0] = 1.$$

20. Show that

$$f[x_0, x_1, \ldots, x_n, x] = \frac{f^{(n+1)}(\xi(x))}{(n+1)!},$$

for some $\xi(x)$ between $x_0, x_1, \ldots, x_n$, and x.

SOLUTION: Equation (3.3) gives

$$f(x) = P_n(x) + \frac{f^{n+1}(\xi(x))}{(n+1)!}(x - x_0) \ldots (x - x_n).$$

Let $x_{n+1} = x$. The interpolation polynomial of degree $n + 1$ on $n + 2$ nodes $x_0, x_1, \ldots, x_{n+1}$ is

$$P_{n+1}(t) = P_n(t) + f[x_0, x_1, \ldots, x_n, x_{n+1}](t - x_0)(t - x_1) \ldots (t - x_n).$$

Since $f(x) = P_{n+1}(x)$ and $x = x_{n+1}$, we have

$$P_{n+1}(x) = P_n(x) + f[x_0, x_1, \ldots, x_n, x](x - x_0)(x - x_1) \ldots (x - x_n)$$

and

$$P_{n+1}(x) = P_n(x) + \frac{f^{n+1}(\xi(x))}{(n+1)!}(x - x_0) \ldots (x - x_n).$$

Hence

$$P_n(x) + \frac{f^{n+1}(\xi(x))}{(n+1)!}(x - x_0) \ldots (x - x_n) = P_n(x) + f[x_0, \ldots, x_n, x](x - x_0) \ldots (x - x_n),$$

which implies that

$$f[x_0, \ldots, x_n, x] = \frac{f^{n+1}(\xi(x))}{(n+1)!}.$$

Exercise Set 3.4, page 142

2. c. Use Theorem 3.9 or Algorithm 3.3 to construct an approximating polynomial for the following data.

x	$f(x)$	$f'(x)$
0.1	−0.29004996	−2.8019975
0.2	−0.56079734	−2.6159201
0.3	−0.81401972	−2.9734038

SOLUTION: We have

$$z_0 = z_1 = 0.1, \quad z_2 = z_3 = 0.2, \quad z_4 = z_5 = 0.3,$$

and

$$f[z_0] = f[z_1] = -0.29004996, \quad f[z_2] = f[z_3] = -0.56079734, \quad f[z_4] = f[z_5] = -0.81401972.$$

This gives

$$f[z_0, z_1] = -2.8019975, \quad f[z_1, z_2] = -2.7074738, \quad f[z_2, z_3] = -2.6159201,$$
$$f[z_3, z_4] = -2.5322238, \quad f[z_4, z_5] = -2.4533949,$$
$$f[z_0, z_1, z_2] = 0.945237, \quad f[z_1, z_2, z_3] = 0.915537, \quad f[z_2, z_3, z_4] = 0.836963,$$
$$f[z_3, z_4, z_5] = 0.788289,$$
$$f[z_0, z_1, z_2, z_3] = -0.29700, \quad f[z_1, z_2, z_3, z_4] = -0.39287, \quad f[z_2, z_3, z_4, z_5] = -0.48674,$$
$$f[z_0, z_1, z_2, z_3, z_4] = -0.47935, \quad f[z_1, z_2, z_3, z_4, z_5] = -0.46935, \text{ and}$$
$$f[z_0, z_1, z_2, z_3, z_4, z_5] = 0.05000.$$

Thus

$$H_5(x) = -0.00985 - 2.80200x + 0.94524(x - 0.1)^2 - 0.29700(x - 0.1)^2(x - 0.2)$$
$$- 0.47935(x - 0.1)^2(x - 0.2)^2 + 0.05000(x - 0.1)^2(x - 0.2)^2(x - 0.3).$$

4. c. The data in Exercise 2(c) were generated using $f(x) = x^2 \cos x - 3x$. Use the polynomials constructed in Exercise 2 for the given value of x to approximate $f(0.18)$, and calculate the absolute error.

SOLUTION: For $f(x) = x^2 \cos(x) - 3x$ and the polynomial $H_5(x)$ in Exercise 2(c) we have, if we keep 10 decimal places,

$$f(0.18) = -0.5081234644, \quad H_5(0.18) = -0.5081234698, \quad \text{and} \quad |f(0.18) - H_5(0.18)| = 5.4 \times 10^{-9}.$$

9. The following table lists data for $f(x) = e^{0.1x^2}$. Approximate $f(1.25)$ with $H_3(1.25)$ and $H_5(1.25)$, and determine error bounds for these approximations.

x	$f(x) = e^{0.1x^2}$	$f'(x) = 0.2xe^{0.1x^2}$
1	1.105170918	0.2210341836
1.5	1.252322716	0.3756968148
2	1.491824698	0.5967298792
3	2.459603111	1.475761867

SOLUTION: The divided-difference table for the Hermite polynomial $H_3(x)$ is given below. It produces

$$H_3(x) = 1.105170918 + 0.221034184(x-1) + 0.146538825(x-1)^2$$
$$+ 0.0324952256(x-1)^2(x-1.5).$$

1	1.105170918			
1	1.105170918	0.2210341836		
1.5	1.252322716	0.294303596	0.146538825	
1.5	1.252322716	0.3756968148	0.162786438	0.0324952256

For $H_5(x)$, we need the additional entries in the table below.

1	1.105170918				
1	1.105170918	0.2210341836			
2	1.491824698	0.386654062	0.165619596		
2	1.491824698	0.5967298792	0.2100760992	0.0444565028	
3	2.459603111	0.967778131	0.37104853	0.0804862173	0.0180148573
3	2.459603111	1.475761867	0.50798345	0.136934920	0.0282243514	0.00510474710

The entries in this table imply that

$$H_5(x) = 1.105170918 + 0.2210341836(x-1) + 0.165619596(x-1)^2$$
$$+ 0.0444565028(x-1)^2(x-2) + 0.0180148573(x-1)^2(x-2)^2$$
$$+ 0.0051047471(x-1)^2(x-2)^2(x-3).$$

Using these polynomials gives $H_3(1.25) = 1.169080403$, with an error bound of 4.81×10^{-5} and $H_5(1.25) = 1.169016064$, with an error bound of 4.43×10^{-4}.

11. a. Show that the Hermite polynomial $H_{2n+1}(x)$ uniquely satisfies the interpolation conditions.

b. Show that the error term for this polynomial has the form

$$f(x) - H_{2n+1}(x) = \frac{(x-x_0)^2 \cdots (x-x_n)^2}{(2n+2)!} f^{(2n+2)}(\zeta),$$

for some ζ between $x_0, x_1, \ldots, x_n$, and x.

SOLUTION: **a.** To show uniqueness, we assume that $P(x)$ is another polynomial with $P(x_k) = f(x_k)$ and $P'(x_k) = f'(x_k)$, for $k = 0, ..., n$, and that the degree of $P(x)$ is at most $2n + 1$. Let

$$D(x) = H_{2n+1}(x) - P(x).$$

Then $D(x)$ is a polynomial of degree at most $2n + 1$ with $D(x_k) = 0$ and $D'(x_k) = 0$, for each $k = 0, 1, ..., n$. Thus, D has zeros of multiplicity 2 at each x_k, so

$$D(x) = (x - x_0)^2 \cdots (x - x_n)^2 Q(x).$$

Either, $D(x)$ is of degree $2n$ or more, which would be a contradiction, or $Q(x) \equiv 0$, which implies that $D(x) \equiv 0$. This implies that $P(x)$ is $H_{2n+1}(x)$, so this polynomial is unique.

b.

To show that the error assumes the form given, we first note that this error formula holds if $x = x_k$ for any choice of ζ. When $x \neq x_k$, for $k = 0, ..., n$, we define

$$g(t) = f(t) - H_{2n+1}(t) - \frac{(t - x_0)^2 \cdots (t - x_n)^2}{(x - x_0)^2 \cdots (x - x_n)^2}[f(x) - \dot{H}_{2n+1}(x)].$$

Note that $g(x_k) = 0$, for $k = 0, ..., n$, and $g(x) = 0$. Thus, g has $n + 2$ distinct zeros in $[a, b]$. By Rolle's Theorem, g' has $n + 1$ distinct zeros, $\xi_0, ..., \xi_n$, which are between the numbers $x_0, ..., x_n, x$. In addition, $g'(x_k) = 0$, for $k = 0, ..., n$, so g' has $2n + 2$ distinct zeros $\xi_0, ..., \xi_n, x_0, ..., x_n$. Since g' is $2n + 1$ times differentiable, the Generalized Rolle's Theorem 1.10 implies that a number ζ in $[a, b]$ exists with $g^{(2n+2)}(\zeta) = 0$.

However,

$$g^{(2n+2)}(t) = f^{(2n+2)}(t) - \frac{d^{2n+2}H_{2n+1}(t)}{dt^{2n+2}}$$
$$- \frac{[f(x) - H_{2n+1}(x)]}{(x - x_0)^2 \cdots (x - x_n)^2} \cdot \frac{d^{2n+2}\left[(t - x_0)^2 \cdots (t - x_n)^2\right]}{dt^{2n+2}}.$$

Since $H_{2n+1}(t)$ is a polynomial of degree $2n + 1$, its derivative of this order is zero. Also,

$$\frac{d^{2n+2}\left[(t - x_0)^2 \cdots (t - x_n)^2\right]}{dt^{2n+2}} = (2n + 2)! \, .$$

As a consequence, a number ζ exists with

$$0 = g^{(2n+2)}(\zeta) = f^{(2n+2)}(\zeta) - \frac{[f(x) - H_{2n+1}(x)]}{(x - x_0)^2 \cdots (x - x_n)^2}(2n + 2)!,$$

so

$$f(x) - H_{2n+1} = \frac{(x - x_0)^2 \cdots (x - x_n)^2}{(2n + 2)!} f^{(2n+2)}(\zeta),$$

for some ζ between $x_0, x_1, \ldots, x_n$, and x.

Exercise Set 3.5, page 161

1. Determine the free cubic spline S that interpolates the data $f(0) = 0$, $f(1) = 1$, and $f(2) = 2$.

SOLUTION: We have $x_0 = 0$, $x_1 = 1$, and $x_2 = 2$ so that $h_0 = h_1 = 1$. Further, $a_0 = 0$, $a_1 = 1$, $a_2 = 2$, $c_0 = 0$, and $c_2 = 0$. The equation for c_1 becomes

$$h_0 c_0 + 2(h_0 + h_1)c_1 + h_1 c_2 = \frac{3}{h_1}(a_2 - a_1) - \frac{3}{h_0}(a_1 - a_0)$$

or

$$2(1 + 1)c_1 = 3(2 - 1) - 3(1 - 0) \quad \text{and} \quad 4c_1 = 0, \quad \text{so} \quad c_1 = 0.$$

The equations for b_0, b_1, d_0, and d_1 are

$$b_1 = \frac{a_2 - a_1}{h_1} - h_1(c_2 + 2c_1)/3 = 1,$$

$$b_0 = \frac{a_1 - a_0}{h_0} - h_0(c_1 + 2c_0)/3 = 1,$$

$$d_1 = \frac{c_2 - c_1}{3h_1} = 0, \quad \text{and} \quad d_0 = \frac{c_1 - c_0}{3h_0} = 0.$$

The natural spline is described by $S_0(x) = x$ and $S_1(x) = 1 + (x - 1) = x$, so that $S(x) = x$ on $[0, 2]$.

3. a. Construct the natural cubic spline for the following data.

x	$f(x)$
8.3	17.56492
8.6	18.50515

SOLUTION: Since $n = 1$ the spline consists of only one cubic polynomial

$$S_0(x) = a_0 + b_0(x - x_0) + c_0(x - x_0)^2 + d_0(x - x_0)^3.$$

With $h = 0.3$, we have

$$a_1 = f(x_1) = a_0 + b_0 h + c_0 h^2 + d_0 h^3, \quad \text{and} \quad S_0''(x_0) = S_0''(x_1) = 0.$$

So

$$S_0''(x) = 2c_0 + 6d_0(x - x_0), \quad 0 = S_0''(x_0) = 2c_0, \quad \text{and} \quad 0 = S_0''(x_1) = 2c_0 + 6d_0 h.$$

This implies that $c_0 = 0$ and $d_0 = 0$. As a consequence,

$$a_1 = f(x_1) = a_0 + b_0 h, \quad \text{which implies that} \quad b_0 = \frac{1}{h}(a_1 - a_0) = 3.1341.$$

The spline is therefore $S_0 = 17.6492 + 3.1342(x - 8.3)$.

5. a. The data in Exercise 3(a) were generated using $f(x) = x \ln x$. Use the cubic splines constructed in Exercise 3(a) for the given value of x to approximate $f(0.9)$ and $f'(0.9)$, and calculate the actual error.

SOLUTION: From Exercise 3(a) we have $S_0 = 17.6492 + 3.1342(x - 8.3)$, so for the function we have

$$f(8.4) = 17.877146, \quad S(8.4) = 17.87833, \quad \text{and} \quad |f(8.4) - S(8.4)| = 0.00118367.$$

and for the derivative we have

$$f'(8.4) = 3.1282317, \quad f'(8.4) = 3.1341, \quad \text{and} \quad |f'(8.4) - S'(8.4)| = 0.0058683.$$

7. a. Construct the clamped cubic spline using the data of Exercise 3(a) and the fact that $f'(8.3) = 3.116256$ and $f'(8.6) = 3.151762$

SOLUTION: Since $n = 1$ the spline consists of only one cubic polynomial

$$s_0(x) = a_0 + b_0(x - x_0) + c_0(x - x_0)^2 + d_0(x - x_0)^3,$$

and

$$s_0'(x) = b_0 + 2c_0(x - x_0) + 3d_0(x - x_0)^2.$$

With $h = 0.3$, we have

$$a_1 = f(x_1) = a_0 + b_0 h + c_0 h^2 + d_0 h^3, \quad b_0 = f'(x_0) = 3.116256,$$

and

$$b_0 + 2c_0 h + 3d_0 h^2 = f'(x_1) = 3.151762.$$

So $b_0 = 3.116256$, and we have two equations involving c_0 and d_0 and known quantities:

$$c_0 + 0.3d_0 = \frac{1}{h^2}(a_1 - a_0) - \frac{1}{h}b_0 \quad \text{and} \quad 2c_0 + 3d_0 h = \frac{1}{h}(f'(x_1) - f'(x_0)).$$

Solving these equations gives $c_0 = 0.06008667$ and $d_0 = -0.00202222$. Hence the clamped spline is

$$s_0(x) = 17.56492 + 3.11626(x - 8.3) + 0.06009(x - 8.3)^2 - 0.002029(x - 8.3)^3.$$

9. a. Repeat Exercise 5(a) using the clamped cubic splines constructed in Exercise 7(a).

SOLUTION: From Exercise 7(a) we have

$$s_0 = 17.56492 + 3.11626(x - 8.3) + 0.06009(x - 8.3)^2 - 0.002029(x - 8.3)^3.$$

So

$$f(8.4) = 17.877146, \quad S(8.4) = 17.877144, \quad \text{and} \quad |f(8.4) - s(8.4)| = 1.6 \times 10^{-6}.$$

For the derivative we have

$$f'(8.4) = 3.1282317, \quad s'(8.4) = 3.182126, \quad \text{and} \quad |f'(8.4) - s'(8.4)| \doteq 1.9 \times 10^{-5}.$$

12. A clamped cubic spline s for a function f is defined on $[1, 3]$ by

$$s(x) = \begin{cases} s_0(x) = 3(x-1) + 2(x-1)^2 - (x-1)^3, & \text{if } 1 \le x < 2, \\ s_1(x) = a + b(x-2) + c(x-2)^2 + d(x-2)^3, & \text{if } 2 \le x \le 3. \end{cases}$$

Given $f'(1) = f'(3)$, find a, b, c, and d.

SOLUTION: We have

$$s_0(x) = 3(x-1) + 2(x-1)^2 - (x-1)^3,$$
$$s_0'(x) = 3 + 4(x-1) - 3(x-1)^2,$$
$$s_0''(x) = 4 - 6(x-1);$$

and

$$s_1(x) = a + b(x-2) + c(x-2)^2 + d(x-2)^3;$$
$$s_1'(x) = b + 2c(x-2) + 3d(x-2)^2;$$
$$s_1''(x) = 2c + 6d(x-2).$$

The properties of splines imply that

$$3 + 2 - 1 = s_0(2) = s_1(2) = a, \quad \text{so } a = 4;$$
$$3 + 4 - 3 = s_0'(2) = s_1'(2) = b, \quad \text{so } b = 4;$$
$$4 - 6 = s_0''(2) = s_1''(2) = 2c, \quad \text{so } c = -1.$$

Since $f'(1) = s'(1) = s_0'(1) = 3$, we have

$$3 = f'(3) = s'(3) = s_1'(3) = b + 2c + 3d.$$

Thus

$$d = \frac{1}{3}[3 - b - 2c] = \frac{1}{3}[3 - 4 + 2] = \frac{1}{3}.$$

Hence

$$a = 4, \quad b = 4, \quad c = -1, \quad \text{and} \quad d = \frac{1}{3}.$$

13. A natural cubic spline S is defined by

$$S(x) = \begin{cases} S_0(x) = 1 + B(x-1) - D(x-1)^3, & \text{if } 1 \le x < 2, \\ S_1(x) = 1 + b(x-2) - \frac{3}{4}(x-2)^2 + d(x-2)^3, & \text{if } 2 \le x \le 3. \end{cases}$$

If S interpolates the data $(1, 1)$, $(2, 1)$, and $(3, 0)$, find B, D, b, and d.

SOLUTION: We have

$$S_0(x) = 1 + B(x-1) - D(x-1)^3,$$
$$S_0'(x) = B - 3D(x-1)^2,$$
$$S_0''(x) = -6D(x-1);$$

and

$$S_1(x) = 1 + b(x - 2) - \frac{3}{4}(x - 2)^2 + d(x - 2)^3,$$

$$S_1'(x) = b - \frac{3}{2}(x - 2) + 3d(x - 2)^2,$$

$$S_1''(x) = -\frac{3}{2} + 6d(x - 2).$$

From the interpolating properties of splines, we have $S_0(1) = 1$, $S_0(2) = S_1(2) = 1$, and $S_1(3) = 0$. The second equation gives

$$1 + B - D = 1 \quad \text{so} \quad B = D.$$

The last equation implies

$$1 + b - \frac{3}{4} + d = 0 \quad \text{so} \quad b + d = -\frac{1}{4}.$$

Since $S_0'(2) = S_1'(2)$, we have $B - 3D = b$. Further,

$$S_0''(2) = S_1''(2) \quad \text{so} \quad -6D = -\frac{3}{2} \quad \text{and} \quad D = \frac{1}{4}.$$

Also,

$$S_1''(3) = 0 \quad \text{so} \quad -\frac{3}{2} + 6d = 0 \quad \text{and} \quad d = \frac{1}{4}.$$

Since $D = \frac{1}{4}$ and $d = \frac{1}{4}$, it follows that $B = \frac{1}{4}$ and $b = -\frac{1}{2}$.

16. Construct a free cubic spline for $f(x) = e^{-x}$ using the values of $f(x)$ when x is 0, 0.25, 0.75, and 1.0.

a. Integrate the spline on $[0, 1]$, and compare this to the value of the integral of $f(x)$ on this interval.

b. Compare the derivatives of the spline at 0.5 to $f'(0.5)$ and $f''(0.5)$.

SOLUTION: **a.** The equation of the spline is

$$S(x) = S_i(x) = a_i + b_i(x - x_i) + c_i(x - x_i)^2 + d_i(x - x_i)^3$$

on the interval $[x_i, x_{i+1}]$, where the results from Algorithm 3.4 are shown in the following table.

x_i	a_i	b_i	c_i	d_i
0	1.00000	−0.923601	0	0.620865
0.25	0.778801	−0.807189	0.465649	−0.154017
0.75	0.472367	−0.457052	0.234624	−0.312832

b. To approximate the integral of $f(x)$, we need to integrate each portion of the spline on the appropriate interval. Notice, however, that, in general,

$$\int_{x_i}^{x_{i+1}} S_i(x)\, dx = \int_{x_i}^{x_{i+1}} a_i + b_i(x - x_i) + c_i(x - x_i)^2 + d_i(x - x_i)^3\, dx$$

$$= \left[a_i(x - x_i) + b_i \frac{(x - x_i)^2}{2} + c_i \frac{(x - x_i)^3}{3} + d_i \frac{(x - x_i)^4}{4} \right]_{x_i}^{x_{i+1}}$$

$$= a_i(x_{i+1} - x_i) + b_i \frac{(x_{i+1} - x_i)^2}{2} + c_i \frac{(x_{i+1} - x_i)^3}{3} + d_i \frac{(x_{i+1} - x_i)^4}{4}$$

$$= a_i h_i + \frac{b_i}{2} h_i^2 + \frac{c_i}{3} h_i^3 + \frac{d_i}{4} h_i^4,$$

so the value of the integral can be easily obtained from the table. Completing the calculations gives

$$\int_0^1 S(x)\, dx = 0.631967$$

compared to the exact value

$$\int_0^1 f(x)\, dx = (1/e)(e - 1) \approx 0.632121.$$

To approximate the derivatives of $f(x)$ is easier than approximating the integral. First note that, in general,

$$S_i'(x) = b_i + 2c_i(x - x_i) + 3d_i(x - x_i)^2 \quad \text{and} \quad S_i''(x) = 2c_i + 6d_i(x - x_i),$$

for x in $[x_i, x_{i+1}]$. As a consequence,

$$S'(0.5) = -0.603243 \quad \text{and} \quad S''(0.5) = 0.700274,$$

compared to

$$f'(0.5) = -e^{0.5} = -0.606531 \quad \text{and} \quad f''(0.5) = e^{0.5} = 0.606531.$$

23. Extend Algorithms 3.4 and 3.5 to include as output the first and second derivative of the spline at the nodes.

    SOLUTION: Insert the following before Step 7 in Algorithm 3.4 and Step 8 in Algorithm 3.5:
    For $j = 0, 1, \ldots, n - 1$ set
    $$l_1 = b_j; \text{ (Note that } l_1 = s'(x_j).)$$
    $$l_2 = 2c_j; \text{ (Note that } l_2 = s''(x_j).)$$
    $$\text{OUTPUT } (l_1, l_2)$$

    Set
    $$l_1 = b_{n-1} + 2c_{n-1}h_{n-1} + 3d_{n-1}h_{n-1}^2; \text{(Note that } l_1 = s'(x_n).)$$
    $$l_2 = 2c_{n-1} + 6d_{n-1}h_{n-1}; \text{(Note that } l_2 = s''(x_n).)$$
    $$\text{OUTPUT } (l_1, l_2).$$

24. Extend Algorithms 3.4 and 3.5 to include as output the integral of the spline over the interval defined by the nodes.

SOLUTION: Insert the following before Step 7 in Algorithm 3.4 and Step 8 in Algorithm 3.5:

Set $I = 0$;
For $j = 0, \ldots, n - 1$ set
$$I = a_j h_j + \frac{b_j}{2} h_j^2 + \frac{c_j}{3} h_j^3 + \frac{d_j}{4} h_j^4 + I. \; \left(\text{Accumulate } \int_{x_j}^{x_{j+1}} S(x)\, dx.\right)$$
OUTPUT (I).

32. Determine the clamped cubic splines for the noble beast.

SOLUTION: The three clamped splines have equations of the form

$$s_i(x) = a_i + b_i(x - x_i) + c_i(x - x_i)^2 + d_i(x - x_i)^3$$

on $[x_i, x_{i+1}]$, where the values of the coefficients are given as follows.

Spline 1

i	x_i	$a_i = f(x_i)$	b_i	c_i	d_i	$f'(x_i)$
0	1	3.0	1.0	−0.347	−0.049	1.0
1	2	3.7	0.447	−0.206	0.027	
2	5	3.9	−0.074	0.033	0.342	
3	6	4.2	1.016	1.058	−0.575	
4	7	5.7	1.409	−0.665	0.156	
5	8	6.6	0.547	−0.196	0.024	
6	10	7.1	0.048	−0.053	−0.003	
7	13	6.7	−0.339	−0.076	0.006	
8	17	4.5				−0.67

Spline 2

i	x_i	$a_i = f(x_i)$	b_i	c_i	d_i	$f'(x_i)$
0	17	4.5	3.0	−1.101	−0.126	3.0
1	20	7.0	−0.198	0.035	−0.023	
2	23	6.1	−0.609	−0.172	0.280	
3	24	5.6	−0.111	0.669	−0.357	
4	25	5.8	0.154	−0.403	0.088	
5	27	5.2	−0.401	0.126	−2.568	
6	27.7	4.1				−4.0

Spline 3

i	x_i	$a_i = f(x_i)$	b_i	c_i	d_i	$f'(x_i)$
0	27.7	4.1	0.330	2.262	−3.800	0.33
1	28	4.3	0.661	−1.157	0.296	
2	29	4.1	−0.765	−0.269	−0.065	
3	30	3.0				−1.5

Exercise Set 3.6, page 161

1. a. Construct a parametric cubic Hermite approximation for the curve with endpoints $(0,0)$ and $(5,2)$ and with guidepoints $(1,1)$ and $(6,1)$, respectively.

SOLUTION: Using the equations (3.22) and (3.23), we have the parametric equations

$$x(t) = -10t^3 + 14t^2 + t \quad \text{and} \quad y(t) = -2t^3 + 3t^2 + t.$$

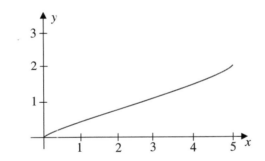

2. a. Construct a parametric cubic Bézier approximation for the curve with endpoints $(0,0)$ and $(5,2)$ and with guidepoints $(1,1)$ and $(6,1)$, respectively.

SOLUTION: Using the Equations (3.24) and (3.25), we have the parametric equations

$$x(t) = -10t^3 + 12t^2 + 3t \quad \text{and} \quad y(t) = 2t^3 - 3t^2 + 3t.$$

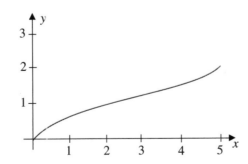

5. a. Suppose a cubic Bézier polynomial is placed through (u_0, v_0) and (u_3, v_3) with guidepoints (u_1, v_1) and (u_2, v_2), respectively. Derive the parametric equations for $u(t)$ and $v(t)$ assuming that

$$u(0) = u_0, \quad u(1) = u_1, \quad u'(0) = u_1 - u_0, \quad u'(1) = u_3 - u_2,$$

and

$$v(0) = v_0, \quad v(1) = v_1, \quad v'(0) = v_1 - v_0, \quad v'(1) = v_3 - v_2.$$

b. Let $f(i/3) = u_i$ and $g(i/3) = v_i$, for $i = 0, 1, 2, 3$. Show that the Bernstein polynomial of degree 3 for f is $u(t)$ and the Bernstein polynomial of degree 3 for g is $v(t)$.

SOLUTION: **a.** The forward divided difference table based on this information is

0	u_0			
0	u_0	$3(u_1 - u_0)$		
1	u_3	$u_3 - u_0$	$u_3 - 3u_1 + 2u_0$	
1	u_3	$3(u_3 - u_2)$	$2u_3 - 3u_2 + u_0$	$u_3 - 3u_2 + 3u_1 - u_0$

So

$$u(t) = u_0 + 3(u_1 - u_0)t + (u_3 - 3u_1 + 2u_0)t^2 + (u_3 - 3u_2 + 3u_1 - u_0)t^2(t-1)$$
$$= u_0 + 3(u_1 - u_0)t + (3u_2 - 6u_1 + 3u_0)t^2 + (u_3 - 3u_2 + 3u_1 - u_0)t^3.$$

Similarly,

$$v(t) = v_0 + 3(v_1 - v_0)t + (3v_2 - 6v_1 + 3v_0)t^2 + (v_3 - 3v_2 + 3v_1 - v_0)t^3.$$

b. The formula for Bernstein polynomial for f is

$$\sum_{k=0}^{3} \binom{3}{k} u_k t^k (1-t)^{3-k} = u_0(1-t)^3 + 3u_1 t(1-t)^2 + 3u_2 t^2(1-t) + u_3 t^3$$

$$= u_0 + 3(u_1 - u_0)t + (3u_2 - 6u_1 + 3u_0)t^2 + (u_3 - 3u_2 + 3u_1 - u_0)t^3$$
$$= u(t).$$

Similarly,

$$\sum_{k=0}^{3} \binom{3}{k} v_k t^k (1-t)^{3-k} = v_0(1-t)^3 + 3v_1 t(1-t)^2 + 3v_2 t^2(1-t) + v_3 t^3$$

$$= v_0 + 3(v_1 - v_0)t + (3v_2 - 6v_1 + 3v_0)t^2 + (v_3 - 3v_2 + 3v_1 - v_0)t^3$$
$$= v(t).$$

is the Bernstein polynomial for g.

Numerical Differentiation and Integration

Exercise Set 4.1, page 182

2. a. Use the forward-difference formulas and backward-difference formulas to determine each missing entry in the table.

x	$f(x)$	$f'(x)$
-0.3	1.9507	
-0.2	2.0421	
-0.1	2.0601	

SOLUTION: We have

$$f'(-0.3) \approx \frac{f(-0.1) - f(-0.3)}{-0.1 - (-0.3)} = \frac{0.0914}{0.1} = 0.914$$

$$f'(-0.2) \approx \frac{f(-0.1) - f(-0.2)}{-0.1 - (-0.2)} = \frac{0.0180}{0.1} = 0.1800$$

$$f'(-0.1) \approx \frac{f(-0.1) - f(-0.2)}{-0.1 - (-0.2)} = \frac{0.0180}{0.1} = 0.1800$$

5. a. Use the most appropriate three-point formula to determine approximations that will complete the following table.

x	$f(x)$	$f'(x)$
1.1	9.025013	
1.2	11.02318	
1.3	13.46374	
1.4	16.44465	

SOLUTION: Using (4.4) for the first and last approximation and (4.5) for the other two gives

$$f'(1.1) \approx \frac{1}{2(0.1)}[-3f(1.1) + 4f(1.2) - f(1.3)] = 17.769705,$$

$$f'(1.2) \approx \frac{1}{2(0.1)}[f(1.3) - f(1.1)] = 22.193635,$$

$$f'(1.3) \approx \frac{1}{2(0.1)}[f(1.4) - f(1.2)] = 27.107350,$$

$$f'(1.4) \approx \frac{1}{2(0.1)}[f(1.2) - 4f(1.3) + 3f(1.4)] = 32.510850.$$

7. a. The data in part (a) of Exercise 5 were taken from the function $f(x) = e^{2x}$. Compute the actual errors in part (a) of Exercise 5, and find error bounds using the error formulas.

SOLUTION: Since $f(x) = e^{2x}$, we have $f'(x) = 2e^{2x}$, $f''(x) = 4e^{2x}$, and $f'''(x) = 8e^{2x}$. The error bound formulas give

$$|f'(1.1) - 17.769705| = \frac{h^2}{3}|f'''(\xi_0)| \leq \frac{(0.1)^2}{3}\left(8e^{2(1.3)}\right) = 0.359033,$$

$$|f'(1.2) - 22.193635| = \frac{h^2}{6}|f'''(\xi_1)| \leq \frac{(0.1)^2}{6}\left(8e^{2(1.3)}\right) = 0.179517,$$

$$|f'(1.3) - 27.107350| = \frac{h^2}{6}|f'''(\xi_2)| \leq \frac{(0.1)^2}{6}\left(8e^{2(1.4)}\right) = 0.219262,$$

$$|f'(1.4) - 32.510850| = \frac{h^2}{3}|f'''(\xi_3)| \leq \frac{(0.1)^2}{3}\left(8e^{2(1.4)}\right) = 0.438524.$$

The actual errors are

$$|f'(1.1) - 17.769705| = |18.050027 - 17.769705| = 0.280322,$$
$$|f'(1.2) - 22.193635| = |22.046353 - 22.193635| = 0.147282,$$
$$|f'(1.3) - 27.107350| = |26.927476 - 27.107350| = 0.179874,$$
$$|f'(1.4) - 32.510850| = |32.889294 - 32.510850| = 0.378444.$$

13. Use the following data and the knowledge that the first five derivatives of f are bounded on $[1, 5]$ by 2, 3, 6, 12 and 23, respectively, to approximate $f'(3)$ as accurately as possible. Find a bound for the error.

x	1	2	3	4	5
$f(x)$	2.4142	2.6734	2.8974	3.0976	3.2804

SOLUTION: We have equally-spaced data on both sides of $x_0 = 3$ so we can use the 5-point centered difference formula with $h = 1$ This gives

$$f(3) \approx \frac{1}{12h}\left(f(x_0 - 2h) - 8f(x_0 - h) + 8f(x_0 + h) - f(x_0 + 2h)\right)$$

$$= \frac{1}{12}(2.4142 - 8(2.6734) + 8(3.0976) - 3.2804) = 0.21062.$$

An error bound is given by

$$\left| \frac{h^4}{30} f^{(5)}(\xi) \right| \leq \frac{1}{30} \max_{1 \leq \xi \leq 5} |f^{(5)}(\xi)| = \frac{23}{30} = 0.7667.$$

16. **a.** Repeat part (a) of Exercise 5 using four-digit chopping arithmetic.

SOLUTION:

$$f'(1.1) \approx \frac{1}{0.2}[-3f(1.1) + 4f(1.2) - f(1.3)] = \frac{1}{0.2}[-3(9.025) + 4(11.02) - 13.46] = 17.75,$$

$$f'(1.2) \approx \frac{1}{0.2}[f(1.3) - f(1.1)] = \frac{1}{0.2}[13.46 - 9.025] = 22.17,$$

$$f'(1.3) \approx \frac{1}{0.2}[f(1.4) - f(1.2)] = \frac{1}{0.2}[16.44 - 11.02] = 27.10,$$

$$f'(1.4) \approx \frac{1}{-0.2}[-3f(1.4) + 4f(1.3) - f(1.2)] = -\frac{1}{0.2}[-3(16.44) + 4(13.46) - 11.02] = 32.50,$$

22. Derive an $O(h^4)$ five-point formula to approximate $f'(x)$ that uses $f(x_0 - h)$, $f(x_0)$, $f(x_0 + h)$, $f(x_0 + 2h)$, and $f(x_0 + 3h)$.

SOLUTION: Expanding the expression

$$Af(x_0 - h) + Bf(x_0 + h) + Cf(x_0 + 2h) + Df(x_0 + 3h)$$

in fourth Taylor polynomials about x_0 and collecting like terms gives

$$Af(x_0 - h) + Bf(x_0 + h) + Cf(x_0 + 2h) + Df(x_0 + 3h)$$
$$= (A + B + C + D)f(x_0) + (-A + B + 2C + 3D)hf'(x_0)$$
$$+ \frac{1}{2}(A + B + 4C + 9D)h^2 f''(x_0) + \frac{1}{6}(-A + B + 8C + 27D)h^3 f'''(x_0)$$
$$+ \frac{1}{24}(A + B + 16C + 81D)h^4 f^{(4)}(x_0)$$
$$+ \frac{1}{120}\left(-Af^{(5)}(\xi_1) + Bf^{(5)}(\xi_2) + 32Cf^{(5)}(\xi_3) + 243Df^{(5)}(\xi_4)\right).$$

To obtain an $O(h^4)$ formula for $f'(x_0)$ we need to have zero coefficients for $f''(x_0)$, $f'''(x_0)$, and $f^{(4)}(x_0)$, and the coefficient of $f(x_0)$ must be 1. Hence, we must choose the constants so that

$$-A + B + 2C + 3D = 1,$$
$$A + B + 4C + 9D = 0,$$
$$-A + B + 8C + 27D = 0,$$
$$A + B + 16C + 81D = 0.$$

Solving this system of equations for $A, B, C,$ and D and simplifying gives

$$f'(x_0) = \frac{1}{12h}[-3f(x_0 - h) + 10f(x_0) + 18f(x_0 + h) - 6f(x_0 + 2h) + f(x_0 + 3h)] - \frac{h^4}{20} f^{(5)}(\xi).$$

The error term is a weighted average of the individual error terms, which must be shown to be valid by alternative means.

24. Analyze the rounding errors as was done in in the Illustration on page 181 for the formula

$$f'(x_0) = \frac{f(x_0 + h) - f(x_0)}{h} - \frac{h}{2}f''(\xi_0).$$

SOLUTION: Assuming that

$$f(x_0 + h) = \tilde{f}(x_0 + h) + e(x_0 + h) \quad \text{and} \quad f(x_0) = \tilde{f}(x_0) + e(x_0),$$

where $\tilde{f}(x_0 + h)$ and $\tilde{f}(x_0)$ are the actual values used in the formula with round-off errors $e(x_0 + h)$ and $e(x_0)$, respectively, gives

$$f'(x_0) - \frac{\tilde{f}(x_0 + h) - \tilde{f}(x_0)}{h} = f'(x_0) - \frac{f(x_0 + h) - f(x_0)}{h} + \frac{(f(x_0 + h) - \tilde{f}(x_0 + h)) - (f(x_0) - \tilde{f}(x_0))}{h}.$$

So

$$f'(x_0) - \frac{\tilde{f}(x_0 + h) - \tilde{f}(x_0)}{h} = -\frac{h}{2}f''(\xi_0) + \frac{e(x_0 + h) - e(x_0)}{h}.$$

If $|f''(x)|$ is bounded by M and the errors are bounded by ε, we have

$$\left| f'(x_0) - \frac{\tilde{f}(x_0 + h) - \tilde{f}(x_0)}{h} \right| < \frac{h}{2}M + \frac{2\varepsilon}{h}.$$

As in the Illustration on page 181, as h approaches zero, the truncation portion of the error, $hM/2$, goes to zero, but the round-off portion, $2\varepsilon/h$, gets large without bound.

27. Choose your favorite differentiable function f and nonzero number x and determine, for various positive integer values of n, the approximations to $f'(x)$ given by

$$f'_n(x) = \frac{f(x + 10^{-n}) - f(x)}{10^{-n}}.$$

Describe what happens and why.

SOLUTION: It is reasonable to state that on a calculator or computer, the derivative of any function at any point is zero. This is because the numerator of the difference quotient defining the approximations $f'_n(x)$ will involve round-off error and eventually go to zero. This will occur before the denominator of $f'_n(x)$ plays any part in the calculations.

29. Show that

$$e(h) = \frac{\varepsilon}{h} + \frac{h^2}{6}M$$

has a minimum at $h = \sqrt[3]{3\varepsilon/M}$.

SOLUTION: Since

$$e'(h) = -\frac{\varepsilon}{h^2} + \frac{hM}{3},$$

we have $e'(h) = 0$ if and only if $h = \sqrt[3]{3\varepsilon/M}$. Also,

$$e'(h) < 0 \quad \text{if} \quad h < \sqrt[3]{3\varepsilon/M}, \quad \text{and} \quad e'(h) > 0 \quad \text{if} \quad h > \sqrt[3]{3\varepsilon/M},$$

so an absolute minimum for $e(h)$ occurs at $h = \sqrt[3]{3\varepsilon/M}$.

Exercise Set 4.2, page 191

1. a. Apply the extrapolation process described in Example 1 to determine $N_3(h)$ to approximate $f'(1)$ for $f(x) = \ln x$ and $h = 0.4$.

SOLUTION: Using the basic formula

$$f'(x_0) = \frac{1}{2h}[f(x_0 + h) - f(x_0 - h)]$$

produces the entries in the first column of the following table. The remaining entries are generated using the averaging formula given in Example 1.

$N_1(0.4) = 1.0591223$		
$N_1(0.2) = 1.0136628$	0.9985096	
$N_1(0.1) = 1.0033535$	0.9999170	1.0000109

The extrapolation produces $f'(1) \approx 1.0000109$.

5. Given that

$$N_1(h) = 1.570796, \quad N_1\left(\frac{h}{2}\right) = 1.896119, \quad N_1\left(\frac{h}{4}\right) = 1.974232, \quad N_1\left(\frac{h}{8}\right) = 1.993570;$$

and that

$$\int_0^\pi \sin x \, dx = N_1(h) + K_1 h^2 + K_2 h^4 + K_3 h^6 + K_4 h^8 + O\left(h^{10}\right),$$

determine $N_4(h)$.

SOLUTION: Since only even powers of h are involved in the error formula, these approximations are generated using the same formula as the derivative problem in Example 1. The calculations are shown in the following table.

$N_1(h) = 1.570796$			
$N_1(h/2) = 1.896119$	2.004560		
$N_1(h/4) = 1.974232$	2.000270	1.999984	
$N_1(h/8) = 1.993570$	2.000016	1.999999	1.999999

Hence $\displaystyle\int_0^\pi \sin x \, dx \approx 1.999999$.

12. a. Show that

$$\lim_{h \to 0} \left(\frac{2 + h}{2 - h}\right)^{1/h} = e.$$

b. Compute approximations to e using the formula $N(h) = \left(\frac{2+h}{2-h}\right)^{1/h}$ for $h = 0.04, 0.02$, and 0.01.

c. Assume that $e = N(h) + K_1 h + K_2 h^2 + K_3 h^3 + \cdots$. Use extrapolation, with at least 16 digits of precision, to compute an $O\left(h^3\right)$ approximation to e with $h = 0.04$. Do you think the assumption is correct?

d. Show that $N(-h) = N(h)$.

e. Use part (d) to show that $K_1 = K_3 = K_5 = \cdots = 0$ in the formula

$$e = N(h) + K_1 h + K_2 h^2 + K_3 h^3 + K_4 h^4 + K_5^5 + \cdots,$$

so that the formula reduces to

$$e = N(h) + K_2 h^2 + K_4 h^4 + K_6 h^6 + \cdots.$$

f. Use the results of part (e) and extrapolation to compute an $O\left(h^6\right)$ approximation to e with $h = 0.04$.

SOLUTION: **a.** By L'Hospital's rule

$$\lim_{h \to 0} \frac{\ln(2+h) - \ln(2-h)}{h} = \lim_{h \to 0} \frac{1}{2+h} + \frac{1}{2-h} = 1,$$

so

$$\lim_{h \to 0} \left(\frac{2+h}{2-h}\right)^{1/h} = \lim_{h \to 0} \exp\left(\frac{1}{h}[\ln(2+h) - \ln(2-h)]\right) = \exp(1) = e.$$

b. We have

$$N(0.04) = \left(\frac{2 + 0.4}{2 - 0.04}\right)^{1/0.04} = \left(\frac{2.04}{1.96}\right)^{25} = 2.718644377221219.$$

Similarly,

$$N(0.02) = 2.718372444800607 \quad \text{and} \quad N(0.01) = 2.718304481241685.$$

c. Let

$$N_2(h) = 2N(h/2) - N(h)$$

and

$$N_3(h) = N_2(h/2) + \frac{1}{3}[N_2(h/2) - N_2(h)].$$

Then

$$N_2(0.04) = 2N(0.02) - N(0.04) = 2.718100512379995,$$
$$N_2(0.02) = 2N(0.01) - N(0.02) = 2.718236517682763,$$

and

$$N_3(0.04) = N_2(0.02) + \frac{1}{3}[N_2(0.02) - N_2(0.04)] = 2.718281852783685.$$

Since

$$|e - N_3(0.04)| \leq 5 \times 10^{-8}$$

and $(0.04)^3 = 6.4 \times 10^{-5}$, the approximation may not be an $O(h^3)$ approximation.

d. We have

$$N(-h) = \left(\frac{2-h}{2+h}\right)^{1/-h} = \left(\frac{2+h}{2-h}\right)^{1/h} = N(h).$$

e. We have

$$e = N(h) + K_1 h + K_2 h^2 + K_3 h^3 + \cdots.$$

Replacing h by $-h$ gives

$$e = N(-h) - K_1 h + K_2 h^2 - K_3 h^3 + \cdots.$$

Since $N(-h) = N(h)$, we have

$$e = N(h) - K_1 h + K_2 h^2 - K_3 h^3 + \cdots.$$

Thus

$$K_1 h + K_3 h^3 + \cdots = -\left(K_1 h + K_3 h^3 + \cdots\right),$$

for all h, which implies that

$$K_1 = K_3 = K_5 = \cdots = 0 \quad \text{and} \quad e = N(h) + K_2 h^2 + K_4 h^4 + \cdots.$$

f. Let

$$N_2(h) = N(h/2) + \frac{1}{3}(N(h/2) - N(h))$$

and

$$N_3(h) = N_2(1/2) + \frac{1}{15}(N_2(h/2) - N_2(h)).$$

Then

$$N_2(0.04) = N(0.02) + \frac{1}{3}(N(0.02) - N(0.04)) = 2.718281800660402,$$

$$N_2(0.02) = N(0.01) + \frac{1}{3}(N(0.01) - N(0.02)) = 2.718281826722043,$$

and

$$N_3(0.04) = N_2(0.02) + \frac{1}{15}(N_2(0.02) - N_2(0.04)) = 2.718281828459487.$$

Since $|e - N_3(0.04)| \le 5 \times 10^{-13}$ and $(0.04)^6 = 4.096 \times 10^{-9}$, $N_3(h)$ might be an $O\left(h^6\right)$ approximation.

13. Suppose that an extrapolation table has been constructed to approximate a number M with $M = N_1(h) + K_1 h^2 + K_2 h^4 + K_3 h^6$. Show that the linear interpolating polynomial $P_{0,1}(x)$ that passes through $\left(h^2, N_1(h)\right)$ and $\left(h^2/4, N_1(h/2)\right)$ satisfies $P_{0,1}(0) = N_2(h)$.

SOLUTION: By the definition of the linear interpolating polynomial we have

$$P_{0,1}(x) = \frac{(x - h^2) N_1\left(\frac{h}{2}\right)}{\frac{h^2}{4} - h^2} + \frac{\left(x - \frac{h^2}{4}\right) N_1(h)}{h^2 - \frac{h^2}{4}}.$$

Evaluating this equation at $x = 0$ gives the extrapolation term

$$P_{0,1}(0) = \frac{4 N_1\left(\frac{h}{2}\right) - N_1(h)}{3}.$$

15. The semiperimeters of the regular polygons that inscribe and circumscribe the unit circle approximate π and are described by

$$p_k = k \sin\left(\frac{\pi}{k}\right) \quad \text{and} \quad P_k = k \tan\left(\frac{\pi}{k}\right),$$

respectively.

a. Show that $p_4 = 2\sqrt{2}$ and $P_4 = 4$.

b. Show that for $k \geq 4$, we have

$$P_{2k} = \frac{2 p_k P_k}{p_k + P_k} \quad \text{and} \quad p_{2k} = \sqrt{p_k P_{2k}}.$$

c. Approximate π to within 10^{-4} by computing p_k and P_k until $P_k - p_k < 10^{-4}$.

d. Show that

$$p_k = \pi - \frac{\pi^3}{3!} \left(\frac{1}{k}\right)^2 + \frac{\pi^5}{5!} \left(\frac{1}{k}\right)^4 + \cdots$$

and

$$P_k = \pi + \frac{\pi^3}{3} \left(\frac{1}{k}\right)^2 + \frac{2\pi^5}{15} \left(\frac{1}{k}\right)^4 + \cdots.$$

e. Use extrapolation to improve the approximation to π.

SOLUTION: **a.** We have

$$p_4 = 4 \sin \frac{\pi}{4} = 4 \frac{\sqrt{2}}{2} = 2\sqrt{2} \quad \text{and} \quad P_4 = 4 \tan \frac{\pi}{4} = 4.$$

b. We have

$$\frac{2 p_k P_k}{p_k + P_k} = \frac{2 \left(k \sin \frac{\pi}{k}\right)\left(k \tan \frac{\pi}{k}\right)}{k \sin \frac{\pi}{k} + k \tan \frac{\pi}{k}} = 2k \left[\frac{\sin\left(\frac{\pi}{k}\right)}{\cos\left(\frac{\pi}{k}\right) + 1}\right]$$

$$= 2k \left[\frac{2 \sin\left(\frac{\pi}{2k}\right) \cos\left(\frac{\pi}{2k}\right)}{2 \left(\cos\left(\frac{\pi}{2k}\right)\right)^2}\right] = 2k \tan\left(\frac{\pi}{2k}\right) = P_{2k}$$

and

$$\sqrt{2 p_k P_{2k}} = \left\{\left(k \sin\left(\frac{\pi}{k}\right)\right)\left(2k \tan\left(\frac{\pi}{2k}\right)\right)\right\}^{\frac{1}{2}} = k \left\{\left(2 \sin\left(\frac{\pi}{2k}\right) \cos\left(\frac{\pi}{2k}\right)\right)\left(2 \tan\left(\frac{\pi}{2k}\right)\right)\right\}^{\frac{1}{2}}$$

$$= 2k \left\{\left(\sin\left(\frac{\pi}{2k}\right)\right)^2\right\}^{\frac{1}{2}} = 2k \sin\left(\frac{\pi}{2k}\right) = p_{2k}.$$

c. The following tables list the appropriate approximations.

k	4	8	16	32
p_k	$2\sqrt{2}$	3.0614675	3.1214452	3.1365485
P_k	4	3.3137085	3.1825979	3.1517249

k	64	128	256	512
p_k	3.1403312	3.1412773	3.1415138	3.1415729
P_k	3.1441184	3.1422236	3.1417504	3.1416321

So $\pi \approx 3.1416$ is accurate to within 10^{-6}.

d. Since

$$\sin x = x - \frac{x^3}{6} + \frac{x^5}{120} \cdots ,$$

we have

$$p_k = k \sin\left(\frac{\pi}{k}\right) = k\left[\frac{\pi}{k} - \frac{\pi^3}{6k^3} + \frac{\pi^5}{120k^5} \cdots\right] = \pi - \frac{\pi^3}{3!}\left(\frac{1}{k}\right)^2 + \frac{\pi^5}{5!}\left(\frac{1}{k}\right)^4 + \cdots .$$

Also,

$$\tan x = x + \frac{x^3}{3} + \frac{2x^5}{15} + \cdots ,$$

so

$$P_k = k \tan\left(\frac{\pi}{k}\right) = k\left[\frac{\pi}{k} + \frac{\pi^3}{3k^3} + \frac{2\pi^5}{15k^5} + \cdots\right] = \pi + \frac{\pi^3}{3}\left(\frac{1}{k}\right)^2 + \frac{2\pi^5}{15}\left(\frac{1}{k}\right)^4 + \cdots .$$

e. Values of P_k and p_k are given in the following tables, together with the extrapolation results:

$P_4 = 4$				
$P_8 = 3.3137085$	3.0849447			
$P_{16} = 3.1825979$	3.1388943	3.1424910		
$P_{32} = 3.1517249$	3.1414339	3.1416032	3.1415891	
$P_{64} = 3.1441184$	3.1415829	3.1415928	3.1415926	3.1415927

$p_4 = 2.8284271$				
$p_8 = 3.0614675$	3.1391476			
$p_{16} = 3.1214452$	3.1414377	3.1415904		
$p_{32} = 3.1365485$	3.1415829	3.1415926	3.1415927	
$p_{64} = 3.1403312$	3.1415920	3.1415927	3.1415927	3.1415927

These results would be even more accurate if additional digits were used in the initial column.

Exercise Set 4.3, page 202

1. c. Use the Trapezoidal rule to approximate

$$\int_1^{1.5} x^2 \ln x \, dx.$$

SOLUTION: The Trapezoidal rule gives

$$\int_1^{1.5} x^2 \ln x \, dx \approx \frac{0.5}{2} \left[1^2 \ln(1) + (1.5)^2 \ln(1.5) \right] = 0.228074.$$

3. c. Find a bound for the error in the Trapezoidal rule approximation of part (c) of Exercise 1.

SOLUTION: Since

$$f'(x) = 2x \ln x + x, \quad \text{we have} \quad f''(x) = 2 \ln x + 3,$$

and the error is bounded on $[1, 1.5]$ by

$$\left| \frac{h^3}{12} f''(x) \right| \le \frac{(0.5)^3}{12} [2 \ln(1.5) + 3] = 0.0396972.$$

The actual error in the approximation is 0.0358146.

5. c. Use Simpson's rule to approximate

$$\int_1^{1.5} x^2 \ln x \, dx.$$

SOLUTION: Simpson's rule gives

$$\int_1^{1.5} x^2 \ln x \, dx \approx \frac{0.25}{3} \left[1^2 \ln(1) + 4(1.25)^2 \ln(1.25) + (1.5)^2 \ln(1.5) \right] = 0.192245.$$

7. c. Find a bound for the error in the Simpson's rule approximation in part (c) of Exercise 3.

SOLUTION: Since

$$f'(x) = 2x \ln x + x, \quad f''(x) = 2 \ln x + 3,$$
$$f'''(x) = \frac{2}{x}, \quad \text{and} \quad f^{(4)}(x) = -\frac{1}{x^2},$$

the error is bounded on $[1, 1.5]$ by

$$\left| \frac{h^5}{90} f^{(4)}(x) \right| \le \frac{(0.25)^5}{90} \frac{2}{1^2} = 2.17014 \times 10^{-5}.$$

The actual error in the approximation is 1.43597×10^{-5}.

9. c. Use the Midpoint rule to approximate

$$\int_1^{1.5} x^2 \ln x \, dx.$$

SOLUTION: The Midpoint rule gives

$$\int_1^{1.5} x^2 \ln x \, dx \approx 2(0.25) \left[(1.25)^2 \ln(1.25) \right] = 0.174331.$$

11. **c.** Find a bound for the error in the Midpoint rule approximation in part (c) of Exercise 5.

SOLUTION: Because of the form of the error term for the Midpoint rule, the error is bounded by half the error bound for the Trapezoidal rule; that is, by 0.0198486. The actual approximation error is 0.0179284.

13. The Trapezoidal rule applied to $\int_0^2 f(x)\,dx$ gives the value 4, and Simpson's rule gives the value 2. What is $f(1)$?

SOLUTION: We have the Trapezoidal rule:

$$4 = [f(0) + f(2)];$$

and Simpson's rule:

$$2 = \frac{1}{3}[f(0) + 4f(1) + f(2)].$$

Substituting $f(0) + f(2) = 4$ into Simpson's rule gives $2 = \frac{1}{3}[4 + 4f(1)]$. Thus, $f(1) = (6-4)/4 = \frac{1}{2}$.

15. Find the degree of precision of the quadrature formula

$$\int_{-1}^{1} f(x)\,dx = f\left(-\frac{\sqrt{3}}{3}\right) + f\left(\frac{\sqrt{3}}{3}\right).$$

SOLUTION: We test the formula with $f(x) = 1, x, x^2, \ldots$ until the integral $\int_{-1}^{1} f(x)\,dx$ does not equal $f(-\sqrt{3}/3) + f(\sqrt{3}/3)$. This gives

$$f(x) = 1: \int_{-1}^{1} dx = 2 \text{ and } f\left(-\frac{\sqrt{3}}{3}\right) + f\left(\frac{\sqrt{3}}{3}\right) = 1 + 1 = 2,$$

$$f(x) = x: \int_{-1}^{1} x\,dx = \frac{1}{2}x^2\Big|_{-1}^{1} = 0; \text{ and } f\left(-\frac{\sqrt{3}}{3}\right) + f\left(\frac{\sqrt{3}}{3}\right) = -\frac{\sqrt{3}}{3} + \frac{\sqrt{3}}{3} = 0,$$

$$f(x) = x^2: \int_{-1}^{1} x^2\,dx = \frac{1}{3}x^3\Big|_{-1}^{1} = \frac{2}{3}; \text{ and } f\left(-\frac{\sqrt{3}}{3}\right) + f\left(\frac{\sqrt{3}}{3}\right) = \left(-\frac{\sqrt{3}}{3}\right)^2 + \left(\frac{\sqrt{3}}{3}\right)^2 = \frac{2}{3},$$

$$f(x) = x^3: \int_{-1}^{1} x^3\,dx = \frac{1}{4}x^4\Big|_{-1}^{1} = 0; \text{ and } f\left(-\frac{\sqrt{3}}{3}\right) + f\left(\frac{\sqrt{3}}{3}\right) = \left(-\frac{\sqrt{3}}{3}\right)^3 + \left(\frac{\sqrt{3}}{3}\right)^3 = 0,$$

$$f(x) = x^4: \int_{-1}^{1} x^4\,dx = \frac{1}{5}x^5\Big|_{-1}^{1} = \frac{2}{5}; \text{ and } f\left(-\frac{\sqrt{3}}{3}\right) + f\left(\frac{\sqrt{3}}{3}\right) = \left(-\frac{\sqrt{3}}{3}\right)^4 + \left(\frac{\sqrt{3}}{3}\right)^4 = \frac{2}{9}.$$

Since the formula is correct for $f(x) = 1, x, x^2$ and x^3, but not exact for $f(x) = x^4$, the degree of precision is 3.

17. The quadrature formula

$$\int_{-1}^{1} f(x)\, dx = c_0 f(-1) + c_1 f(0) + c_2 f(1)$$

is exact for all polynomials of degree less than or equal to 2. Determine c_0, c_1, and c_2.

SOLUTION: Since the formula is exact for $f(x) = 1$, x and x^2, we have

$$\int_{-1}^{1} dx = 2 = c_0 + c_1 + c_2, \quad \int_{-1}^{1} x\, dx = 0 = -c_0 + c_2, \quad \text{and} \quad \int_{-1}^{1} x^2\, dx = \frac{2}{3} = c_0 + c_2.$$

Solving the 3 equations we find that $c_0 = c_2 = \frac{1}{3}$ and $c_1 = \frac{4}{3}$.

19. Find the constants c_0, c_1, and x_1 so that the quadrature formula

$$\int_{0}^{1} f(x)\, dx = c_0 f(0) + c_1 f(x_1)$$

has the highest possible degree of precision.

SOLUTION: Since there are three unknowns c_0, c_1, and x_1, we would expect the formula to be exact for $f(x) = 1$, x and x^2. Thus,

$$f(x) = 1 : \int_{0}^{1} f(x)\, dx = 1 = c_0 + c_1,$$

$$f(x) = x : \int_{0}^{1} f(x)\, dx = \frac{1}{2} = c_1 x_1,$$

and

$$f(x) = x^2 : \int_{0}^{1} f(x)\, dx = \frac{1}{3} = c_1 x_1^2.$$

The last two equations yield $x_1 = \frac{2}{3}$ and $c_1 = \frac{3}{4}$. The first equation gives $c_0 = 1 - c_1 = \frac{1}{4}$. The formula is $\int_{0}^{1} f(x)\, dx = \frac{1}{4} f(0) + \frac{3}{4} f(2/3)$. We check the formula for

$$f(x) = x^3 : \int_{0}^{1} x^3\, dx = \frac{1}{4}$$

and

$$\frac{1}{4}(0)^3 + \frac{3}{4}\left(\frac{2}{3}\right)^3 = \frac{2}{9}.$$

So the degree of precision is 2.

24. Derive Simpson's rule with its $O\left(h^5\right)$ error term by using the fact that the rule is exact for x^n when $n = 1, 2$, and 3 to determine the appropriate constants in the formula

$$\int_{x_0}^{x_2} f(x)\, dx = a_0 f(x_0) + a_1 f(x_1) + a_2 f(x_2) + k f^{(4)}(\xi).$$

SOLUTION: We have

$$f(x) = x : a_0 x_0 + a_1(x_0 + h) + a_2(x_0 + 2h) = 2x_0 h + 2h^2;$$

$$f(x) = x^2 : a_0 x_0^2 + a_1(x_0 + h)^2 + a_2(x_0 + 2h)^2 = 2x_0^2 h + 4x_0 h^2 + \frac{8h^3}{3};$$

$$f(x) = x^3 : a_0 x_0^3 + a_1(x_0 + h)^3 + a_2(x_0 + 2h)^3 = 2x_0^3 h + 6x_0^2 h^2 + 8x_0 h^3 + 4h^4.$$

Solving this linear system for a_0, a_1, and a_2 gives $a_0 = \frac{h}{3}$, $a_1 = \frac{4h}{3}$, and $a_2 = \frac{h}{3}$, so

$$\int_{x_0}^{x_2} f(x)\, dx = \frac{h}{3} f(x_0) + \frac{4h}{3} f(x_1) + \frac{h}{3} f(x_2) + k f^{(4)}(\xi).$$

For the function $f(x) = x^4$, we have $f^{(4)}(\xi) = 24$, for any number ξ, so integrating the left side of the equation and using the values of a_0, a_1, and a_2 determined earlier produces

$$\frac{1}{5}(x_2^5 - x_0^5) = \frac{h}{3}(x_0^4 + 4x_1^4 + x_2^4) + 24k.$$

Hence

$$k = \frac{1}{120}(x_2^5 - x_0^5) - \frac{h}{72}(x_0^4 + 4x_1^4 + x_2^4)$$

$$= \frac{1}{120}\left((x_0 + 2h)^5 - x_0^5\right) - \frac{h}{72}\left(x_0^4 + 4(x_0 + h)^4 + (x_0 + 2h)^4\right)$$

$$= \frac{1}{120}\left(x_0^5 + 10hx_0^4 + 40h^2 x_0^3 + 80h^3 x_0^2 + 80h^4 x_0 + 32h^5 - x_0^5\right)$$

$$- \frac{h}{72}\left[x_0^4 + 4\left(x_0^4 + 4hx_0^3 + 6h^2 x_0^2 + 4h^3 x_0 + h^4\right) \right.$$

$$\left. + \left(x_0^4 + 8hx_0^3 + 24h^2 x_0^2 + 32h^3 x_0 + 16h^4\right) \right].$$

Gathering like terms, we find that all the coefficients of powers of x_0 are zero and $k = -h^5/90$.

26. Derive Simpson's three-eighths rule with error term.

SOLUTION: Using $n = 3$ in Theorem 4.2 gives

$$\int_a^b f(x)\, dx = \sum_{i=0}^{3} a_i f(x_i) + \frac{h^5 f^{(4)}(\xi)}{24} \int_0^3 t(t-1)(t-2)(t-3)\, dt.$$

Since

$$\int_0^3 t(t-1)(t-2)(t-3)\, dt = -\frac{9}{10},$$

the error term is

$$-\frac{3}{80} h^5 f^{(4)}(\xi).$$

Also,

$$a_i = \int_{x_0}^{x_3} \prod_{\substack{j=0 \\ j \neq i}}^{3} \frac{x - x_j}{x_i - x_j}\, dx, \quad \text{for each } i = 0, 1, 2, 3.$$

Using the change of variables $x = x_0 + th$ gives

$$a_i = h \int_0^3 \prod_{\substack{j=0 \\ j \neq i}}^3 \frac{t - j}{i - j}\, dt, \quad \text{for each } i = 0, 1, 2, 3.$$

Evaluating the integrals gives $a_0 = \dfrac{3h}{8}, a_1 = \dfrac{9h}{8}, a_2 = \dfrac{9h}{8}$, and $a_3 = \dfrac{3h}{8}$.

Exercise Set 4.4, page 210

2. b. Use the Composite Trapezoidal rule with $n = 6$ to approximate

$$\int_{-0.5}^{0.5} x \ln(x + 1)\, dx = 0.08802039175$$

SOLUTION: For $n = 6$ the Composite Trapezoidal rule has $h = (0.05 - (-0.05))/6 = 1/6$. So

$$\int_{-0.5}^{0.5} x \ln(x + 1)\, dx \approx \frac{1}{12} \left(f\left(-\tfrac{1}{2}\right) + 2 \left(f\left(-\tfrac{1}{3}\right) + f\left(-\tfrac{1}{6}\right) + f(0) + f\left(\tfrac{1}{6}\right) + f\left(\tfrac{1}{3}\right) \right) + f\left(\tfrac{1}{2}\right) \right)$$

$$= \frac{1}{12} \left(-\tfrac{1}{2} \ln \tfrac{1}{2} + 2 \left(-\tfrac{1}{3} \ln \tfrac{2}{3} - \tfrac{1}{6} \ln \tfrac{5}{6} + 0 + \tfrac{1}{6} \ln \tfrac{7}{6} + \tfrac{1}{3} \ln \tfrac{4}{3} \right) + \tfrac{1}{2} \ln \tfrac{3}{2} \right)$$

$$= 0.09363013976.$$

4. b. Use the Composite Simpson's rule to approximate the integrals in Exercise 2(b).

SOLUTION: For $n = 6$ the Composite Simpson's rule has $h = (0.05 - (-0.05)/6 = 1/6$. So

$$\int_{-0.5}^{0.5} x \ln(x + 1)\, dx \approx \frac{1}{18} \left(f\left(-\tfrac{1}{2}\right) + 4 \left(f\left(-\tfrac{1}{3}\right) + f(0) + f\left(\tfrac{1}{3}\right) \right) + 2 \left(f\left(-\tfrac{1}{6}\right) + f\left(\tfrac{1}{6}\right) \right) + f\left(\tfrac{1}{2}\right) \right)$$

$$= \frac{1}{18} \left(\tfrac{1}{2} \ln \tfrac{1}{2} + 4 \left(-\tfrac{1}{3} \ln \tfrac{2}{3} + \tfrac{1}{3} \ln \tfrac{4}{3} \right) + 2 \left(-\tfrac{1}{6} \ln \tfrac{5}{6} + \tfrac{1}{6} \ln \tfrac{7}{6} \right) + \tfrac{1}{2} \ln \tfrac{3}{2} \right)$$

$$= 0.08809221094.$$

6. b. Use the Composite Midpoint rule with $n + 2$ subintervals to approximate the integrals in Exercise 2(b).

SOLUTION: For $n = 6$ the Composite Midpoint rule has $h = (0.05 - (-0.05)/(6 + 2) = 1/8$. So

$$\int_{-0.5}^{0.5} x \ln(x + 1)\, dx \approx \frac{2}{8} \left(f\left(-\tfrac{3}{8}\right) + f\left(-\tfrac{1}{8}\right) + f\left(\tfrac{1}{8}\right) + f\left(\tfrac{3}{8}\right) \right)$$

$$= \frac{1}{4} \left(-\tfrac{3}{8} \ln \tfrac{5}{8} - \tfrac{1}{8} \ln \tfrac{7}{8} + \tfrac{1}{8} \ln \tfrac{9}{8} + \tfrac{3}{8} \ln \tfrac{11}{8} \right)$$

$$= 0.08177145342.$$

8. Approximate the integral

$$\int_0^2 x^2 e^{-x^2}\, dx.$$

a. Use the Composite Trapezoidal rule with $n = 8$.

b. Use the Composite Simpson's rule with $n = 8$.

c. Use the Composite Midpoint rule with $n = 6$.

SOLUTION: **a.** For the Composite Trapezoidal rule we have

$$\int_0^2 x^2 e^{-x^2}\, dx \approx \frac{0.25}{2}\Big[0^2 e^0 + 2\Big((0.25)^2 e^{-(0.25)^2} + (0.5)^2 e^{-(0.5)^2} + (0.75)^2 e^{-(0.75)^2} + (1)^2 e^{-(1)^2}$$

$$+ (1.25)^2 e^{-(1.25)^2} + (1.5)^2 e^{-(1.5)^2} + (1.75)^2 e^{-(1.75)^2}\Big) + 2^2 e^{-(2)^2}\Big] = 0.42158204.$$

b. For the the Composite Simpson's rule we have

$$\int_0^2 x^2 e^{-x^2}\, dx \approx \frac{0.25}{3}\Big[0^2 e^0 + 4\Big((0.25)^2 e^{-(0.25)^2} + (0.75)^2 e^{-(0.75)^2} + (1.25)^2 e^{-(1.25)^2} + (1.75)^2 e^{-(1.75)^2}\Big)$$

$$+ 2\Big((0.5)^2 e^{-(0.5)^2} + (1)^2 e^{-(1)^2} + (1.5)^2 e^{-(1.5)^2}\Big) + 2^2 e^{-(2)^2}\Big] = 0.42271619.$$

c. For the Composite Midpoint rule we have

$$\int_0^2 x^2 e^{-x^2}\, dx \approx 2(0.25)\Big[(0.25)^2 e^{-(0.25)^2} + (0.75)^2 e^{-(0.75)^2} + (1.25)^2 e^{-(1.25)^2} + (1.75)^2 e^{-(1.75)^2}\Big]$$

$$= 0.42498448.$$

9. Suppose that $f(0.25) = f(0.75) = \alpha$. Find α if the Composite Trapezoidal rule with $n = 2$ gives the value 2, for $\int_0^1 f(x)\, dx$, and with $n = 4$ gives the value 1.75.

SOLUTION: With $n = 2$, the Composite Trapezoidal rule gives

$$h = \frac{b - a}{n} = \frac{1 - 0}{2} = \frac{1}{2}$$

and

$$\int_0^1 f(x)\, dx \approx \frac{1}{4}[f(0) + 2f(1/2) + f(1)],$$

so

$$2 = \frac{1}{4}[f(0) + 2f(1/2) + f(1)].$$

For $n = 4$, the Composite Trapezoidal rule gives $h = 1/4$ and

$$\int_0^1 f(x)\, dx \approx \frac{1}{8}[f(0) + 2f(0.25) + 2f(0.5) + 2f(0.75) + f(1)],$$

so

$$1.75 = \frac{1}{8}[f(0) + 2f(0.25) + 2f(0.5) + 2f(0.75) + f(1)].$$

Substituting

$$f(0) + 2f(0.5) + f(1) = 8 \quad \text{and} \quad f(0.25) = f(0.75) = \alpha \quad \text{gives} \quad 1.75 = \frac{1}{8}(8 + 4\alpha).$$

So $\alpha = 1.5$.

16. Show that the error for Composite Simpson's rule can be approximated by

$$-\frac{h^4}{180}[f'''(b) - f'''(a)].$$

SOLUTION: We will first show that

$$\sum_{j=1}^{n/2} f^{(4)}(\xi_j) 2h$$

is a Riemann Sum for $f^{(4)}$ on $[a, b]$. Let $y_i = x_{2i}$, for $i = 0, 1, \ldots n/2$. Then $\Delta y_i = y_{i+1} - y_i = 2h$ and $y_{i-1} \le \xi_i \le y_i$. Thus,

$$\sum_{j=1}^{n/2} f^{(4)}(\xi_j) \Delta y_j = \sum_{j=1}^{n/2} f^{(4)}(\xi_j) 2h$$

is a Riemann Sum for $f^{(4)}$ on $[a, b]$. Hence, the error is

$$E(f) = -\frac{h^5}{90} \sum_{j=1}^{n/2} f^{(4)}(\xi_j) = -\frac{h^4}{180}\left[\sum_{j=1}^{n/2} f^{(4)}(\xi_j) 2h\right] \approx -\frac{h^4}{180}\int_a^b f^{(4)}(x)\,dx = -\frac{h^4}{180}[f'''(b) - f'''(a)].$$

21. Determine to within 10^{-6} the length of the graph of the ellipse with equation $4x^2 + 9y^2 = 36$.

SOLUTION: To find the length of the ellipse we will compute the length of $y = \frac{2}{3}\sqrt{9 - x^2}$, for $0 \le x \le 3$, and multiply this by 4. Hence,

$$L = 4\int_0^3 \sqrt{1 + \left[D_x\left(\frac{2}{3}\sqrt{9 - x^2}\right)\right]^2}\,dx = 4\int_0^3 \sqrt{\frac{81 - 5x^2}{81 - 9x^2}}\,dx.$$

Applying Composite Simpson's rule to this integral gives the length as approximately 15.8655.

22. A car laps a race track in 84 seconds, and the speed of the car is timed according to the entries in the following table

Time	0	6	12	18	24	30	36
Speed	124	134	148	156	147	133	121

Time	42	48	54	60	66	72	78	84
Speed	109	99	85	78	89	104	116	123

How long is the track?

SOLUTION: Since the velocity, $v(t)$, is the derivative of the distance function, $s(t)$, the total distance traveled in the 84 second interval is

$$s(84) = \int_0^{84} v(t)\,dt.$$

However, we do not have an explicit representation for the velocity, only its values at each 6-second interval. We can approximate the distance by doing numerical integration on the velocity. Using the Composite Simpson's rule gives the approximate length of the track as

$$s(84) \approx \int_0^{84} v(t)\, dt \approx \frac{6}{3}[124 + 4(134 + 156 + 133 + 109 + 85 + 89 + 116)$$
$$+ 2(148 + 147 + 121 + 99 + 78 + 104) + 123] = 9858 \text{ feet}.$$

The average lap speed of the car is $\dfrac{9858}{84} \approx 117$ ft/sec, or about 80 mi/hr.

26. The equation

$$\int_0^x \frac{1}{\sqrt{2\pi}} e^{-t^2/2}\, dx = 0.45$$

can be solved using Newton's method on

$$f(x) = \int_0^x \frac{1}{\sqrt{2\pi}} e^{-t^2/2}\, dx - 0.45.$$

To evaluate f at the approximation p_k, we need a quadrature formula to approximate

$$\int_0^{p_k} \frac{1}{\sqrt{2\pi}} e^{-t^2/2}\, dx.$$

Find a solution to $f(x) = 0$ accurate to within 10^{-5} using Newton's method with $p_0 = 0.5$ and the Composite Simpson's rule.

SOLUTION: Since

$$f(x) = \int_0^x \frac{1}{\sqrt{2\pi}} e^{-t^2/2}\, dt - 0.45,$$

the Fundamental Theorem of Calculus implies that

$$f'(x) = \frac{1}{\sqrt{2\pi}} e^{-x^2/2},$$

and Newton's method generates the sequence of iterates

$$p_k = p_{k-1} - \frac{\int_0^{p_{k-1}} e^{-t^2/2}\, dt - 0.45\sqrt{2\pi}}{e^{-p_{k-1}^2/2}}.$$

We approximate the integral in this formula using the Composite Simpson's rule to obtain

$$\int_0^{p_{k-1}} e^{-t^2/2}\, dt \approx \frac{h}{3}\left[e^{-0^2/2} + e^{-p_{k-1}^2/2} \right.$$
$$+ 2\left(e^{-(2h)^2/2} + e^{-(4h)^2/2} + \cdots + e^{-(p_{k-1}-2h)^2/2} \right)$$
$$\left. + 4\left(e^{-(h)^2/2} + e^{-(3h)^2/2} + \cdots + e^{-(p_{k-1}-h)^2/2} \right) \right],$$

where $h = p_{k-1}/n$ and n is the number of steps used in Composite Simpson's rule.

With $n = 20$ and $p_0 = 0.5$, we have $p_6 = 1.644854$.

Exercise Set 4.5, page 218

1. a. Use Romberg integration to compute R_{33} for $\int_{1}^{1.5} x^2 \ln x \, dx$.

SOLUTION: Romberg integration gives the entries in the following table.

$$
\begin{aligned}
R_{11} &= 0.2280741 \\
R_{21} &= 0.2012025 & R_{22} &= 0.1922453 \\
R_{31} &= 0.1944945 & R_{32} &= 0.1922585 & R_{33} &= 0.1922593
\end{aligned}
$$

3. a. Calculate $R_{4,4}$ for the integral in part (a) of Exercise 1.

SOLUTION: This adds a new line to the bottom of the table.

$$R_{41} = 0.1928181 \quad R_{42} = 0.1922593 \quad R_{43} = 0.1922594 \quad R_{44} = 0.1922594$$

8. Romberg integration is used to approximate

$$\int_{0}^{1} \frac{x^2}{1+x^3} \, dx.$$

If $R_{11} = 0.250$ and $R_{22} = 0.2315$, what is R_{21}?

SOLUTION: Since

$$R_{22} = R_{21} + \frac{1}{3}(R_{21} - R_{11}),$$

we have

$$R_{21} = \frac{1}{4}(3R_{22} + R_{11}) = \frac{1}{4}(3(0.2315) + 0.250)) = 0.2361.$$

Notice that no additional information about the integral was needed to solve the problem, only the values of R_{11} and R_{22}.

9. Romberg integration is used to approximate

$$\int_{2}^{3} f(x) \, dx.$$

If $f(2) = 0.51342$, $f(3) = 0.36788$, $R_{31} = 0.43687$, and $R_{33} = 0.43662$, find $f(2.5)$.

SOLUTION: We first use the Trapezoidal rule to calculate

$$R_{11} = \frac{3-2}{2}[f(2) + f(3)] = 0.44065.$$

Then using the following Romberg table,

$$
\begin{aligned}
R_{11} &= 0.44065 \\
R_{21} &= \ ? & R_{22} &= \ ? \\
R_{31} &= 0.43687 & R_{32} &= \ ? & R_{33} &= 0.43662
\end{aligned}
$$

we know that

$$R_{21} = \frac{1}{2}[R_{11} + f(2.5)] = 0.22033 + \frac{1}{2}f(2.5).$$

Now

$$R_{22} = R_{21} + \frac{1}{3}(R_{21} - R_{11})$$

$$= 0.22033 + \frac{1}{2}f(2.5) + \frac{1}{3}\left[0.22033 + \frac{1}{2}f(2.5) - 0.44065\right] = 0.14689 + \frac{2}{3}f(2.5),$$

$$R_{32} = R_{31} + \frac{1}{3}(R_{31} - R_{21})$$

$$= 0.43687 + \frac{1}{3}\left[0.43687 - 0.22033 - \frac{1}{2}f(2.5)\right] = 0.50905 - \frac{1}{6}f(2.5),$$

and

$$R_{33} = R_{32} + \frac{1}{15}(R_{32} - R_{22}).$$

Thus

$$0.43662 = 0.50905 - \frac{1}{6}f(2.5) + \frac{1}{15}\left[0.50905 - \frac{1}{6}f(2.5) - 0.14689 - \frac{2}{3}f(2.5)\right] = 0.53319 - \frac{2}{9}f(2.5)$$

and $f(2.5) = \frac{9}{2}(0.53319 - 0.43662) = 0.43457.$

13. Show that the approximation obtained from $R_{k,2}$ is the same as the Composite Simpson's rule with $h = h_k$.

SOLUTION: We have

$$R_{k,2} = \frac{4R_{k,1} - R_{k-1,1}}{3} = \frac{1}{3}\left[R_{k-1,1} + 2h_{k-1}\sum_{i=1}^{2^{k-2}} f\left(a + \left(i - \frac{1}{2}\right)h_{k-1}\right)\right].$$

Replacing $R_{k-1,1}$ with the appropriate Composite Trapezoidal formula gives

$$R_{k,2} = \frac{1}{3}\left[\frac{h_{k-1}}{2}(f(a) + f(b)) + h_{k-1}\sum_{i=1}^{2^{k-2}-1} f(a + ih_{k-1}) + 2h_{k-1}\sum_{i=1}^{2^{k-2}} f\left(a + \left(i - \frac{1}{2}\right)h_{k-1}\right)\right]$$

$$= \frac{1}{3}\left[h_k(f(a) + f(b)) + 2h_k\sum_{i=1}^{2^{k-2}-1} f(a + 2ih_k) + 4h_k\sum_{i=1}^{2^{k-2}} f(a + (2i - 1)h_k)\right]$$

$$= \frac{h}{3}\left[f(a) + f(b) + 2\sum_{i=1}^{M-1} f(a + 2ih) + 4\sum_{i=1}^{M} f(a + (2i - 1)h)\right],$$

where $h = h_k$ and $M = 2^{k-2}$. This is Composite Simpson's rule with $h = h_k$.

Exercise Set 4.6, page 227

1. a. Compute the Simpson's rule approximations $S(1, 1.5)$, $S(1, 1.25)$, and $S(1.25, 1.5)$ for

$$\int_1^{1.5} x^2 \ln x \, dx,$$

and verify the error estimate given in the approximation formula.

SOLUTION: The Simpson's rule approximations are

$$S(1, 1.5) = 0.19224530, \quad S(1, 1.25) = 0.039372434 \quad \text{and} \quad S(1.25, 1.5) = 0.15288602,$$

and the actual value is 0.19225935. So

$$|S(1, 1.25) + S(1.25, 1.5) - S(1, 1.5)| = |0.039372434 + 0.15288602 - 0.19224530| = 0.000013154$$

and

$$\left| S(1, 1.25) + S(1.25, 1.5) - \int_1^{1.5} x^2 \ln x \, dx \right| = |0.039372434 + 0.15288602 - 0.19225935| = 9.0 \times 10^{-7}.$$

As a consequence, $S(1, 1.25) + S(1.25, 1.5)$ agrees with $\int_1^{1.5} x^2 \ln x \, dx$ about 14.6 times better than it agrees with $S(1, 1.5)$.

2. a. Use Adaptive quadrature to approximate

$$\int_1^{1.5} x^2 \ln x \, dx$$

to within 10^{-3}.

SOLUTION: Initially Simpson's method with $h = 0.25$ gives

$$S(1, 1.5) = 0.19224530.$$

With $h = 0.125$, Simpson's method gives

$$S(1, 1.25) = 0.039372434 \quad \text{and} \quad S(1.25, 1.5) = 0.15288602.$$

Thus

$$S(1, 1.25) + S(1.25, 1.5) = 0.19225845.$$

But

$$|S(1, 1.25) + S(1.25, 1.5) - S(1, 1.5)| = 1.315 \times 10^{-5} < 10(0.001),$$

so we accept the value 0.19225845 as accurate to 10^{-3}.

6. Sketch the graphs of $\sin(1/x)$ and $\cos(1/x)$, and use Adaptive quadrature to approximate the integrals

$$\int_{0.1}^2 \sin \frac{1}{x} \, dx \quad \text{and} \quad \int_{0.1}^2 \cos \frac{1}{x} \, dx$$

to within 10^{-3}

SOLUTION: The graphs of the functions are shown.

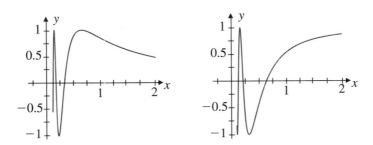

Adaptive quadrature for $f(x) = \sin(1/x)$ uses function evaluations at the 37 points listed below:

$0.1^*, 0.1037109375, 0.107421875^*, 0.1111328125, 0.11484375^*, 0.1185546875, 0.122265625^*,$

$0.1259765625, 0.1296875^*, 0.137109375^*, 0.14453125^*, 0.151953125^*, 0.159375^*, 0.17421875^*,$

$0.1890625^*, 0.20390625^*, 0.21875^*, 0.23359375, 0.2484375^*, 0.26328125, 0.278125^*,$

$0.29296875, 0.3078125^*, 0.32265625, 0.3375^*, 0.396875^*, 0.45625^*, 0.515625^*, 0.575^*, 0.69375^*,$

$0.8125^*, 0.93125^*, 1.05^*, 1.2875^*, 1.525^*, 1.7625^*, 2^*.$

This gives

$$\int_{0.1}^{2} \sin \frac{1}{x} \, dx \approx 1.1454.$$

The nodes used in the evaluation of the integral for $f(x) = \cos(1/x)$ is a subset of those used for the sine function. They are indicated by the asterisk. The approximation for this integral is

$$\int_{0.1}^{2} \cos \frac{1}{x} \, dx \approx 0.67378.$$

The unequal spacing for the evaluations is due to the wide variation in the values of the functions, more for the sine function than the cosine.

Exercise Set 4.7, page 234

1. b. Use Gaussian quadrature with $n = 2$ to approximate

$$\int_{0}^{1} x^2 e^{-x} \, dx.$$

SOLUTION: We first need to convert the integral into one whose integrand has $[-1, 1]$ as its domain. We do this with formula (4.42), which is given by

$$\int_{0}^{1} x^2 e^{-x} \, dx = \frac{1}{2} \int_{-1}^{1} \left(\frac{t+1}{2} \right)^2 e^{-(t+1)/2} \, dt.$$

Using the Gaussian roots given in Table 4.11 (reduced for convenience of representation) gives

$$\int_0^1 x^2 e^{-x}\, dx \approx \frac{1}{2}\left[\left(\frac{-0.5773503+1}{2}\right)^2 e^{-(-0.5773503+1)/2}\right.$$
$$\left.+\left(\frac{0.5773503+1}{2}\right)^2 e^{-(0.5773503+1)/2}\right]$$
$$= 0.1594104.$$

2. b. Use Gaussian quadrature with $n = 3$ to approximate

$$\int_0^1 x^2 e^{-x}\, dx.$$

SOLUTION: This problem is similar to part (b) of Exercise 1, but now we will need to multiply the evaluations at the Gaussian roots by appropriate Gaussian coefficients. As in Exercise 1, the first step is to convert the integral into one whose integrand has $[-1, 1]$ as its domain with the formula

$$\int_0^1 x^2 e^{-x}\, dx = \frac{1}{2}\int_{-1}^1 \left(\frac{t+1}{2}\right)^2 e^{-(t+1)/2}\, dt.$$

Now, using the Gaussian roots in Table 4.12 gives

$$\int_0^1 x^2 e^{-x}\, dx \approx \frac{1}{2}\left[0.5555556\left(\frac{-0.7745967+1}{2}\right)^2 e^{-(-0.7745967+1)/2} + 0.8888889\left(\frac{0+1}{2}\right)^2 e^{-(0+1)/2}\right.$$
$$\left.+0.5555556\left(\frac{0.7745967+1}{2}\right)^2 e^{-(0.7745967+1)/2}\right] = 0.1605954.$$

5. Determine constants a, b, c, and d that will produce a quadrature formula

$$\int_{-1}^1 f(x)\, dx = af(-1) + bf(1) + cf'(-1) + df'(1)$$

that has degree of precision 3.

SOLUTION: The formula must be exact for $f(x) = 1$, x, x^2 and x^3. Thus,

$$f(x) = 1,\ f'(x) = 0: \int_{-1}^1 dx = 2 = a + b;$$

$$f(x) = x,\ f'(x) = 1: \int_{-1}^1 x\, dx = 0 = -a + b + c + d;$$

$$f(x) = x^2,\ f'(x) = 2x: \int_{-1}^1 x^2\, dx = \frac{2}{3} = a + b - 2c + 2d;$$

$$f(x) = x^3,\ f'(x) = 3x^2: \int_{-1}^1 x^3\, dx = 0 = -a + b + 3c + 3d.$$

Solving the four equations gives $a = 1$, $b = 1$, $c = \frac{1}{3}$, and $d = -\frac{1}{3}$.

7. Find the roots of the Legendre polynomials that give the root entries for $n = 2$ and $n = 3$ in Table 4.11. Then use the equations preceeding this table to show that the coefficients in Table 4.11 are correct when $n = 2$ and $n = 3$.

   SOLUTION: The Legendre polynomials P_2 and P_3 are given by

   $$P_2(x) = x^2 - \frac{1}{3} \quad \text{and} \quad P_3(x) = x^3 - \frac{3}{5}x,$$

   so it is easy to verify that the roots in Table 4.11 are correct for $n = 1$ and $n = 2$.
   To verify that the coefficients are correct: For $n = 2$,

   $$c_1 = \int_{-1}^{1} \frac{x + 0.5773502692}{1.1547005} \, dx = 1 \quad \text{and} \quad c_2 = \int_{-1}^{1} \frac{x - 0.5773502692}{-1.1547005} \, dx = 1.$$

   And for $n = 3$:

   $$c_1 = \int_{-1}^{1} \frac{x(x + 0.7745966692)}{1.2} \, dx = \frac{5}{9},$$

   $$c_2 = \int_{-1}^{1} \frac{(x + 0.7745966692)(x - 0.7745966692)}{-0.6} \, dx = \frac{8}{9},$$

   and

   $$c_3 = \int_{-1}^{1} \frac{x(x - 0.7745966692)}{1.2} \, dx = \frac{5}{9}.$$

9. Apply Maple's Composite Gaussian Quadrature routine to approximate $\int_{-1}^{1} x^2 e^x \, dx$ in the following manner.

   a. Use Gaussian Quadrature with $n = 8$ on the single interval $[-1, 1]$.
   b. Use Gaussian Quadrature with $n = 4$ on the intervals $[-1, 0]$ and $[0, 1]$.
   c. Use Gaussian Quadrature with $n = 2$ on the intervals $[-1, -0.5]$, $[-0.5, 0]$, $[0, 0.5]$ and $[0.5, 1]$.
   d. Give an explanation for the accuracy of the results.

   SOLUTION:

   a. The approximations and errors using Maple's routine with

   $f := x^2 \cdot exp(x); a := -1; b := 1;$
   $Quadrature(f(x), x = a..b, method = gaussian[n], partition = p, output = information)$
   give the following approximations to the exact 10-digit value 0.878884623.

n	p	Number of Function Evaluations	Approximation	Error
8	1	8	0.878884623	0.0
4	2	8	0.878884546	8×10^{-8}
2	8	8	0.878387796	5×10^{-4}

   b. Gaussian quadrature chooses the evaluation points in an optimal way for the given interval. If the interval is partitioned it uses then uses points that are not optimal in the Gaussian sense, and less accuracy is to be expected.

Exercise Set 4.8, page 248

1. c. Use Composite Simpson's rule with $n = m = 4$ to approximate

$$\int_{2}^{2.2} \int_{x}^{2x} \left(x^2 + y^3\right) \, dy \, dx.$$

SOLUTION: To gain a better appreciation for the method we will give the details of the implementation of the method. First we need to apply Simpson's rule to the inner integral. This gives

$$\int_{x}^{2x} \left(x^2 + y^3\right) \, dy = \frac{\frac{2x-x}{2}}{3}\left[\left(x^2 + x^3\right) + 4\left(x^2 + \left(x + \frac{x}{2}\right)^3 \right) + \left(x^2 + (2x)^3\right) \right]$$

$$= \frac{x}{6}\left[\left(x^2 + x^3\right) + 4\left(x^2 + \frac{27}{8}x^3 \right) + \left(x^2 + 8x^3\right) \right].$$

As a consequence, we have

$$\int_{2}^{2.2}\left[\int_{x}^{2x} \left(x^2 + y^3\right) \, dy \, dx\right] \approx \int_{2}^{2.2} \frac{x}{6}\left[\left(x^2 + x^3\right) + 4\left(x^2 + \frac{27}{8}x^3 \right) + \left(x^2 + 8x^3\right) \right] \, dx.$$

We now use Simpson's rule on this integral to obtain

$$\int_{2}^{2.2}\left[\int_{x}^{2x} \left(x^2 + y^3\right) \, dy \, dx\right] \approx \frac{0.1}{3}\left[\frac{2}{6}\left[\left(2^2 + 2^3\right) + 4\left(2^2 + \frac{27}{8}(2)^3 \right) + \left(2^2 + 8(2^3)\right) \right] \right.$$

$$+ 4\left\{ \frac{2.1}{6}\left[\left(2.1^2 + 2.1^3\right) + 4\left(2.1^2 + \frac{27}{8}(2.1)^3 \right) + \left(2.1^2 + 8(2.1)^3\right) \right] \right\}$$

$$\left. + \frac{2.2}{6}\left[\left(2.2^2 + 2.2^3\right) + 4\left(2.2^2 + \frac{27}{8}(2.2)^3 \right) + \left(2.2^2 + 8(2.2)^3\right) \right] \right]$$

$$= 16.50865.$$

5. c. Use Gaussian quadrature with $n = m = 2$ to approximate

$$\int_{2}^{2.2} \int_{x}^{2x} \left(x^2 + y^3\right) \, dy \, dx.$$

SOLUTION: This is the same integral as we approximated using Composite Simpson's rule in Exercise 1. Again, to gain a better appreciation for the method we will give the details of the implementation of the method.

First transform the inner integral to an integral on the interval $[-1, 1]$. We do this by mapping the variable y in $[x, 2x]$ onto the variable t in $[-1, 1]$ using the linear transformation

$$t = \frac{1}{x}[2y - 3x]; \quad \text{that is,} \quad y = \frac{1}{2}x(t + 3).$$

Thus,

$$\int_{x}^{2x} \left(x^2 + y^3\right) \, dy = \int_{-1}^{1} \frac{x}{2}\left(x^2 + \frac{x^3}{8}(t + 3)^3 \right) \, dt.$$

Using Gaussian quadrature with $m = 2$ on this integral gives

$$\int_x^{2x} (x^2 + y^3) \, dy \approx \frac{x}{2}\left[x^2 + \frac{x^3}{8}(3.577350269)^3\right] + \frac{x}{2}\left[x^2 + \frac{x^3}{8}(2.422649731)^3\right] = x^3 + 3.75x^4.$$

We now use Gaussian Quadrature with $n = 2$ on the integral with respect to x. This requires transforming x in $[2, 2.2]$ to s in $[-1, 1]$. It is accomplished using the the mapping

$$s = 10x - 21; \quad \text{that is,} \quad x = 0.1s + 2.1.$$

The final approximation using Gaussian quadrature with $n = m = 2$ is

$$\int_2^{2.2} \int_x^{2x} x^2 + y^3 \, dy \, dx \approx \int_2^{2.2} (x^3 + 3.75x^4) \, dx = 0.1 \int_{-1}^1 \left[(0.1s + 2.1)^3 + 3.75(0.1s + 2.1)^4\right] ds$$

$$\approx 0.1\left[\left((0.1)(0.577350269) + 2.1\right)^3 + 3.75\left((0.1)(0.577350269) + 2.1\right)^4\right.$$

$$\left. + \left((0.1)(-0.577350269) + 2.1\right)^3 + 3.75\left((0.1)(-0.577350269) + 2.1\right)^4\right]$$

$$= 16.508633.$$

10. Use the Composite Simpson's Rule Algorithm 4.4 with $n = m = 6$ to approximate

$$\iint_R \sqrt{xy + y^2} \, dA,$$

where R is the region bounded by the graphs of $x + y = 6$, $3y - x = 2$, and $3x - y = 2$.

SOLUTION: The region R is shown in the following figure. To evaluate the integral we need to rewrite it as

$$\iint_R \sqrt{xy + y^2} \, dA = \int_1^2 \int_{(2+x)/3}^{3x-2} \sqrt{xy + y^2} \, dy \, dx + \int_2^4 \int_{(2+x)/3}^{6-x} \sqrt{xy + y^2} \, dy \, dx.$$

We can now apply Algorithm 4.4 to these two double integrals to find that the approximation to the given integral is 13.15229.

13. The surface area of the surface described by $z = f(x, y)$, for (x, y), in the region R is given by

$$\iint_R \sqrt{[f_x(x, y)]^2 + [f_y(x, y)]^2 + 1} \, dA.$$

Use Algorithm 4.4 with $n = m = 8$ to approximate the area of the surface on the hemisphere with equation $x^2 + y^2 + z^2 = 9$, for $z \geq 0$, that lies above the rectangle that is described by $R = \{(x, y) \mid 0 \leq x \leq 1 \text{ and } 0 \leq y \leq 1\}$.

SOLUTION: First note that the surface is described by

$$z = f(x, y) = \sqrt{9 - x^2 - y^2}$$

and that

$$f_x(x,y) = \frac{-x}{\sqrt{9 - x^2 - y^2}} \quad \text{and} \quad f_y(x,y) = \frac{-y}{\sqrt{9 - x^2 - y^2}},$$

so

$$[f_x(x,y)]^2 + [f_y(x,y)]^2 + 1 = \left[\frac{-x}{\sqrt{9 - x^2 - y^2}}\right]^2 + \left[\frac{-y}{\sqrt{9 - x^2 - y^2}}\right]^2 + 1$$

$$= \frac{x^2 + y^2 + 9 - x^2 - y^2}{9 - x^2 - y^2} = \frac{9}{9 - x^2 - y^2}.$$

The area of the surface is consequently

$$\int_0^1 \int_0^1 \frac{3}{\sqrt{9 - x^2 - y^2}} \, dy \, dx.$$

Applying Algorithm 4.4 to this integral gives the approximation 1.040253.

Exercise Set 4.9, page 254

1. a. Use the Composite Simpson's rule to approximate the value of the improper integral

$$\int_0^1 x^{-1/4} \sin x \, dx.$$

SOLUTION: For the sine function, the fourth Taylor polynomial about zero is

$$P_4(x) = x - \frac{x^3}{6}$$

and

$$\int_0^1 x^{-1/4} P_4(x) \, dx = \int_0^1 x^{3/4} - \frac{1}{6} x^{11/4} \, dx = \left[\frac{4}{7} x^{7/4} - \frac{2}{45} x^{15/4}\right]_0^1 = \frac{4}{7} - \frac{2}{45} = 0.52698412.$$

Now define

$$G(x) = \begin{cases} x^{-1/4}(\sin x - P_4(x)), & \text{if } 0 < x \le 1 \\ 0, & \text{if } x = 0, \end{cases}$$

and apply Composite Simpson's rule to $G(x)$. This gives

$$\int_0^1 G(x) \, dx = \frac{0.25}{3}[G(0) + 4G(0.25) + 2G(0.5) + 4G(0.75) + G(1)]$$

$$= \frac{0.25}{3}[0 + 4(0.00001149) + 2(0.00030785) + 4(0.00209676) + (0.00813765)]$$

$$= 0.00143220.$$

Hence

$$\int_0^1 x^{-1/4} \sin x \, dx \approx 0.52698412 + 0.00143220 = 0.52841632.$$

3. a. Use the Composite Simpson's rule to approximate the value of the improper integral

$$\int_1^\infty \frac{1}{x^2 + 9}\, dx.$$

SOLUTION: First transform the integral using the substitution $t = x^{-1}$ and $dx = -t^{-2}\, dt$. This produces

$$\int_1^\infty \frac{1}{x^2 + 9}\, dx = \int_1^0 \frac{1}{(1/t)^2 + 9}\left(-t^{-2}\right) dt = \int_0^1 \frac{1}{1 + 9t^2}\, dt.$$

Now we apply the Composite Simpson's rule to this integral to obtain

$$\int_1^\infty \frac{1}{x^2 + 9}\, dx \approx 0.4112649.$$

4. a. Approximate, to within 10^{-6}, the value of the improper integral $\int_0^\infty \frac{1}{1 + x^4}\, dx$.

SOLUTION: First we break up the integral into

$$\int_0^\infty \frac{1}{1 + x^4}\, dx = \int_0^1 \frac{1}{1 + x^4}\, dx + \int_1^\infty \frac{1}{1 + x^4}\, dx.$$

The first integral on the right can be approximated using Composite Simpson's rule with $n = 16$ to obtain 0.8669735. The second integral on the right is converted using the substitution in Exercise 2:

$$t = x^{-1} \quad \text{and} \quad dx = -t^{-2}\, dt.$$

This produces

$$\int_1^\infty \frac{1}{1 + x^4}\, dx = \int_1^0 \frac{1}{1 + (1/t)^4}\left(-t^{-2}\right) dt = \int_0^1 \frac{t^2}{t^4 + 1}\, dt.$$

This integral can be approximated directly using the Composite Simpson's rule with $n = 16$ to give 0.2437483. Summing these approximations gives

$$\int_0^\infty \frac{1}{1 + x^4}\, dx \approx 1.1107218.$$

6. Show that the Laguerre polynomials can be used to give a formula with degree of precision at least five for approximating improper integrals of the form

$$\int_0^\infty e^{-x} f(x)\, dx.$$

SOLUTION: The polynomial $L_n(x)$ has n distinct zeros in $[0, \infty)$.

Let $x_1, ..., x_n$ be the n distinct zeros of L_n and define, for each $i = 1, ..., n$,

$$c_i = \int_0^\infty e^{-x} \prod_{\substack{j=1 \\ j \neq i}}^n \frac{(x - x_j)}{(x_i - x_j)}\, dx.$$

Let $P(x)$ be any polynomial of degree $n-1$ or less, and let $P_{n-1}(x)$ be the $(n-1)$st Lagrange polynomial for P on the nodes $x_1, ..., x_n$. As in the proof of Theorem 4.7,

$$\int_0^\infty P(x)e^{-x}\, dx = \int_0^\infty P_{n-1}(x)e^{-x}\, dx = \sum_{i=1}^n c_i P(x_i),$$

so the quadrature formula is exact for polynomials of degree $n-1$ or less.

If $P(x)$ has degree $2n-1$ or less, $P(x)$ can be divided by the nth Laguerre polynomial $L_n(x)$ to obtain

$$P(x) = Q(x)L_n(x) + R(x),$$

where $Q(x)$ and $R(x)$ are both polynomials of degree less than n. As in the proof of Theorem 4.7, the orthogonality of the Laguerre polynomials on $[0, \infty)$ implies that

$$Q(x) = \sum_{i=0}^{n-1} d_i L_i(x),$$

for some constants d_i. Thus

$$\int_0^\infty e^{-x} P(x)\, dx = \int_0^\infty \sum_{i=0}^{n-1} d_i L_i(x) L_n(x) e^{-x}\, dx + \int_0^\infty e^{-x} R(x)\, dx$$

$$= \sum_{i=0}^{n-1} d_i \int_0^\infty L_i(x) L_n(x) e^{-x}\, dx + \sum_{i=1}^n c_i R(x_i)$$

$$= 0 + \sum_{i=1}^n c_i R(x_i) = \sum_{i=1}^n c_i R(x_i).$$

But

$$P(x_i) = Q(x_i)L_n(x_i) + R(x_i) = 0 + R(x_i) = R(x_i),$$

so

$$\int_0^\infty e^{-x} P(x)\, dx = \sum_{i=1}^n c_i P(x_i).$$

Hence, the quadrature formula has degree of precision $2n-1$.

Initial-Value Problems for Ordinary Differential Equations

Exercise Set 5.1, page 264

1. a. Use Theorem 5.4 to show that the initial-value problem

$$y' = y \cos t, \quad 0 \leq t \leq 1, \quad y(0) = 1$$

has a unique solution, and find the solution.

SOLUTION: Since $f(t, y) = y \cos t$, we have $\dfrac{\partial f}{\partial y}(t, y) = \cos t$, and f satisfies a Lipschitz condition in y with $L = 1$ on

$$D = \{(t, y) \mid 0 \leq t \leq 1, -\infty < y < \infty\}.$$

Also, f is continuous on D, so there exists a unique solution.

To determine this unique solution, we can rewrite the equation as follows and integrate with respect to t. This gives

$$y' = y \cos t \quad \Rightarrow \quad \frac{y'}{y} = \cos t \quad \Rightarrow \quad \ln|y(t)| = \sin t + C,$$

for some constant C. This implies that

$$|y(t)| = e^{\sin t + C} = e^C e^{\sin t}, \quad \text{so} \quad y(t) = \hat{C} e^{\sin t},$$

for some constant $\hat{C}$. Using the initial condition gives $\hat{C} = 1$, and therefore, the solution is $y(t) = e^{\sin t}$.

5. a. Show that the equation $y^3 t + yt = 2$ implicitly defines a solution to the initial-value problem

$$y' = -\frac{y^3 + y}{(3y^2 + 1)t}, \quad 1 \leq t \leq 2, \quad y(1) = 1,$$

and use Newton's method to approximate $y(2)$.

SOLUTION: Differentiating the implicit equation $y^3 t + yt = 2$ with respect to t gives

$$3y^2 y' t + y^3 + y' t + y = 0,$$

and solving for y' gives the original differential equation. Setting $t = 1$ and $y = 1$ verifies the initial condition.

89

To approximate $y(2)$, we will use Newton's method to solve the equation $g(y) = y^3 + y - 1 = 0$. This produces the sequence of approximations

$$y_{n+1} = y_n - \frac{g(y_n)}{g'(y_n)} = y_n - \frac{y_n^3 + y_n - 1}{3y_n^2 + 1}$$

and gives $y(2) \approx 0.6823278$.

6. Prove Theorem 5.3 by applying the Mean Value Theorem to $f(t, y)$.

SOLUTION: Let (t, y_1) and (t, y_2) be in D with $y_1 < y_2$. Hold t fixed, and define $g(y) = f(t, y)$. Since the line joining (t, y_1) to (t, y_2) lies in D and f is continuous on D, we have $g \in C[y_1, y_2]$. Further, $g'(y) = \frac{\partial f}{\partial y}(t, y)$. Using the Mean Value Theorem on g, a number ξ, with $y_1 < \xi < y_2$, exists for which

$$g(y_2) - g(y_1) = g'(\xi)(y_2 - y_1).$$

Thus

$$f(t, y_2) - f(t, y_1) = \frac{\partial f(t, \xi)}{\partial y}(y_2 - y_1)$$

and

$$|f(t, y_2) - f(t, y_1)| \le L|y_2 - y_1|.$$

So f satisfies a Lipschitz condition on D in the variable y with Lipschitz constant L.

9. Picard's method for solving the initial-value problem

$$y' = f(t, y), \quad a \le t \le b, \quad y(a) = \alpha$$

is described by the sequence $y_0(t) = \alpha$, and for $k \ge 1$,

$$y_k(t) = \alpha + \int_a^t f(\tau, y_{k-1}(\tau)) \, d\tau.$$

a. Integrate $y' = f(t, y(t))$, and use the initial-condition to derive Picard's method.

b. Generate $y_0(t), y_1(t), y_2(t)$, and $y_3(t)$ for the initial-value problem

$$y' = -y + t + 1, \quad 0 \le t \le 1, \quad y(0) = 1.$$

c. Compare the results in part (b) to the Maclaurin series for the exact solution $y(t) = t + e^{-t}$.

SOLUTION: **a.** Since $y' = f(t, y(t))$, we have

$$\int_a^t y'(z) \, dz = \int_a^t f(z, y(z)) \, dz.$$

So

$$y(t) - y(a) = \int_a^t f(z, y(z)) \, dz \quad \text{and} \quad y(t) = \alpha + \int_a^t f(z, y(z)) \, dz.$$

The iterative method follows from this equation.

b. We have $y_0(t) = 1$, since this is the initial condition. The formula implies that

$$y_1(t) = y_0(t) + \int_0^t (-y_0(\tau) + \tau + 1)\, d\tau = 1 + \int_0^t \tau\, d\tau = 1 + \frac{t^2}{2}.$$

Proceeding in the same manner produces

$$y_2(t) = 1 + \frac{1}{2}t^2 - \frac{1}{6}t^3 \quad \text{and} \quad y_3(t) = 1 + \frac{1}{2}t^2 - \frac{1}{6}t^3 + \frac{1}{24}t^4.$$

c. The Maclaurin series for $y(t)$ is simply t added to the Maclaurin series for e^{-t}, so

$$y(t) = 1 + \frac{1}{2}t^2 - \frac{1}{6}t^3 + \frac{1}{24}t^4 - \frac{1}{120}t^5 + \cdots .$$

The nth term of Picard's method is the nth partial sum of the Maclaurin series for the solution.

Exercise Set 5.2, page 273

1. a. Use Euler's method with $h = 0.5$ to approximate the solution to

$$y' = te^{3t} - 2y, \quad 0 \le t \le 1, \quad y(0) = 0.$$

SOLUTION: Since $w_0 = \alpha = y(0) = 0$ and $h = 0.5$, the first step gives

$$y(0.5) \approx w_1 = 0 + 0.5[0 \cdot e^{3 \cdot 0} - 0] = 0,$$

and the second step gives

$$y(1.0) \approx w_2 = 0 + 0.5 \left[0.5 \cdot e^{3(0.5)} - 0 \right] = 1.1204223.$$

3. a. The exact solution to the initial-value problem

$$y' = te^{3t} - 2y, \quad 0 \le t \le 1, \quad y(0) = 0$$

is

$$y(t) = \frac{1}{5}te^{3t} - \frac{1}{25}e^{3t} + \frac{1}{25}e^{-2t}.$$

Determine an error bound for the approximation obtained in Exercise 1(a).

SOLUTION: The first two derivatives of $y(t)$ are

$$y'(t) = \frac{2}{25}e^{3t} + \frac{3}{5}te^{3t} - \frac{2}{25}e^{-2t} \quad \text{and} \quad y''(t) = \frac{21}{25}e^{3t} + \frac{9}{5}te^{3t} + \frac{4}{25}e^{-2t}.$$

Since $y''(t)$ is increasing on $[0, 1]$, the maximum value on $[0, 1]$ is $M = y''(1) = 53.047$. The partial derivative of f with respect to y is

$$\frac{\partial f}{\partial y}(t, y) = -2, \quad \text{so} \quad L = \left| \frac{\partial f}{\partial y}(t, y) \right| = 2$$

is a Lipschitz constant for f on $[0, 1]$. Hence, an error bound is given by

$$\frac{hM}{2L} \left[e^{L(t_i - a)} - 1 \right] = \frac{0.5(53.047)}{2(2)} \left(e^{2t_i} - 1 \right).$$

At $t_1 = 0.5$, this bound is 11.3938, and at $t_2 = 1.0$, it is 42.3654.

5. b. Use Euler's method to approximate the solutions for the initial-value problem

$$y' = 1 + \frac{y}{t} + \left(\frac{y}{t}\right)^2, \quad 1 \le t \le 3, \quad y(1) = 0, \text{ with } h = 0.2.$$

SOLUTION: We have $N = (3 - 1)/0.2 = 10$. The values of t to approximate $y(t)$ are $t_0 = 1.0$, $t_1 = 1.2, \ldots t_9 = 2.8$, and $t_{10} = 3.0$. Using Euler's method gives

$$w_0 = y(1) = 0,$$

$$w_1 = w_0 + hf(t_0, w_0) = 0 + 0.2 \left(1 + \frac{0}{1}\left(\frac{0}{1}\right)^2\right) = 0.2 \approx y(1.2),$$

$$w_2 = w_1 + hf(t_1, w_1) = 0.2 + 0.2 \left(1 + \frac{0.2}{1.2} + \left(\frac{0.2}{1.2}\right)^2\right) = 0.4388889 \approx y(1.4).$$

The remaining entries are in the table.

i	t_i	w_i	$y(t_i)$
3	1.6	0.7212428	0.8127527
4	1.8	1.0520380	1.1994386
5	2.0	1.4372511	1.6612818
6	2.2	1.8842608	2.2135018
7	2.4	2.4022696	2.8765514
8	2.6	3.0028372	3.6784753
9	2.8	3.7006007	4.6586651
10	3.0	4.5142774	5.8741000

9. The initial-value problem

$$y' = \frac{2}{t}y + t^2 e^t, \quad 1 \le t \le 2, \quad y(1) = 0$$

has the exact solution $y(t) = t^2 (e^t - e)$.

a. Use Euler's method with $h = 0.05$ to approximate the solution.

b. Use linear interpolation to approximate $y(1.04)$, $y(1.55)$, and $y(1.97)$.

c. Use Eq. (5.10) to compute the value of h needed to ensure that $|y(t_i) - w_i| \le 0.1$.

SOLUTION: Euler's method gives the approximations in the following table. These are the results we will need for part (b).

i	t_i	w_i	$|y(t_i) - w_i|$
0	1.0	0	0
1	1.1	0.271828	0.07409
5	1.5	3.18744	0.7802
6	1.6	4.62080	1.100
9	1.9	11.7480	2.575
10	2.0	15.3982	3.285

b. Linear interpolation on the most appropriate data gives

$$y(1.04) \approx 0.6w_0 + 0.4w_1 = 0.108731, \quad y(1.55) \approx 0.5w_5 + 0.5w_6 = 3.90412,$$

and

$$y(1.97) \approx 0.3w_9 + 0.7w_{10} = 14.3031.$$

c. Since

$$y'(t) = 2t\left(e^t - e\right) + t^2 e^t \quad \text{and} \quad y''(t) = \left(2 + 4t + t^2\right)e^t - 2e,$$

the function $y''(t)$ is increasing on $[1, 2]$ and

$$|y''(t)| \le y''(2) = 14e^2 - 2e = 98.0102.$$

Since a Lipschitz constant for $f(t, y)$ is $L = 2$, to have

$$|y(t_i) - w_i| \le \frac{hM}{2L}\left[e^{L(t_i - a)} - 1\right] < 0.1,$$

we need to choose h so that

$$\frac{98.0102h}{4}\left[e^{2(2-1)} - 1\right] < 0.1 \quad \Rightarrow \quad h < \frac{0.4}{98.0102\left(e^2 - 1\right)} = 0.00064.$$

12. Apply Euler's method to the initial-value problem

$$y' = -10y, \quad 0 \le t \le 2, \quad y(0) = 1,$$

which has exact solution $y(t) = e^{-10t}$, and explain why the behavior does not violate Theorem 5.9.

SOLUTION: For $t > 0$, the approximations are zero and are not adequate approximations until the solution becomes close to zero. This behavior does not violate Theorem 5.9.

13. b. Use the results of Exercise 5(b) and linear interpolation with $y(2.1)$ and $y(2.75)$ to approximate the following values of $y(t)$. Compare the approximations obtained to the actual values obtained using the functions given in Exercise 7(b).

SOLUTION: We have $y(2.0) \approx w_5 = 1.4372511$ and $y(2.2) \approx w_6 = 1.8842608$. So

$$y(2.1) \approx \frac{2.1 - 2.2}{2.0 - 2.2}\, 1.4372511 + \frac{2.1 - 2.0}{2.2 - 2.0}\, 1.8842608 = 1.6607560.$$

The actual value is $y(2.1) = 1.9249617$ so, the absolute error is 0.2642057.

We have $y(2.6) \approx w_8 = 3.0028372$ and $y(2.8) \approx w_9 = 3.7006007$. So

$$y(2.1) \approx \frac{2.75 - 2.6}{2.6 - 2.8}\, 3.0028372 + \frac{2.75 - 2.8}{2.8 - 2.6}\, 3.7006007 = 3.5261598.$$

The actual value is $y(2.75) = 4.3941697$, so the absolute error is 0.8680099.

Exercise Set 5.3, page 281

1. a. Use Taylor's method of order two to approximate the solution for the initial-value problem $y' = te^{3t} - 2y$, $0 \le t \le 1$, $y(0) = 0$, with $h = 0.5$.

SOLUTION: The first approximation is

$$w_1 = w_0 + h\left(t_0 e^{3t_0} - 2w_0\right) + \frac{h^2}{2}\left(t_0 e^{3t_0} + e^{3t_0} + 4w_0\right)$$

$$= 0 + 0.5(0 - 0) + \frac{(0.5)^2}{2}(0 + 1 + 0) = 0.125,$$

and the second is

$$w_2 = w_1 + h\left(t_1 e^{3t_1} - 2w_1\right) + \frac{h^2}{2}\left(t_1 e^{3t_1} + e^{3t_1} + 4w_1\right)$$

$$= 0.125 + 0.5\left(0.5e^{1.5} - 2(0.125)\right) + \frac{(0.5)^2}{2}\left(0.5e^{1.5} + e^{1.5} + 4(0.125)\right)$$

$$= 2.02323897.$$

3. a. Use Taylor's method of order four to approximate the solution to $y' = te^{3t} - 2y$, $0 \le t \le 1$, $y(0) = 0$, with $h = 0.5$.

SOLUTION: The first approximation is

$$w_1 = w_0 + h\left(t_0 e^{3t_0} - 2w_0\right) + \frac{h^2}{2}\left(t_0 e^{3t_0} + e^{3t_0} + 4w_0\right)$$

$$+ \frac{h^3}{6}\left(7t_0 e^{3t_0} + 4e^{3t_0} - 8w_0\right) + \frac{h^4}{24}\left(13t_0 e^{3t_0} + 19e^{3t_0} + 16w_0\right)$$

$$= 0 + 0.5(0 - 0) + \frac{(0.5)^2}{2}(0 + 1 + 0) + \frac{(0.5)^3}{6}(0 + 4 - 0) + \frac{(0.5)^4}{24}(0 + 19 + 0)$$

$$= 0.2578125,$$

and the second is

$$w_2 = w_1 + h\left(t_1 e^{3t_1} - 2w_1\right) + \frac{h^2}{2}\left(t_1 e^{3t_1} + e^{3t_1} + 4w_1\right)$$

$$+ \frac{h^3}{6}\left(7t_1 e^{3t_1} + 4e^{3t_1} - 8w_1\right) + \frac{h^4}{24}\left(13t_1 e^{3t_1} + 19e^{3t_1} + 16w_1\right)$$

$$= 0.2578125 + 0.5(1.72521954) + \frac{(0.5)^2}{2}(7.75378361)$$

$$+ \frac{(0.5)^3}{6}(31.55016803) + \frac{(0.5)^4}{24}(118.4080713)$$

$$= 3.05529474.$$

5. c. Use Taylor's method of order two to approximate the solution for initial-value problem

$$y' = \frac{1}{t}(y^2 + y), \quad 1 \le t \le 3, \quad y(1) = -2, \text{ with } h = 0.5.$$

SOLUTION: We have $t_0 = 1.0$, $w_0 = -2$, $h = 0.5$

For $f(t, y) = \dfrac{y^2 + y}{t}$ we have

$$
\begin{aligned}
D_t f(t, y(t)) &= \frac{t(2y(t)y'(t) + y'(t)) - y(t)^2 - y(t)}{t^2} \\
&= \frac{t \left(2y(t) \left(y(t)^2 + y(t) \right) /t + \left(y(t)^2 + y(t) \right) /t \right) - y(t)^2 - y(t)}{t^2} \\
&= \frac{2y(t)^2(y(t) + 1)}{t^2}.
\end{aligned}
$$

Taylor's method of order 2 uses the formula

$$
w_{i+1} = w_i + hf(t_i, w_i) + \frac{h^2}{2} D_t f(t_i, w_i).
$$

In this case we have $h = 0.5$ and $t_0 = 1.0$ which gives

$$
w_1 = (-2.0) + (0.5)(2.0) + \frac{(0.5)^2}{2}(-8.0) = -2.0,
$$

$$
w_2 = (-2.0) + (0.5)(1.\overline{3}) + \frac{(0.5)^2}{2}(-3.\overline{5}) = -1.\overline{7},
$$

$$
w_3 = -1.585733883 \quad \text{and} \quad w_4 = -1.458884601
$$

7. c. Repeat Exercise 5(c) using Taylor's method of order four.

SOLUTION: In Exercise 5(c) we found that

$$
D_t f(t, y(t)) = \frac{2y(t)^3 + 2y(t)^2}{t^2}.
$$

Differentiating $f'(t, y(t))$ with respect to t and then substituting $f(t, y)$ for $y'(t)$ gives

$$
D_t^2 f(t, y(t)) = \frac{6y(t)^3(y(t) + 1)}{t^3}.
$$

Continuing the differentiation and substitution process produces

$$
D_t^3 f(t, y(t)) = \frac{24y(t)^4(y(t) + 1)}{t^4}.
$$

This gives

$$
w_1 = (-2.0) + (0.5)(2.0) + \frac{(0.5)^2}{2}(-8.0) + \frac{(0.5)^3}{6}(48) + \frac{(0.5)^4}{24}(-384) = -2.0,
$$

$$
w_2 = (-2.0) + (0.5)(1.\overline{3}) + \frac{(0.5)^2}{2}(-3.\overline{5}) + \frac{(0.5)^3}{6}(14.\overline{2}) + \frac{(0.5)^4}{24}(-75.\overline{851}) = -1.679012346,
$$

$$
w_3 = -1.484492968 \quad \text{and} \quad w_4 = -1.374439974.
$$

9. Consider the following initial-value problem, which has the exact solution $y(t) = t^2 (e^t - e)$:

$$
y' = \frac{2}{t}y + t^2 e^t, \quad 1 \le t \le 2, \quad y(1) = 0.
$$

a. Use Taylor's method of order two with $h = 0.1$ to approximate the solution.

b. Use the results of part (a) and linear interpolation to approximate $y(1.04)$, $y(1.55)$, and $y(1.97)$.

c. Use Taylor's method of order four with $h = 0.1$ to approximate the solution.

d. Use the results of part (c) and cubic Hermite interpolation to approximate $y(1.04)$, $y(1.55)$, and $y(1.97)$.

SOLUTION: **a.** The results we will need in part (b) are given in the following table.

i	t_i	w_i
1	1.1	0.3397852
5	1.5	3.910985
6	1.6	5.643081
9	1.9	14.15268
10	2.0	18.46999

b. Linear interpolation on the most appropriate data gives

$$y(1.04) \approx 0.6w_0 + 0.4w_1 = 0.1359139,$$
$$y(1.55) \approx 0.5w_5 + 0.5w_6 = 4.777033,$$

and

$$y(1.97) \approx 0.3w_9 + 0.7w_{10} = 17.17480.$$

c. The results we will need in part (d) are given in the following table.

i	t_i	w_i
1	1.1	0.3459127
5	1.5	3.967603
6	1.6	5.720875
9	1.9	14.32290
10	2.0	18.68287

d. The difference table needed for the $y(1.04)$ approximation is given below.

1.0	0			
1.0	0	2.718282		
1.1	0.3459127	3.459127	7.40845	
1.1	0.3459127	4.263973	8.04846	6.40009

This gives

$$H(t) = 2.718282(t-1) + 7.40845(t-1)^2 + 6.40009(t-1)^2(t-1.1)$$

and $H(1.04) = 0.1199704$.

To approximate $y(1.55)$, we use the following difference table.

1.5	3.967603			
1.5	3.967603	15.373937		
1.6	5.720875	17.53272	21.58782	
1.6	5.720875	19.83086	22.98137	13.93545

This gives

$$H(t) = 3.967603 + 15.373937(t-1.5) + 21.58782(t-1.5)^2 + 13.93545(t-1.5)^2(t-1.6)$$

and $H(1.55) = 4.788527$.

Finally, to approximate $y(1.97)$ we use the following table.

1.9	14.32290			
1.9	14.32290	39.21282		
2.0	18.68287	43.59970	43.86884	
2.0	18.68287	48.23909	46.39394	25.25102

This gives

$$H(t) = 14.32290 + 39.21282(t-1.9) + 43.86884(t-1.9)^2 + 25.25102(t-1.9)^2(t-2.0)$$

and $H(1.97) = 17.27904$.

The exact values to the digits listed are

$$y(1.04) = 0.1199875, \quad y(1.55) = 4.788635, \quad \text{and} \quad y(1.97) = 17.27930.$$

If you compare the approximations in this part of the problem with those in part (b) and the corresponding results in Exercise 9 of Section 5.2, you will find that the results here are at least 100 times more accurate.

Exercise Set 5.4, page 291

3. b. Use the Modified Euler method to approximate the solutions to the following initial-value problem, and compare the results to the actual solution $y(t) = t\tan(\ln t)$.

$$y' = 1 + \frac{y}{t} + \left(\frac{y}{t}\right)^2, \quad 1 \le t \le 3, \quad y(1) = 0, \text{ with } h = 0.2.$$

SOLUTION: We have $h = 0.2$, $t_0 = 1.0$, $w_0 = 0$ and, for each i,

$$w_{i+1} = w_i + \frac{h}{2}\left(f(t_{i+1}, w_i + hf(t_i, w_i))\right).$$

So

$$w_1 = 0.0 + \frac{0.2}{2}(1.0 + f(1.2, 0.2)) = 0.1(1.0 + 1.194444445) = 0.219\overline{4},$$

and

$$w_2 = 0.219\overline{4} + \frac{0.2}{2}(1.216311924 + 1.439738357) = 0.4850495.$$

The complete entries are shown in the table.

| | Modified Euler Method | | |
t_i	w_i	$y(t_i)$	Error
1.2	0.2194444	0.2212428	1.8×10^{-3}
1.4	0.4850495	0.4896817	4.6×10^{-3}
1.6	0.8040116	0.8127527	8.7×10^{-3}
1.8	1.1848560	1.1994386	1.5×10^{-2}
2.0	1.6384229	1.6612818	2.3×10^{-2}
2.2	2.1788772	2.2135018	3.4×10^{-2}
2.4	2.8250651	2.8765514	5.1×10^{-2}
2.6	3.6025247	3.6784753	7.6×10^{-2}
2.8	4.5466136	4.6586651	1.1×10^{-1}
3.0	5.7075699	5.8741000	1.6×10^{-1}

7. b. Repeat Exercise 3(b) using the Midpoint method.

SOLUTION: We have $h = 0.2$, $t_0 = 1.0$, $w_0 = 0$ and, for each i,

$$w_{i+1} = w_i + h\left(f\left(t_i + \frac{h}{2}, w_i + \frac{h}{2}f(t_i, w_i)\right)\right).$$

So

$$w_1 = 0.0 + 0.2f(1.1, 0.0 + 0.1f(1.0, 0.0)) = 0.2(1.099173554) = 0.219\overline{8}347,$$

and

$$w_2 = 0.2198347 + 0.2(1.331711680) = 0.4861770.$$

The complete entries are shown in the table.

| | Midpoint Method | | |
t_i	w_i	$y(t_i)$	Error
1.2	0.2198347	0.2212428	1.4×10^{-3}
1.4	0.4861770	0.4896817	3.5×10^{-3}
1.6	0.8061849	0.8127527	6.7×10^{-3}
1.8	1.1884393	1.1994386	1.1×10^{-2}
2.0	1.6438889	1.6612818	1.7×10^{-2}
2.2	2.1868609	2.2135018	2.7×10^{-2}
2.4	2.8364357	2.8765514	4.0×10^{-2}
2.6	3.6184926	3.6784753	6.0×10^{-2}
2.8	4.5688944	4.6586651	9.0×10^{-2}
3.0	5.7386475	5.8741000	1.4×10^{-1}

11. b. Repeat Exercise 3(b) using Heun's method.

SOLUTION: Heun's method uses the difference equation, for each $i = 0, 1, \ldots, N - 1$,

$$k_1 = hf(t_i, w_i),$$
$$k_2 = hf\left(t_i + \frac{1}{3}h, w_i + \frac{1}{3}k_1\right),$$
$$k_3 = hf\left(t_i + \frac{2}{3}h, w_i + \frac{2}{3}k_2\right),$$
$$w_{i+1} = w_i + \frac{1}{4}(k_1 + 3k_3)$$

With $t_0 = 1.10$, $h = 0.2$, $w_0 = y(1) = 0$, $N = 10$, and the given value of $f(t, y(t))$ we have

$$k_1 = 0.2f(0.1, 0.0) = 0.2(1.0) = 0.2$$
$$k_2 = 0.2f\left(1.0 + \frac{1}{3}(0.2), 0.0 + \frac{1}{3}(0.2)\right) = 0.2(1.0664063) = 0.2132813$$
$$k_3 = 0.2f\left(1.0 + \frac{2}{3}(0.2), 0.0 + \frac{2}{3}(0.2132813)\right) = 0.2(1.1411997) = 0.2282399$$
$$w_1 = w_0 + \frac{1}{4}(k_1 + 3k_3) = 0.0 + 0.25(0.2 + 3(0.2282399)) = 0.2211799.$$

All the approximations and the actual values are given in the table.

| | Heun's Method | | |
t_i	w_i	$y(t_i)$	Error
1.2	0.2211799	0.2212428	6.3×10^{-5}
1.4	0.4895074	0.4896817	1.7×10^{-4}
1.6	0.8124061	0.8127527	3.5×10^{-4}
1.8	1.1988343	1.1994386	6.0×10^{-4}
2.0	1.6438889	1.6612818	9.7×10^{-4}
2.2	2.2119484	2.2135018	1.6×10^{-3}
2.4	2.8364357	2.8765514	2.4×10^{-3}
2.6	3.6747836	3.6784753	3.7×10^{-3}
2.8	4.6529736	4.6586651	5.7×10^{-3}
3.0	5.8652189	5.8741000	8.9×10^{-3}

15. b. Repeat Exercise 3(b) using the Runge-Kutta method of order four.

The Runge-Kutta Order Four method uses the difference equation, for each $i = 0, 1, \ldots, N-1$,

$$k_1 = hf(t_i, w_i),$$
$$k_2 = hf\left(t_i + \frac{1}{2}h, w_i + \frac{1}{2}k_1\right),$$
$$k_3 = hf\left(t_i + \frac{1}{2}h, w_i + \frac{1}{2}k_2\right),$$
$$k_4 = hf(t_i + h, w_i + k_3),$$
$$w_{i+1} = w_i + \frac{1}{6}(k_1 + 2k_2 + 2k_3 + k_4)$$

With $t_0 = 1.10$, $h = 0.2$, $w_0 = y(1) = 0$, $N = 10$, and the given value of $f(t, y(t))$ we have

$$k_1 = 0.2f(0.1, 0.0) = 0.2(1.0) = 0.2$$
$$k_2 = 0.2f\left(1.0 + \frac{1}{2}(0.2), 0.0 + \frac{1}{2}(0.2)\right) = 0.2(1.0991736) = 0.2198347$$
$$k_3 = 0.2f\left(1.0 + \frac{1}{2}(0.2), 0.0 + \frac{1}{2}(0.2198347)\right) = 0.2(1.1099098) = 0.2219820$$
$$k_4 = 0.2f(1.2, 0.0 + 0.2219820) = 0.2(1.2192044) = 0.2438409$$
$$w_1 = w_0 + \frac{1}{6}(k_1 + 2k_2 + 2k_3 + k_4)$$
$$= 0.0 + \frac{1}{6}(0.2 + 2(0.2198347) + 2(0.2219820) + 0.2438409) = 0.2212457.$$

All the approximations and the actual values are given in the table.

| | Runge-Kutta Order Four | | |
t_i	w_i	$y(t_i)$	Error
1.2	0.2212457	0.2212428	2.9×10^{-6}
1.4	0.4896842	0.4896817	2.5×10^{-6}
1.6	0.8127522	0.8127527	5.8×10^{-7}
1.8	1.1994320	1.1994386	6.6×10^{-6}
2.0	1.6612651	1.6612818	1.7×10^{-5}
2.2	2.2134693	2.2135018	3.2×10^{-5}
2.4	2.8764941	2.8765514	5.7×10^{-5}
2.6	3.6783790	3.6784753	9.6×10^{-5}
2.8	4.6585063	4.6586651	1.6×10^{-4}
3.0	5.8738386	5.8741000	2.6×10^{-4}

25. b. Use the results of Exercise 15 and Cubic Hermite interpolation with $y(2.1)$ and $y(2.75)$ to approximate values of $y(t)$, and compare the approximations to the actual values.

SOLUTION: To approximate $y(2.1)$ we use $z_0 = z_1 = x_0 = 2.0$ and $z_2 = z_3 = x_1 = 2.2$ and construct the divided difference table

z	$f(z)$	First divided differences	Second divided differences	Third divided difference
$z_0 = 2.0$ $f[z_0] = f(2.0) = 1.6612651$				
		$f[z_0, z_1] = f'(2.0) = 2.5205830$		
$z_1 = 2.0$ $f[z_1] = f(2.0) = 1.6612651$			$f[z_0, z_1, z_2] = 1.2021900$	
		$f[z_1, z_2] = 2.7610210$		$f[z_0, z_1, z_2, x_3] = 0.4236432$
$z_2 = 2.2$ $f[z_2] = f(2.2) = 2.2134693$			$f[z_1, z_2, z_3] = 1.2869187$	
		$f[z_2, z_3] = f'(2.2) = 3.0184047$		
$z_3 = 2.2$ $f[z_3] = f(2.2) = 2.2134693$				

The cubic Hermite polynomial is

$$H_3(x) = 1.6612651 + 2.5205830(x - 2.0) + 1.2021900(x - 2.0)^2 + 0.4236432((x - 2.0)^2(x - 2.2),$$

so $H_3(2.1) = 1.949217 \approx y(2.1) = 1.924962$.

To approximate $y(2.75)$ we use $z_0 = z_1 = x_0 = 2.6$ and $z_2 = z_3 = x_1 = 2.8$ and construct the divided difference table

z	$f(z)$	First divided differences	Second divided differences	Third divided difference
$z_0 = 2.6 \ f[z_0] = f(2.0) = 3.6783790$				
		$f[z_0, z_1] = f'(2.6) = 4.4163103$		
$z_1 = 2.6 \ f[z_1] = f(2.0) = 3.6783790$			$f[z_0, z_1, z_2] = 2.4216309$	
		$f[z_1, z_2] = 4.9006364$		$f[z_0, z_1, z_2, x_3] = 1.1715291$
$z_2 = 2.8 \ f[z_2] = f(2.2) = 4.6585063$			$f[z_1, z_2, z_3] = 2.6559367$	
		$f[z_2, z_3] = f'(2.8) = 5.4318238$		
$z_3 = 2.8 \ f[z_3] = f(2.2) = 4.6585063$				

The cubic Hermite polynomial is

$$H_3(x) = 3.6783790 + 4.4163103(x - 2.6) + 2.4216309(x - 2.6)^2 + 1.1715291((x - 2.6)^2(x - 2.8),$$

so $H_3(2.75) = 4.393994 \approx y(2.75) = 4.394170$.

30. Show that an $O(h^3)$ local truncation error method cannot be obtained from a difference equation in the form

$$w_{i+1} = w_i + a_1 f(t_i, w_i) + a_2 f(t_i + \alpha_2, w_1 + \gamma_2 f(t_i, w_i)).$$

SOLUTION: Using the notation $y_{i+1} = y(t_{i+1})$, $y_i = y(t_i)$, and $f_i = f(t_i, y(t_i))$, we have

$$h\tau_{i+1} = y_{i+1} - y_i - a_1 f_i - a_2 f(t_i + \alpha_2, y_i + \delta_2 f_i).$$

Expanding y_{i+1} and $f(t_i + \alpha_2, y_i + \delta_2 f_i)$ in Taylor series about t_i and $f(t_i, y_i)$, gives

$$h\tau_{i+1} = (h - a_1 - a_2)f_i + \frac{h^2}{2}f_i' - a_2\alpha_2 f_t(t_i, y_i) - a_2\delta_2 f_i f_y(t_i, y_i) + \frac{h^3}{6}f_i'' - a_2\frac{\alpha_2^2}{2}f_{tt}(t_i, y_i)$$

$$- a_2\alpha_2\delta_2 f_i f_{ty}(t_i, y_i) - a_2\frac{\delta_2^2}{2}f_i^2 f_{yy}(t_i, y_i) + \cdots$$

$$= (h - a_1 - a_2)f_i + \left(\frac{h^2}{2} - a_2\alpha_2\right)f_t(t_i, y_i) + \left(\frac{h^2}{2} - a_2\delta_2\right)f_i f_y(t_i, y_i) + \left(\frac{h^3}{6} - a_2\frac{\alpha_2^2}{2}\right)f_{tt}(t_i, y_i)$$

$$+ \left(\frac{h^3}{3} - a_2\alpha_2\delta_2\right)f_i f_{ty}(t_i, y_i) + \left(\frac{h^3}{6} - a_2\frac{\delta_2^2}{2}\right)f_i^2 f_{tt}(t_i, y_i)$$

$$+ \frac{h^3}{6}\left[f_t(t_i, y_i)f_y(t_i, y_i) + f_i f_y^2(t_i, y_i)\right] + \cdots.$$

The local truncation error τ_{i+1} is $O(h^2)$ because regardless of the choice of a_1, a_2, α_2, and δ_2 we cannot cancel the term

$$\frac{h^3}{6}\left[f_t(y_i, t_i)f_y(t_i, y_i) + f_i f_y^2(t_i, y_i)\right].$$

Exercise Set 5.5, page 300

2. a. Use the Runge-Kutta-Fehlberg method with *hmax* $= 0.05$ to approximate the solution to

$$y' = \left(\frac{y}{t}\right)^2 + \left(\frac{y}{t}\right), \quad 1 \le t \le 1.2, \quad y(1) = 1.$$

SOLUTION: To determine the first approximation with $w_0 = 1$, we need

$$k_1 = 0.05f(1, 1) = 0.1,$$

$$k_2 = 0.05f\left(1 + \frac{1}{4}(0.05), 1 + \frac{1}{4}(0.1)\right) = 0.1018595,$$

$$k_3 = 0.05f\left(1 + \frac{3}{8}(0.05), 1 + \frac{3}{32}(0.1) + \frac{9}{32}(0.1018595)\right) = 0.1028556,$$

$$k_4 = 0.05f\left(1 + \frac{12}{13}(0.05), 1 + \frac{1932}{2197}(0.1) - \frac{7200}{2197}(0.1018595) + \frac{7296}{2197}(0.1028556)\right)$$
$$= 0.1072158,$$

$$k_5 = 0.05f\left(1.05, 1 + \frac{439}{216}(0.1) - 8(0.1018595) - \frac{3680}{513}(0.1028556) - \frac{845}{4104}(0.1072158)\right)$$
$$= 0.1078648,$$

$$k_6 = 0.05f\left(1 + \frac{1}{2}(0.05), 1 - \frac{8}{27}(0.1) + 2(0.1018595) - \frac{3544}{2565}(0.1028556) + \frac{1859}{4104}(0.1072158)\right.$$
$$\left. - \frac{11}{40}(0.1078648)\right) = 0.1038191,$$

and

$$R = \frac{1}{h}\left|\frac{1}{360}(0.1) - \frac{128}{4275}(0.1028556) - \frac{2197}{75240}(0.1072158) + \frac{1}{50}(0.1078648) + \frac{2}{55}(0.1038191)\right|$$
$$\approx 6.27858 \times 10^{-7}.$$

Since $R \le 10^{-4}$, we have a sufficiently accurate approximation

$$w_1 = 1 + \frac{25}{216}(0.1) + \frac{1408}{2565}(0.1028556) + \frac{2197}{4104}(0.1072158) - \frac{1}{5}(0.1078648)$$
$$= 1.103857,$$

and we have a new value of

$$\delta = 0.84\left(\frac{10^{-4}}{6.277091 \times 10^{-7}}\right)^{1/4} = 2.984180.$$

The new value of h would be $2.984180(0.05)$, but this would exceed *hmax* $= 0.05$, so the stepsize is unchanged.

The remaining calculations follow in a similar manner and are given in the following table.

i	t_i	w_i	h_i
1	1.0500000	1.1038574	0.0500000
2	1.1000000	1.2158864	0.0500000
3	1.1500000	1.3368393	0.0500000
4	1.2000000	1.4675697	0.0500000

4. Construct an algorithm to implement the Runge-Kutta-Verner method, and repeat the calculations in Exercise 3(a) with this technique.

SOLUTION: The Runge-Kutta-Verner method is frequently used in general purpose software packages. To modify the Runge-Kutta-Fehlberg Algorithm in implementing this new technique, we must first change Steps 3 and 6 to use the new equations. Then change Step 4 to use the new difference

$$R = \frac{1}{h}\left| -\frac{1}{160}K_1 - \frac{125}{17952}K_3 + \frac{1}{144}K_4 - \frac{12}{1955}K_5 - \frac{3}{44}K_6 + \frac{125}{11592}K_7 + \frac{43}{616}K_8 \right|.$$

Finally, in Step 8 we change the tolerance specification to $\delta = 0.871(TOL/R)^{\frac{1}{5}}$ to reflect the higher order of the method.

Repeating Exercise 3 part (a) using the Runge-Kutta-Verner method gives the results in the following table.

i	t_i	w_i	h_i	y_i
1	1.42087564	1.05149775	0.42087564	1.05150868
3	2.28874724	1.25203709	0.50000000	1.25204675
5	3.28874724	1.50135401	0.50000000	1.50136369
7	4.00000000	1.67622922	0.21125276	1.67623914

Exercise Set 5.6, page 314

1. a. Use the Adams-Bashforth methods with $h = 0.2$ to approximate the solution to

$$y' = te^{3t} - 2y, \quad 0 \le t \le 1, \quad y(0) = 0,$$

and compare the results to those of the exact solution $y(t) = \frac{1}{5}te^{3t} - \frac{1}{25}e^{3t} + \frac{1}{25}e^{-2t}$.

SOLUTION: The first steps of the Adams-Bashforth Two-Step method are

$w_0 = \alpha = 0,$

$w_1 = \alpha_1 = y(0.2) = 0.0268128,$

$w_2 = w_1 + \dfrac{h}{2}[3f(t_1, w_1) - f(t_0, w_0)] = 0.0268128 - 0.1[3(0.3107982) - 0] = 0.1200522,$

$w_3 = w_2 + \dfrac{h}{2}[3f(t_2, w_2) - f(t_1, w_1)] = 0.4153551.$

The approximations w_4 and w_5 are found in a similar manner.

The first steps of the Adams-Bashforth Three-Step method are

$w_0 = \alpha = 0,$

$w_1 = \alpha_1 = y(0.2) = 0.0268128,$

$w_2 = \alpha_2 = y(0.4) = 0.1507778,$

$w_3 = w_2 + \dfrac{h}{12}[23f(t_2, w_2) - 16f(t_1, w_1) + 5f(t_0, w_0)]$

$\qquad = 0.1507778 - \dfrac{0.2}{12}[23f(0.4, 0.1507778) - 16f(0.2, 0.0268128) + 5f(0, 0)]$

$\qquad = 0.1507778 + \dfrac{0.2}{12}[23(1.026491) - 16(0.3107982) + 0] = 0.4613866.$

The approximations w_4 and w_5 are found in a similar manner. The results for all the Adams-Bashforth methods are summarized in the following table.

t_i	2 step	3 step	4 step	5 step	$y(t_i)$
0.2	0.0268128	0.0268128	0.0268128	0.0268128	0.0268128
0.4	0.1200522	0.1507778	0.1507778	0.1507778	0.1507778
0.6	0.4153551	0.4613866	0.4960196	0.4960196	0.4960196
0.8	1.1462844	1.2512447	1.2961260	1.3308570	1.3308570
1.0	2.8241683	3.0360680	3.1461400	3.1854002	3.2190993

5. **a.** Use the Adams-Bashforth Predictor-Corrector Algorithm with $h = 0.2$ to approximate the solution to the problem in Exercise 1(a).

SOLUTION: The Runge-Kutta method of order four is first used to generate

$$w_1 = 0.0269059, \quad w_2 = 0.1510468, \quad \text{and} \quad w_3 = 0.4966479.$$

Then the Adams-Bashforth Four-Step method is used as predictor to generate

$w_4^{(0)} = w_3 + \dfrac{h}{24}[55f(t_3, w_3) - 59f(t_2, w_2) + 37f(t_1, w_1) - 9f(t_0, w_0)]$

$\qquad = 0.4966479 + \dfrac{0.2}{24}[55f(0.6, 0.4966479) - 59f(0.4, 0.1510468) + 37f(0.2, 0.0269059) - 9f(0, 0)]$

$\qquad = 1.296385.$

Finally, the Adams-Moulton Three-Step method is now used as corrector to generate

$$w_4 = w_3 + \frac{h}{24}\left[9f\left(t_4, w_4^{(0)}\right) + 19f(t_3, w_3) - 5f(t_2, w_2) + f(t_1, w_1)\right]$$

$$= 0.4966479 + \frac{0.2}{24}[9(6.225770) + 19(2.636493) - 5(1.025953) + 0.3106120]$$

$$= 1.340866.$$

The following table summarizes these results.

t_i	w_i	$y(t_i)$	Error
0.0	0.0000000	0.0000000	0
0.2	0.0269059	0.0268128	0.0000931
0.4	0.1510468	0.1507778	0.0002690
0.6	0.4966479	0.4960196	0.0006283
0.8	1.3408657	1.3308570	0.0100087
1.0	3.2450881	3.2190993	0.0259888

11. a. Use the Lagrange interpolation formula to derive the Adams-Bashforth two-step method.

SOLUTION: For some ξ_i in (t_{i-1}, t_i), we have

$$f(t, y(t)) = P_1(t) + \frac{f''(\xi_i, y(\xi_i))}{2}(t - t_i)(t - t_{i-1}),$$

where $P_1(t)$ is the linear Lagrange polynomial

$$P_1(t) = \frac{(t - t_{i-1})}{(t_i - t_{i-1})}f(t_i, y(t_i)) + \frac{(t - t_i)}{(t_{i-1} - t_i)}f(t_{i-1}, y(t_{i-1})).$$

Thus

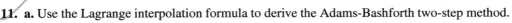

$$\int_{t_i}^{t_{i+1}} P_1(t)\, dt = \frac{f(t_i, y(t_i))}{t_i - t_{i-1}}\int_{t_i}^{t_{i+1}}(t - t_{i-1})\, dt + \frac{f(t_{i-1}, y(t_{i-1}))}{t_{i-1} - t_i}\int_{t_i}^{t_{i+1}}(t - t_i)\, dt$$

$$= \frac{3h}{2}f(t_i, y(t_i)) - \frac{h}{2}f(t_{i-1}, y(t_{i-1})).$$

Since $(t - t_i)(t - t_{i-1})$ does not change sign on (t_i, t_{i+1}), the Mean Value Theorem for Integrals gives

$$\int_{t_i}^{t_{i+1}} \frac{f''(\xi_i, y(\xi_i))(t - t_i)(t - t_{i-1})}{2}\, dt = \frac{f''(\mu, y(\mu))}{2}\int_{t_i}^{t_{i+1}}(t-t_i)(t-t_{i-1})\, dt = \frac{5h^2 f''(\mu, y(\mu))}{12}.$$

Replacing $y(t_j)$ with w_j, for $j = i - 1, i$, and $i + 1$, in the formula

$$y(t_{i+1}) = y(t_i) + \int_{t_i}^{t_{i+1}} f(t, y(t))\, dt$$

gives

$$w_{i+1} = w_i + \frac{h[3f(t_i, w_i) - f(t_{i-1}, w_{i-1})]}{2},$$

and the local truncation error is

$$\tau_{i+1}(h) = \frac{5h^2 y'''(\mu)}{12},$$

for some μ in (t_{i-1}, t_{i+1}).

12. Derive the Adams-Bashforth three-step method by setting

$$y(t_{i+1}) = y(t_i) + ahf(t_i, y(t_i)) + bhf(t_{i-1}, y(t_{i-1})) + chf(t_{i-2}, y(t_{i-2}))$$

and expanding $y(t_{i+1})$, $f(t_{i-1}, y(t_{i-1}))$, and $f(t_{i-2}, y(t_{i-2}))$ in a Taylor series about $y(t_i)$.

SOLUTION: To simplify the notation we define $y = y(t_i)$, $f = f(t_i, y(t_i))$, $f_t = f_t(t_i, y(t_i))$, ..., suppressing the arguments of the various functions. Then we have

$$y + hf + \frac{h^2}{2}(f_t + ff_y) + \frac{h^3}{6}\left(f_{tt} + f_t f_y + 2ff_{yt} + ff_y^2 + f^2 f_{yy}\right)$$

$$= y + ahf + bh\left[f - h(f_t + ff_y) + \frac{h^2}{2}\left(f_{tt} + f_t f_y + 2ff_{yt} + ff_y^2 + f^2 f_{yy}\right)\right]$$

$$+ ch\left[f - 2h(f_t + ff_y) + 2h^2\left(f_{tt} + f_t f_y + 2ff_{yt} + ff_y^2 + f^2 f_{yy}\right)\right]$$

$$= y + (a + b + c)hf + (-b - 2c)h^2(f_t + ff_y)$$

$$+ \left(\frac{1}{2}b + 2c\right)h^3\left(f_{tt} + f_t f_y + 2ff_{yt} + ff_y^2 + f^2 f_{yy}\right).$$

Matching coefficients produces the system of equations

$$a + b + c = 1, \quad -b - 2c = \frac{1}{2}, \quad \text{and} \quad \frac{1}{2}b + 2c = \frac{1}{6}.$$

This system has the solution $a = \dfrac{23}{12}$, $b = -\dfrac{16}{12}$, and $c = \dfrac{5}{12}$.

14. Derive Simpson's method by applying Simpson's quadrature rule to the integral

$$y(t_{i+1}) - y(t_{i-1}) = \int_{t_{i-1}}^{t_{i+1}} f(t, y(t))\, dt.$$

SOLUTION: Applying Simpson's quadrature rule to the integral gives

$$y(t_{i+1}) - y(t_{i-1}) = \int_{t_{i-1}}^{t_{i+1}} f(t, y(t))\, dt$$

$$= \frac{h}{3}[f(t_{i-1}, y(t_{i-1})) + 4f(t_i, y(t_i)) + f(t_{i+1}, y(t_{i+1}))] - \frac{h^5}{90} f^{(4)}(\xi, y(\xi)).$$

This leads to the difference equation

$$w_{i+1} = w_{i-1} + \frac{h}{3}[f(t_{i-1}, w_{i-1}) + 4f(t_i, w_i) + f(t_{i+1}, w_{i+1})],$$

with local truncation error

$$\tau_{i+1}(h) = -\frac{h^4}{90} y^{(5)}(\xi).$$

16. Verify the entries in Table 5.12.

SOLUTION: The entries are generated by evaluating the following integrals:

$$k = 0 : (-1)^k \int_0^1 \binom{-s}{k} \, ds = \int_0^1 ds = 1,$$

$$k = 1 : (-1)^k \int_0^1 \binom{-s}{k} \, ds = -\int_0^1 -s \, ds = \frac{1}{2},$$

$$k = 2 : (-1)^k \int_0^1 \binom{-s}{k} \, ds = \int_0^1 \frac{s(s+1)}{2} \, ds = \frac{5}{12},$$

$$k = 3 : (-1)^k \int_0^1 \binom{-s}{k} \, ds = -\int_0^1 \frac{-s(s+1)(s+2)}{6} \, ds = \frac{3}{8},$$

$$k = 4 : (-1)^k \int_0^1 \binom{-s}{k} \, ds = \int_0^1 \frac{s(s+1)(s+2)(s+3)}{24} \, ds = \frac{251}{720}, \quad \text{and}$$

$$k = 5 : (-1)^k \int_0^1 \binom{-s}{k} \, ds = -\int_0^1 -\frac{s(s+1)(s+2)(s+3)(s+4)}{120} \, ds = \frac{95}{288}.$$

Exercise Set 5.7, page 320

2. a. Use the Adams Variable Step-Size Predictor-Corrector Algorithm with *hmax* = 0.05 to approximate the solution to

$$y' = \left(\frac{y}{t}\right)^2 + \left(\frac{y}{t}\right), \quad 1 \le t \le 1.2, \quad y(1) = 1.$$

SOLUTION: The Runge-Kutta method of order four is first used to generate

$$w_1 = 1.103857, \quad w_2 = 1.215886, \quad \text{and} \quad w_3 = 1.336838.$$

The Adams-Bashforth Four-Step method is used as predictor to give $w_4^{(0)} = 1.4675626$, and the Adams-Moulton Three-Step method as corrector improves this to $w_4 = 1.4675688$. Since

$$\sigma = \frac{19}{270h} \left| w_4 - w_4^{(0)} \right| = \frac{19}{270(0.05)} |1.4675633 - 1.4675688| = 7.7407 \times 10^{-6} < 10^{-4},$$

the result is accepted as w_4.

4. Construct an Adams-Variable Step-Size Predictor-Corrector Algorithm based on the Adams-Bashforth Five-Step method and the Adams-Moulton Four-Step method.

SOLUTION: Using the Adams-Variable Step-Size Predictor-Corrector Algorithm as a basis for the new algorithm we need to change the following steps:

Step 1   Set up an algorithm, denoted *RK5*, for the Runge Kutta Method of Order 5.

Step 3   Call $RK5(h, w_0, t_0, w_1, t_1, w_2, t_2, w_3, t_3, w_4, t_4)$:

Set $NFLAG = 1$;
        $i = 5$;
        $t = t_4 + h$.

Step 5 Set

$$WP = w_{i-1} + \tfrac{h}{720}\Bigg[1901f(t_{i-1}, w_{i-1}) - 2774f(t_{i-2}, w_{i-2})$$

$$+2616f(t_{i-3}, w_{i-3}) - 1274f(t_{i-4}, w_{i-4}) + 251f(t_{i-5}, w_{i-5})\Bigg];$$

$$WC = w_{i-1} + \tfrac{h}{720}\Bigg[251f(t, WP) + 646f(t_{i-1}, w_{i-1}) - 264f(t_{i-2}, w_{i-2})$$

$$+106f(t_{i-3}, w_{i-3}) - 19f(t_{i-4}, w_{i-4})\Bigg];$$

$$\sigma = 27|WC - WP|/(502)h.$$

Step 8 If $NFLAG = 1$ then for $j = i - 4, i - 3, i - 2, i - 1, i$

Step 12 Set $q = (0.5TOL/\sigma)^{\frac{1}{5}}$

Step 15 If $t_{i-1} + 5h > b$, then set $h = (b - t_{i-1})/5$

Step 16 Call $RK5(h, w_{i-1}, t_{i-1}, w_i, t_i, w_{i+1}, t_{i+1}, w_{i+2}, t_{i+2}, w_{i+3}, t_{i+3})$;

    Set $NFLAG = 1$;

    $i = i + 4$.

Step 17 Set $q = (0.5TOL/\sigma)^{\frac{1}{5}}$.

Step 19 else

    if $NFLAG = 1$ then set $i = i - 4$;

    Call $RK5(h, w_{i-1}, t_{i-1}, w_i, t_i, w_{i+1}, t_{i+1}, w_{i+2}, t_{i+2}, w_{i+3}, t_{i+3})$;

    set $i = i + 4$;

      $NFLAG = 1$.

The following are partial results obtained by applying the new algorithm to the problem in part (a) of Exercise 3.

i	t_i	w_i	h_i	y_i
5	1.17529879	1.01186066	0.03505976	1.01186063
15	1.56737794	1.08139480	0.05580055	1.08139470
25	2.25808774	1.24445586	0.08897663	1.24445574
35	3.51328927	1.55692781	0.14118166	1.55692763
40	4.00000000	1.67623932	0.09734215	1.67623914

Exercise Set 5.8, page 327

2. a. Use the Extrapolation Algorithm with $TOL = 10^{-4}$ and $hmax = 0.05$ to approximate the solution to

$$y' = \left(\frac{y}{t}\right)^2 + \frac{y}{t}, \quad 1 \le t \le 1.2, \quad y(1) = 1.$$

SOLUTION: The initial condition gives $w_0 = 1.0$. With $h = 0.05$ and $h_0 = 0.025$, Euler's method gives

$$y(1.025) \approx 1.05,$$

and the Midpoint method yields

$$y(1.05) \approx 1.0 + 2(0.025)f(1.025, 1.05) = 1.1036883.$$

The correction produces

$$y_{1,1} = \frac{1}{2}[1.1036883 + 1.05 + 0.025f(1.05, 1.1036883)] = 1.1037943.$$

The process is repeated with $h = 0.05/4 = 0.0125$. We now have $w_0 = 1.0$, from the initial condition, and

$$y(1.0125) \approx w_1 = 1 + 0.0125f(1,1) = 1.025,$$

from Euler's method. Using the Midpoint method gives

$$y(1.025) \approx w_2 = w_0 + 2h_1f(1 + h_1, w_1) = 1 + 0.025f(1.0125, 1.025) = 1.0509297,$$

$$y(1.0375) \approx w_3 = w_1 + 2h_1f(1 + 2h_1, w_2) = 1.025 + 0.025f(1.025, 1.0509297) = 1.0769133,$$

and

$$y(1.05) \approx w_4 = 1.0509297 + 0.025f(1.0375, 1.0769133) = 1.1038150.$$

The endpoint correction is

$$y_{2,1} = \frac{1}{2}[w_4 + w_3 + h_1f(1 + h, w_4)]$$

$$= \frac{1}{2}[1.1038150 + 1.0769133 + 0.0125f(1.05, 1.1038150)] = 1.1038415.$$

Extrapolation on $y_{1,1}$ and $y_{2,1}$ gives

$$y_{2,2} = y_{2,1} + \frac{h_1^2(y_{2,1} - y_{1,1})}{h_0^2 - h_1^2} = y_{2,1} + \frac{y_{2,1} - y_{1,1}}{3}$$

$$= 1.1038415 + \frac{(1.1038415 - 1.1037943)}{3} = 1.1038573.$$

Since $|y_{2,2} - y_{1,1}| = |1.1038573 - 1.1037943| = 6.3 \times 10^{-5} < 10^{-4}$, we accept $y(1.05) \approx y_{2,2} = 1.1038573$. We now need to approximate $y(1.1)$ using $h = 0.05$. The extrapolation for this approximation is

$$y_{1,1} = 1.2158192, \quad y_{2,1} = 1.2158694, \quad \text{and} \quad y_{2,2} = 1.2158861,$$

and since $|1.2158861 - 1.2158192| < 10^{-4}$, we accept $y(1.1) \approx y_{2,2} = 1.2158861$. The remaining results are computed in a similar manner.

A summary is given in the following table.

i	t_i	w_i	h_i	k
1	1.05	1.10385729	0.05	2
2	1.10	1.21588614	0.05	2
3	1.15	1.33683891	0.05	2
4	1.20	1.46756907	0.05	2

4. Approximate the solution to the logistic population equation

$$P'(t) = bP(t) - k[P(t)]^2$$

at $t = 5$ with $P(0) = 50,976$, $b = 2.9 \times 10^{-2}$, and $k = 1.4 \times 10^{-7}$.

SOLUTION: The logistic equation is a first-order differential equation of Bernoulli type that can be transformed into a linear equation by use of the substitutions

$$Q(t) = (P(t))^{-1} \quad \text{and} \quad Q'(t) = -(P(t))^{-2}P'(t).$$

Making these changes produces the linear equation in $Q(t)$:

$$Q'(t) = -bQ(t) + k \quad \text{with solution} \quad Q(t) = \frac{k}{b} + Ce^{-bt}.$$

Transforming back to $P(t)$ gives

$$P(t) = \frac{1}{k/b + Ce^{-bt}}.$$

Since $P(0) = 50,976$, we can solve for C and use the values of k and b in the exercise to determine that

$$P(t) = \frac{1}{k/b + (1/P(0) - k/b)e^{-bt}} = \frac{1}{4.82759 \times 10^{-6} + \left(1.4789 \times 10^{-5}\right)e^{-(2.9 \times 10^{-2})t}}.$$

When $t = 5$, this gives the population that agrees with the approximation that is produced using the Extrapolation Algorithm, $P(5) = 56{,}751$.

Exercise Set 5.9, page 337

1. a. Use the Runge-Kutta for Systems Algorithm with $h = 0.2$ to approximate the solution to the system

$$u_1' = 3u_1 + 2u_2 - \left(2t^2 + 1\right)e^{2t}, \quad u_1(0) = 1$$

and

$$u_2' = 4u_1 + u_2 + \left(t^2 + 2t - 4\right)e^{2t}, \quad u_2(0) = 1,$$

for $0 \le t \le 1$.

SOLUTION: The initial conditions give $w_{1,0} = 1$ and $w_{2,0} = 1$. Then

$$\begin{aligned}
k_{1,1} &= 0.2f_1(0, 1, 1) = 0.8, \\
k_{1,2} &= 0.2f_2(0, 1, 1) = 0.2, \\
k_{2,1} &= 0.2f_1(0.1, 1 + 0.4, 1 + 0.1) = 0.2f_1(0.1, 1.4, 1.1) = 1.03083384, \\
k_{2,2} &= 0.2f_2(0.1, 1.4, 1.1) = 0.41417671 \\
k_{3,1} &= 0.2f_1(0.1, 1.51541692, 1.20708836) = 1.14291933, \\
k_{3,2} &= 0.2f_2(0.1, 1.51541692, 1.20708836) = 0.52792792, \\
k_{4,1} &= 0.2f_1(0.2, 2.14291933, 1.52792792) = 1.57468863, \\
k_{4,2} &= 0.2f_2(0.2, 2.14291933, 1.52792792) = 0.95774186,
\end{aligned}$$

which gives

$$w_{1,1} = w_{1,0} + \frac{1}{6}(k_{1,1} + 2k_{2,1} + 2k_{3,1} + k_{4,1}) = -1 + \frac{1}{6}(6.72219497) = 2.12036583$$

and

$$w_{2,1} = w_{2,0} + \frac{1}{6}(k_{1,2} + 2k_{2,2} + 2k_{3,2} + k_{4,2}) = 1 + \frac{1}{6}(3.04195112) = 1.50699185.$$

We then compute new values of $k_{1,1}$, $k_{1,2}$, $k_{2,1}$, $k_{2,2}$, $k_{3,1}$, $k_{3,2}$, $k_{4,1}$, and $k_{4,2}$ to provide the approximations $w_{1,2}$ and $w_{2,2}$.

The results are summarized in the following table.

t_i	w_{1i}	u_{1i}	w_{2i}	u_{2i}
0.200	2.12036583	2.12500839	1.50699185	1.51158743
0.400	4.44122776	4.46511961	3.24224021	3.26598528
0.600	9.73913329	9.83235869	8.16341700	8.25629549
0.800	22.67655977	23.00263945	21.34352778	21.66887674
1.000	55.66118088	56.73748265	56.03050296	57.10536209

3. a. Use the Runge-Kutta for Systems Algorithm with $h = 0.1$ to approximate the solution to the second order initial-value problem

$$y'' - 2y' + y = te^t - t, \quad 0 \le t \le 1 \quad y(0) = y'(0) = 0.$$

SOLUTION: We first convert the second-order equation to a system of first-order differential equations using

$$u_1(t) = y(t) \quad \text{and} \quad u_2(t) = y'(t).$$

This gives the system

$$u_1'(t) = u_2(t),$$
$$u_2'(t) = -u_1(t) + 2u_2(t) + te^t - t.$$

The setup is now similar to that used in Exercise 1 except we use

$$f_1(t, w_1, w_2) = w_2,$$
$$f_2(t, w_1, w_2) = -w_1 + 2w_2 + te^t - t.$$

The results are summarized in the following table.

i	t_i	w_{1i}	w_{2i}
2	0.200	0.0001535	0.0001535
5	0.500	0.0074297	0.0074303
7	0.700	0.0329962	0.0329980
10	1.000	0.1713222	0.1713288

5. Modify the Adams Fourth-Order Predictor-Corrector Algorithm to approximate solutions of systems of equations.

SOLUTION: The following algorithm can be used to approximate the solution of the mth-order system of the first-order initial-value problems using Adams Fourth-Order Predictor-Corrector method

$$u'_j(t) = f_j(t, u_1, u_2, \ldots, u_m),$$

for $j = 1, 2, \ldots, m$ and $a \le t \le b$, with $u_j(a) = \alpha_j$, $j = 1, 2, \ldots, m$, at $(n + 1)$ equally spaced numbers in the interval $[a, b]$:

INPUT endpoints a, b; number of equations m; integer N; initial conditions $\alpha_1, \ldots, \alpha_m$.
OUTPUT approximations $w_{i,j}$ to $u_j(t_i)$.

Step 1   Set $h = (b - a)/N$;
Step 2   For $j = 1, 2, \ldots, m$ set $w_{0,j} = \alpha_j$.
Step 3   OUTPUT $(t_0, w_{0,1}, w_{0,2}, \ldots, w_{0,m})$.
Step 4   For $i = 1, 2, 3$ do Steps 5–11.
    Step 5   For $j = 1, 2, \ldots, m$ set
        $k_{1,j} = h f_j(t_{i-1}, w_{i-1,1}, \ldots, w_{i-1,m})$.
    Step 6   For $j = 1, 2, \ldots, m$ set
        $k_{2,j} = h f_j\left(t_{i-1} + \frac{h}{2}, w_{i-1,1} + \frac{1}{2}k_{1,1}, w_{i-1,2} + \frac{1}{2}k_{1,2}, \ldots, w_{i-1,m} + \frac{1}{2}k_{1,m}\right)$.
    Step 7   For $j = 1, 2, \ldots, m$ set
        $k_{3,j} = h f_j\left(t_{i-1} + \frac{h}{2}, w_{i-1,1} + \frac{1}{2}k_{2,1}, w_{i-1,2} + \frac{1}{2}k_{2,2}, \ldots, w_{i-1,m} + \frac{1}{2}k_{2,m}\right)$.
    Step 8   For $j = 1, 2, \ldots, m$ set
        $k_{4,j} = h f_j(t_{i-1} + h, w_{i-1,1} + k_{3,1}, w_{i-1,2} + k_{3,2}, \ldots, w_{i-1,m} + k_{3,m})$.
    Step 9   For $j = 1, 2, \ldots, m$ set
        $w_{i,j} = w_{i-1,j} + (k_{1,j} + 2k_{2,j} + 2k_{3,j} + k_{4,j})/6$.
    Step 10 Set $t_i = a + ih$.
    Step 11 OUTPUT $(t_i, w_{i,1}, w_{i,2}, \ldots, w_{i,m})$.
Step 12 For $i = 4, \ldots, N$ do Steps 13–16.
    Step 13 Set $t_i = a + ih$.
    Step 14 For $j = 1, 2, \ldots, m$ set

$$w_{i,j}^{(0)} = w_{i-1,j} + \frac{h}{24}\Bigg[55 f_j(t_{i-1}, w_{i-1,1}, \ldots, w_{i-1,m}) - 59 f_j(t_{i-2}, w_{i-2,1}, \ldots, w_{i-2,m})$$

$$+ 37 f_j(t_{i-3}, w_{i-3,1}, \ldots, w_{i-3,m}) - 9 f_j(t_{i-4}, w_{i-4,1}, \ldots, w_{i-4,m})\Bigg].$$

    Step 15 For $j = 1, 2, \ldots, m$ set

$$w_{i,j} = w_{i-1,j} + \frac{h}{24}\Bigg[9 f_j\left(t_i, w_{i,1}^{(0)}, \ldots, w_{i,m}^{(0)}\right) + 19 f_j(t_{i-1}, w_{i-1,1}, \ldots, w_{i-1,m})$$

$$- 5 f_j(t_{i-2}, w_{i-2,1}, \ldots, w_{i-2,m}) + f_j(t_{i-3}, w_{i-3,1}, \ldots, w_{i-3,m})\Bigg].$$

    Step 16   OUTPUT $(t_i, w_{i,1}, w_{i,2}, \ldots, w_{i,m})$.
Step 17   STOP

Exercise Set 5.10, page 347

2. a. Show that for the Adams-Bashforth method of order four, when $f \equiv 0$, we also have
$F(t_i, h, w_{i+1}, \ldots, w_{i+1-m}) = 0$.

SOLUTION: For the Adams-Bashforth method,

$$F(t_i, h, w_{i+1}, w_i, w_{i-1}, w_{i-2}, w_{i-3}) = \frac{1}{24}\bigg[55f(t_i, w_i) - 59f(t_{i-1}, w_{i-1})$$
$$+ 37f(t_{i-2}, w_{i-2}) - 9f(t_{i-3}, w_{i-3})\bigg],$$

so if $f \equiv 0$, then $F \equiv 0$.

4. a. Derive the equation

$$y'(t_i) = \frac{-3y(t_i) + 4y(t_{i+1}) - y(t_{t+2})}{2h} + \frac{h^2}{3}y'''(\xi_i).$$

b. Use the method suggested by part (a) with $h = 0.1$ and $w_1 = 1 - e^{-0.01}$ to approximate the solution to
$$y' = 1 - y, \quad 0 \leq t \leq 1 \quad y(0) = 0.$$

c. Repeat part (b) using instead $h = 0.01$ and $w_1 = 1 - e^{-0.01}$.

d. Discuss the consistency, convergence, and stability of the method.

SOLUTION: **a.** First expand $y(t_{i+1})$ and $y(t_{i+2})$ in second Taylor polynomials about $y(t_i)$. This gives

$$y(t_{i+1}) = y(t_i) + hy'(t_i) + \frac{h^2}{2}y''(t_i) + \frac{h^3}{6}y'''(\xi_1)$$

and

$$y(t_{i+2}) = y(t_i) + 2hy'(t_i) + \frac{4h^2}{2}y''(t_i) + \frac{8h^3}{6}y'''(\xi_2).$$

Subtracting the second equation from 4 times the first gives

$$4y(t_{i+1}) - y(t_{i+2}) = 3y(t_i) + 2hy'(t_i) + \frac{h^3}{6}(4y'''(\xi_1) - 8y'''(\xi_2)).$$

Solving this equation for $y'(t_i)$ gives

$$y'(t_i) = \frac{1}{2h}(4y(t_{i+1}) - y(t_{i+2}) - 3y(t_i)) - \frac{h^2}{12}(4y'''(\xi_1) - 8y'''(\xi_2)).$$

The error term is the weighted average of the errors in this equation, but it must be verified by an alternative technique.

b. The method gives $w_2 = 0.18065 \approx y(0.2), w_5 = 0.35785 \approx y(0.5)$, $w_7 = 0.15340 \approx y(0.7)$, and $w_{10} = -9.7822 \approx y(1.0)$.

c. With $h = 0.01$, we have the very poor approximations $w_{20} = -60.402 \approx y(0.2)$, $w_{50} = -1.37 \times 10^{17} \approx y(0.5)$, $w_{70} = -5.11 \times 10^{26} \approx y(0.7)$, and $w_{100} = -1.16 \times 10^{41} \approx y(1.0)$. The solution to the problem is $y(t) = 1 - e^{-t}$, so the exact results are $y(0.2) = 0.18127$, $y(0.5) = 0.39347$, $y(0.7) = 0.50341$, and $y(1) = 0.63212$.

d. The difference method

$$w_{i+2} = 4w_{i+1} - 3w_i - 2hf(t_i, w_i), \quad \text{for } i = 0, 1, \ldots, N-2,$$

is a multistep method with local truncation error $\dfrac{h^2}{3} y'''(\xi_i)$.

Suppose

$$|y'''(t)| \leq M, \quad \text{for } a \leq x \leq b.$$

Then

$$|\tau_i(h)| \leq \frac{h^2}{3} M \quad \text{and} \quad \lim_{h \to 0} |\tau_i(h)| = 0.$$

So the method is consistent. The characteristic polynomial of the method is

$$p(\lambda) = \lambda^2 - 4\lambda + 3 = (\lambda - 3)(\lambda - 1).$$

Since one of the roots of this polynomial is $\lambda = 3$, the method is not stable, so it is also not convergent.

8. Suppose that the method in Exercise 4 is applied to the problem

$$y' = 0, \quad 0 \leq t \leq 10 \quad y(0) = 0,$$

which has exact solution $y \equiv 0$, and that an error ε is introduced in the first step. Find how this error propagates.

SOLUTION: The next five terms are

$$w_2 = 4\varepsilon = (3+1)\varepsilon,$$
$$w_3 = [4(3+1) - 3(1)]\varepsilon = [(4-1)3 + 4]\varepsilon = \left(3^2 + 3 + 1\right)\varepsilon = 13\varepsilon,$$
$$w_4 = \left[4\left(3^2 + 3 + 1\right) - 3(3+1)\right]\varepsilon = \left[4\left(3^2 + 3\right) - \left(3^2 + 3\right) + 4\right]\varepsilon$$
$$= \left[3\left(3^2 + 3\right) + 4\right]\varepsilon = \left(3^3 + 3^2 + 3 + 1\right)\varepsilon = 40\varepsilon,$$

and, in general,

$$w_n = \left(3^{n-1} + 3^{n-2} + \cdots 3 + 1\right)\varepsilon = \frac{1 - 3^n}{1 - 3}\varepsilon = \frac{1}{2}(3^n - 1)\varepsilon.$$

This error grows exponentially with n.

Exercise Set 5.11, page 354

1. a. Use Euler's method $h = 0.1$ to approximate the solution to the stiff initial-value problem

$$y' = -9y, \quad 0 \leq t \leq 1, \quad y(0) = e.$$

SOLUTION: Euler's method applied to this differential equation gives

$$w_0 = e \quad \text{and} \quad w_{i+1} = w_i + hf(t_i, w_i) = (1 - 9h)w_i,$$

for $i \geq 1$. Since $w_i = (1 - 9h)w_{i-1}$, for each i, we have

$$w_{i+1} = (1 - 9h)w_i = (1 - 9h)^2 w_{i-1} = \cdots = (1 - 9h)^{i+1}w_0.$$

Since $w_0 = e$ and $h = 0.1$, we have $w_{i+1} = (0.1)^{i+1}e$, for each $i = 0, 1, \ldots, N - 1$, and the following values are generated.

i	t_i	w_i	$y(t_i)$
2	0.2	0.0271828	0.4493290
5	0.5	0.0000272	0.0301974
7	0.7	0.0000003	0.0049916
10	1.0	0.0000000	0.0003355

7. a. Use the Trapezoidal method with $h = 0.1$ and $TOL = 10^{-5}$ to approximate the solution to the stiff initial-value problem

$$y' = -9y, \ 0 \leq t \leq 1, \ y(0) = e.$$

SOLUTION: We have $w_0 = e$, so $w_1^{(0)} = w_0 = e$ and

$$w_1^{(1)} = w_1^{(0)} - \frac{w_1^{(0)} - w_0 - h/2\left[f(t_0, w_0) + f\left(t_1, w_1^{(0)}\right)\right]}{1 - h/2\left(f_y\left(t_1, w_1^{(0)}\right)\right)}$$

$$= e - \frac{e - e - 0.1/2[f(0, e) + f(0.1, e)]}{1 + 0.1/2(9)} = 1.03107242.$$

Further,

$$w_1^{(2)} = w_1^{(1)} - \frac{w_1^{(1)} - w_0 - 0.1/2\left[f(t_0, w_0) + f\left(t_1, w_1^{(1)}\right)\right]}{1 + 0.1/2(9)} = 1.03107242.$$

Since $\left|w_1^{(2)} - w_1^{(1)}\right| < 10^{-5}$, we set

$$w_1 = w_1^{(2)} = 1.03107242.$$

We use $w_2^{(0)} = w_1 = 1.03107242$ and

$$w_2^{(1)} = w_2^{(0)} - \frac{w_2^{(0)} - w_1 - \frac{h}{2}\left[f(t_1, w_1) + f\left(t_2, w_2^{(0)}\right)\right]}{1 - \frac{h}{2}\left(f_y\left(t_2, w_2^{(0)}\right)\right)}.$$

Continuing in this manner gives the following table.

t_i	w_i	number of iterations	$y(t_i)$
0.100	1.03107242	2	1.10517092
0.200	0.39109643	2	0.44932896
0.300	0.14834692	2	0.18268352
0.400	0.05626952	2	0.07427358
0.500	0.02134361	2	0.03019738
0.600	0.00809585	2	0.01227734
0.700	0.00307084	2	0.00499159
0.800	0.00116480	2	0.00202943
0.900	0.00044182	2	0.00082510
1.000	0.00016759	2	0.00033546

11. Discuss the consistency, stability, and convergence of the Trapezoidal method

$$w_{i+1} = w_i + \frac{h}{2}[f(t_i, w_i) + f(t_{i+1}, w_{i+1})].$$

SOLUTION: Using Eq. (4.25) gives $\tau_{i+1} = \frac{-y'''(\xi_i)}{12}h^2$, for some $t_i < \xi_i < t_{i+1}$, and by Definition 5.18, the Trapezoidal method is consistent. Once again using Eq. (4.25) gives

$$y(t_{i+1}) = y(t_i) + \frac{h}{2}[f(t_i, y(t_i)) + f(t_{i+1}, y(t_{i+1}))] - \frac{y'''(\xi_i)}{12}h^3.$$

Subtracting the difference equation and using the Lipschitz constant L for f gives

$$|y(t_{i+1}) - w_{i+1}| \le |y(t_i) - w_i| + \frac{hL}{2}|y(t_i) - w_i| + \frac{hL}{2}|y(t_{i+1}) - w_{i+1}| + \frac{h^3}{12}|y'''(\xi_i)|.$$

Let $M = \max_{a \le x \le b}|y'''(x)|$. Then, assuming $hL \neq 2$,

$$|y(t_{i+1}) - w_{i+1}| \le \frac{2 + hL}{2 - hL}|y(t_i) - w_i| + \frac{h^3}{6(2 - hL)}M.$$

Using Lemma 5.8, we have

$$|y(t_{i+1}) - w_{i+1}| \le e^{2(b-a)L/(2-hL)}\left[\frac{Mh^2}{12L} + |\alpha - w_0|\right] - \frac{Mh^2}{12L}.$$

So if $hL \neq 2$, the Trapezoidal method is convergent, and consequently stable.

Direct Methods for Solving Linear Systems

Exercise Set 6.1, page 368

1. For each of the following linear systems, obtain a solution by graphical methods, if possible. Explain the results from a geometrical standpoint.

   a. $x_1 + 2x_2 = 3,$
      $x_1 - \ \ x_2 = 0.$

   d. $2x_1 + \ \ x_2 = -1,$
      $4x_1 + 2x_2 = -2,$
      $x_1 - 3x_2 = 5.$

SOLUTION: The graphs are shown. Notice that although (d) has three equations, two of them produce the same line. The intersection of the lines in part (a) clearly occurs when $x_1 = x_2 = 1$. The intersection point of the lines in part (d) is not as obvious, but it occurs when $x_1 = \frac{2}{7}$ and $x_2 = -\frac{11}{7}$.

a.

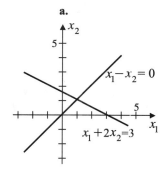

d.

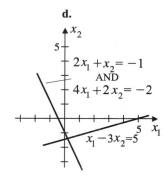

5. a. Use the Gaussian Elimination Algorithm to solve

$$x_1 - x_2 + 3x_3 = 2,$$
$$3x_1 - 3x_2 + x_3 = -1,$$
$$x_1 + \quad x_2 = 3.$$

SOLUTION: We first perform $(E_2 - 3E_1) \rightarrow E_2$ and $(E_3 - E_1) \rightarrow E_3$ to obtain

$$x_1 - x_2 + 3x_3 = 2,$$
$$-8x_3 = -7,$$
$$2x_2 - 3x_3 = 1.$$

A second to third row interchange is now required. Using backward substitution gives

$$x_3 = \frac{-7}{-8} = 0.875, \quad x_2 = \frac{1}{2}(1 + 3x_3) = 1.8125,$$

and $x_1 = 2 + x_2 - 3x_3 = 1.1875$.

9. Given the linear system

$$2x_1 - 6\alpha x_2 = 3 \quad \text{and} \quad 3\alpha x_1 - x_2 = \frac{3}{2},$$

find value(s) of α for which the system has

a. no solutions,

b. an infinite number of solutions,

c. a unique solution.

SOLUTION: **a.** When $\alpha = -1/3$, there is no solution since the equations describe parallel lines.

b. When $\alpha = 1/3$, there is an infinite number of solutions with $x_1 = x_2 + 1.5$, and x_2 arbitrary, because the two equations describe the same line.

c. If $\alpha \neq \pm 1/3$, then the unique solution is

$$x_1 = \frac{3}{2(1 + 3\alpha)} \quad \text{and} \quad x_2 = \frac{-3}{2(1 + 3\alpha)},$$

since the equations describe two intersecting lines.

12. The Gauss-Jordan method uses the ith equation to eliminate not only x_i from the equations E_{i+1}, $E_{i+2}, \ldots, E_n$, as was done in the Gaussian elimination method, but also from $E_1, E_2, \ldots, E_{i-1}$. Construct an algorithm for the procedure.

SOLUTION: Make the following changes in Algorithm 6.1:

Step 1 For $i = 1, \ldots, n$ do Steps 2, 3, and 4.

Step 4 For $j = 1, \ldots, i - 1, i + 1, \ldots, n$ do Steps 5 and 6.

Step 8 For $i = 1, \ldots, n$ set $x_i = a_{i,n+1}/a_{ii}$.

In addition, delete Step 9.

15. a. Show that the Gauss-Jordan method requires

$$\frac{n^3}{2} + n^2 - \frac{n}{2} \quad \text{Multiplications/Divisions} \quad \text{and} \quad \frac{n^3}{2} - \frac{n}{2} \quad \text{Additions/Subtractions.}$$

b. Compare the number of operations needed for Gauss-Jordan and Gaussian Elimination when $n = 3, 10, 50$ and 100.

SOLUTION: Refer to Exercise 8 for the algorithm for the Gauss-Jordan method. Using this algorithm, we have the following operation counts:

	Multiplications/Divisions	Additions/Subtractions
Step 5	$n - 1$ for each i	0
Step 6	$(n - 1)(n - i + 1)$ for each i	$(n - 1)(n - i + 1)$ for each i
Step 8	1 for each i	0

The totals are

Multiplications/Divisions:

$$\sum_{i=1}^{n} [(n - 1) + (n - 1)(n - i + 1) + 1] = \frac{n^3}{2} + n^2 - \frac{n}{2};$$

Additions/Subtractions:

$$\sum_{i=1}^{n} (n - 1)(n - i + 1) = \frac{n^3 - n}{2}.$$

b. The results for this exercise are listed in the following table. In this table the abbreviations M/D and A/S are used for Multiplications/Divisions and for Additions/Subtractions, respectively.

	Gaussian Elimination		Gauss-Jordan	
n	M/D	A/S	M/D	A/S
3	17	11	21	12
10	430	375	595	495
50	44150	42875	64975	62475
100	343300	338250	509950	499950

16. a. Apply the Gaussian-elimination technique to reduce a system to triangular form. Then use the nth equation to eliminate the coefficients of x_n in each of the first $n - 1$ rows. After this is completed, use the $(n - 1)$st equation to eliminate the coefficients of x_{n-1} in the first $n - 2$ rows, etc. The system will eventually appear as the reduced system in Exercise 8. Determine the number of calculations needed for this method.

b. Compare the number of operations needed for Gaussian Elimination, Gauss-Jordan, and the hybrid method when $n = 3, 10, 50$ and 100.

SOLUTION: The Gaussian elimination procedure requires

$$\frac{2n^3 + 3n^2 - 5n}{6} \text{ Multiplications/Divisions} \quad \text{and} \quad \frac{n^3 - n}{3} \text{ Additions/Subtractions.}$$

The additional elimination steps are:

For $i = n, n - 1, ..., 2$

    for $j = 1, ..., i - 1$,

$$\text{set} \quad a_{j,n+1} = a_{j,n+1} - \frac{a_{ji} a_{i,n+1}}{a_{ii}}.$$

This requires

$$n(n-1) \text{ Multiplications/Divisions} \quad \text{and} \quad \frac{n(n-1)}{2} \text{ Additions/Subtractions.}$$

Solving for

$$x_i = \frac{a_{i,n+1}}{a_{ii}}$$

requires n divisions. Thus, the totals are

$$\frac{n^3}{3} + \frac{3n^2}{2} - \frac{5n}{6} \text{ Multiplications/Divisions} \quad \text{and} \quad \frac{n^3}{3} + \frac{n^2}{2} - \frac{5n}{6} \text{ Additions/Subtractions.}$$

b. In the following table the abbreviations M/D and A/S are used for Multiplications/Divisions and Additions/Subtractions, respectively.

	Gaussian Elimination		Gauss-Jordan		Hybrid	
n	M/D	A/S	M/D	A/S	M/D	A/S
3	17	11	21	12	20	11
10	430	375	595	495	475	375
50	44150	42875	64975	62475	45375	42875
100	343300	338250	509950	499950	348250	338250

20. A Fredholm integral equation of the second kind is an equation of the form

$$u(x) = f(x) + \int_a^b K(x,t)u(t)\, dt,$$

where a and b and the functions f and K are given. To approximate the function u on the interval $[a, b]$, a partition $x_0 = a \leq x_1 \leq \cdots \leq x_{m-1} \leq x_m = b$ is selected and the equations

$$u(x_i) = f(x_i) + \int_a^b K(x_i,t)u(t)\, dt, \quad \text{for each} \quad i = 0, \ldots, m,$$

are solved for $u(x_0), u(x_1), \ldots, u(x_m)$. The integrals are approximated using quadrature formulas based on the nodes $x_0, \ldots, x_m$. In our problem, $a = 0$, $b = 1$, $f(x) = x^2$, and $K(x, t) = e^{|k-t|}$.

a. Show that the linear system

$$u(0) = f(0) + \frac{1}{2}[K(0,0)u(0) + K(0,1)u(1)],$$

$$u(1) = f(1) + \frac{1}{2}[K(1,0)u(0) + K(1,1)u(1)]$$

must be solved when the Trapezoidal rule is used.

b. Set up and solve the linear system that results when the Composite Trapezoidal rule is used with $n = 4$.

c. Repeat part (b) using the Composite Simpson's rule.

SOLUTION: **a.** For the Trapezoidal rule, $m = n = 1$, $x_0 = 0$, $x_1 = 1$, so for $i = 0$ and $i = 1$, we have

$$u(x_i) = f(x_i) + \int_0^1 K(x_i, t)u(t)\, dt = f(x_i) + \frac{1}{2}[K(x_i,0)u(0) + K(x_i,1)u(1)].$$

Substituting for x_i gives the desired equations.

b. We have

$$n = 4, \quad h = \frac{1}{4}, \quad x_0 = 0, \quad x_1 = \frac{1}{4}, \quad x_2 = \frac{1}{2}, \quad x_3 = \frac{3}{4}, \quad \text{and} \quad x_4 = 1,$$

so

$$u(x_i) = f(x_i) + \frac{h}{2}\left[K(x_i,0)u(0) + K(x_i,1)u(1) + 2K\left(x_i, \frac{1}{4}\right)u\left(\frac{1}{4}\right) \right.$$
$$\left. + 2K\left(x_i, \frac{1}{2}\right)u\left(\frac{1}{2}\right) + 2K\left(x_i, \frac{3}{4}\right)u\left(\frac{3}{4}\right) \right],$$

for $i = 0, 1, 2, 3, 4$. This gives

$$u(x_i) = x_i^2 + \frac{1}{8}\left[e^{x_i}u(0) + e^{|x_i-1|}u(1) + 2e^{|x_i-\frac{1}{4}|}u\left(\frac{1}{4}\right) + 2e^{|x_i-\frac{1}{2}|}u\left(\frac{1}{2}\right) + 2e^{|x_i-\frac{3}{4}|}u\left(\frac{3}{4}\right) \right],$$

for each $i = 1, \ldots, 4$.

This 5×5 linear system has solution

$$u(0) = -1.154255, \quad u\left(\frac{1}{4}\right) = -0.9093298, \quad u\left(\frac{1}{2}\right) = -0.7153145, \quad u\left(\frac{3}{4}\right) = -0.5472949,$$

and $u(1) = -0.3931261$.

c. The Composite Simpson's rule gives

$$\int_0^1 K(x_i, t)u(t)\, dt = \frac{h}{3}\left[K(x_i,0)u(0) + 2K\left(x_i, \frac{1}{2}\right)u\left(\frac{1}{2}\right) \right.$$
$$\left. + 4K\left(x_i, \frac{1}{4}\right)u\left(\frac{1}{4}\right) + 4K\left(x_i, \frac{3}{4}\right)u\left(\frac{3}{4}\right) + K(x_i,1)u(1) \right],$$

which results in the linear equations given, for each $i = 1, \ldots, 4$, by

$$u(x_i) = x_i^2 + \frac{1}{12}\left[e^{x_i}u(0) + e^{|x_i-1|}u(1) + 2e^{|x_i-\frac{1}{2}|}u\left(\frac{1}{2}\right) + 4e^{|x_i-\frac{1}{4}|}u\left(\frac{1}{4}\right) + 4e^{|x_i-\frac{3}{4}|}u\left(\frac{3}{4}\right) \right].$$

This 5×5 linear system has solutions

$$u(0) = -1.234286, \quad u\left(\frac{1}{4}\right) = -0.9507292, \quad u\left(\frac{1}{2}\right) = -0.7659400, \quad u\left(\frac{3}{4}\right) = -0.5844737,$$

and $u(1) = -0.4484975$.

Exercise Set 6.2, page 379

1. b. Find the row interchanges that are required to solve the following linear system using Algorithm 6.1.

$$\begin{aligned}
x_1 + x_2 - x_3 &= 1, \\
x_1 + x_2 + 4x_3 &= 2, \\
2x_1 - x_2 + 2x_3 &= 3.
\end{aligned}$$

SOLUTION: We form the augmented matrix

$$\begin{bmatrix}
1 & 1 & -1 & \vdots & 1 \\
1 & 1 & 4 & \vdots & 2 \\
2 & -1 & 2 & \vdots & 3
\end{bmatrix}.$$

Performing the operations $E_2 - E_1 \rightarrow E_2$ and $E_3 - 2E_1 \rightarrow E_3$ gives

$$\begin{bmatrix}
1 & 1 & -1 & \vdots & 1 \\
0 & 0 & 5 & \vdots & 1 \\
0 & -3 & 4 & \vdots & 1
\end{bmatrix}.$$

Since a_{22} is now 0, we need to interchange rows 2 and 3. No further interchanges are needed.

3. b. Repeat part (b) of Exercise 1 using Algorithm 6.2.

SOLUTION: We form the augmented matrix

$$\begin{bmatrix}
1 & 1 & -1 & \vdots & 1 \\
1 & 1 & 4 & \vdots & 2 \\
2 & -1 & 2 & \vdots & 3
\end{bmatrix}.$$

Since $|a_{31}| = 2 > 1 = |A_{11}| = |a_{21}|$, we first interchange rows 1 and 3. Thus, we have

$$\begin{bmatrix}
2 & -1 & 2 & \vdots & 3 \\
1 & 1 & 4 & \vdots & 2 \\
1 & 1 & -1 & \vdots & 1
\end{bmatrix}.$$

Performing the operations $E_2 - \frac{1}{2}E_1 \to E_2$ and $E_3 - \frac{1}{2}E_1 \to E_3$ gives

$$\begin{bmatrix} 2 & -1 & 2 & \vdots & 3 \\ 0 & \frac{3}{2} & 3 & \vdots & \frac{1}{2} \\ 0 & \frac{3}{2} & -2 & \vdots & -\frac{1}{2} \end{bmatrix}.$$

No further interchanges are needed.

5. b. Repeat part (b) of Exercise 1 using Algorithm 6.3.

SOLUTION: With $S_1 = 1$, $S_2 = 4$ and $S_3 = 2$, we form the augmented matrix

$$\begin{bmatrix} 1 & 1 & -1 & \vdots & 1 \\ 1 & 1 & 4 & \vdots & 2 \\ 2 & -1 & 2 & \vdots & 3 \end{bmatrix}.$$

Since $|a_{11}|/S_1 = 1$, $|a_{12}|/S_2 = 1/4$ and $|a_{13}|/S_3 = 1$, no interchange is needed. Performing the operations $E_2 - E_1 \to E_2$ and $E_3 - 2E_1 \to E_3$ gives

$$\begin{bmatrix} 1 & 1 & -1 & \vdots & 1 \\ 0 & 0 & 5 & \vdots & 1 \\ 0 & -3 & 4 & \vdots & 1 \end{bmatrix}.$$

Since $a_{22} = 0$, we need to interchange rows 2 and 3. No further interchange is needed.

7(b)

7. b. Repeat Exercise 1(b) using complete pivoting.

SOLUTION: Complete pivoting first requires that columns 1 and 3 be interchanged, and rows 1 and 2.

$$\begin{bmatrix} 1 & 1 & -1 & \vdots & 1 \\ 1 & 1 & 4 & \vdots & 2 \\ 2 & -1 & 2 & \vdots & 3 \end{bmatrix} \; C1 \leftrightarrow C3 \; \begin{bmatrix} -1 & 1 & 1 & \vdots & 1 \\ 4 & 1 & 1 & \vdots & 2 \\ 2 & -1 & 2 & \vdots & 3 \end{bmatrix} \; R1 \leftrightarrow R2 \; \begin{bmatrix} 4 & 1 & 1 & \vdots & 2 \\ -1 & 1 & 1 & \vdots & 1 \\ 2 & -1 & 2 & \vdots & 3 \end{bmatrix}.$$

Then we perform $R2 + \frac{1}{4}R1 \to R2$ followed by $R2 - \frac{1}{2}R1 \to R2$ to obtain

$$\begin{bmatrix} 4 & 1 & 1 & \vdots & 2 \\ 0 & \frac{5}{4} & \frac{5}{4} & \vdots & \frac{3}{2} \\ 0 & -\frac{3}{2} & \frac{3}{2} & \vdots & 2 \end{bmatrix}.$$

Complete pivoting requires the interchange of rows 2 and 3. After that we perform $R3 + \frac{5}{6}R2 \to R3$ to reduce the matrix to the upper triangular form

$$\begin{bmatrix} 4 & 1 & 1 & \vdots & 2 \\ 0 & -\dfrac{3}{2} & \dfrac{3}{2} & \vdots & 2 \\ 0 & 0 & \dfrac{5}{2} & \vdots & \dfrac{19}{6} \end{bmatrix}.$$

Because columns 1 and 3 were interchanged, backward substitution gives x_1, then x_2, and finally x_3.

$$x_1 = \frac{19}{15}, \quad x_2 = -\frac{1}{15}, \quad \text{and} \quad x_3 = \frac{1}{5}.$$

9. a. Solve the linear system

$$0.03x_1 + 58.9x_2 = 59.2,$$
$$5.31x_1 - 6.10x_2 = 47.0$$

using Gaussian elimination and three-digit chopping arithmetic.

SOLUTION: Using the pivot

$$\frac{5.31}{0.03} = 177$$

gives the reduced system

$$0.03x_1 + 58.9x_2 = 59.2,$$
$$-10400x_2 = -10300,$$

so

$$x_2 = \frac{-10300}{-10400} = 0.990 \quad \text{and} \quad x_1 = \frac{1}{0.03}(59.2 - 58.9(0.990)) = 30.00.$$

The exact solution is $x_1 = 10.0$ and $x_2 = 1.00$, so the roundoff error here is devastating.

13. a. Repeat part (a) of Exercise 9 using Gaussian elimination with partial pivoting.

SOLUTION: The coefficient of x_1 in the first row is smaller than the coefficient of x_1 in the second row, so partial pivoting calls for a row interchange to

$$5.31x_1 - 6.10x_2 = 47.0,$$
$$0.03x_1 + 58.9x_2 = 59.2,$$

which produces the pivot

$$\frac{0.03}{5.31} = 0.00565.$$

The reduced system is now

$$5.31x_1 - 6.10x_2 = 47.0,$$
$$58.9x_2 = 58.9,$$

which produces the exact the solution

$$x_2 = \frac{58.9}{58.9} = 1.00 \quad \text{and} \quad x_1 = \frac{1}{5.31}(47.0 + 6.10(1.00)) = 10.00.$$

17. a. Repeat part (a) of Exercise 9 using Gaussian elimination with scaled-partial pivoting.

SOLUTION: The scaling factors for the rows are 58.9 and 6.10, respectively. Since

$$\frac{0.03}{58.9} < \frac{5.31}{6.10},$$

a row interchange is required. The calculations then proceed the same as those in part (a) of Exercise 7.

31. Suppose that

$$2x_1 + x_2 + 3x_3 = 1,$$
$$4x_1 + 6x_2 + 8x_3 = 5,$$
$$6x_1 + \alpha x_2 + 10x_3 = 5,$$

with $|\alpha| < 10$. For which of the following values of α will there be no row interchange required when solving this system using scaled partial pivoting?

a. $\alpha = 6$ **b.** $\alpha = 9$ **c.** $\alpha = -3$

SOLUTION: Regardless of whether $\alpha = 6, 9$, or -3, we have $S_1 = 3$, $S_2 = 8$ and $S_3 = 10$. Since

$$\frac{|a_{11}|}{S_1} = \frac{2}{3}, \quad \frac{|a_{21}|}{S_2} = \frac{1}{2} \quad \text{and} \quad \frac{|a_{31}|}{S_3} = \frac{3}{5},$$

we do not need a row interchange yet. Using the augmented matrix

$$\begin{bmatrix} 2 & 1 & 3 & \vdots & 1 \\ 4 & 6 & 8 & \vdots & 5 \\ 6 & \alpha & 10 & \vdots & 5 \end{bmatrix}$$

and the row operations $E_2 - 2E_1 \to E_2$ and $E_3 - 3E_1 \to E_3$ gives

$$\begin{bmatrix} 2 & 1 & 3 & \vdots & 1 \\ 0 & 4 & 2 & \vdots & 3 \\ 0 & \alpha - 3 & 1 & \vdots & 2 \end{bmatrix}.$$

No row interchange is required if

$$\frac{|a_{32}|}{S_3} \leq \frac{|a_{22}|}{S_2},$$

which implies that

$$\frac{|\alpha - 3|}{10} \leq \frac{4}{8}, \quad \text{or} \quad |\alpha - 3| \leq 5.$$

Thus, no interchange is required if $\alpha = 6$. When $\alpha = 9$ or $\alpha = -3$, an interchange of rows 2 and 3 is required.

32. Construct an algorithm for implementing complete pivoting.

SOLUTION: Change Algorithm 6.2 as follows:

Add to Step 1.
$$NCOL(i) = i$$

Replace Step 3 with the following.

Let p and q be the smallest integers with $i \leq p, q \leq n$ and

$$|a(NROW(p), NCOL(q))| = \max_{i \leq k, j \leq n} |a(NROW(k), NCOL(j))|.$$

Add to Step 4.
$$A(NROW(p), NCOL(q)) = 0$$

Add to Step 5.

If $NCOL(q) \neq NCOL(i)$ then set

$$NCOPY = NCOL(i);$$
$$NCOL(i) = NCOL(q);$$
$$NCOL(q) = NCOPY.$$

Replace Step 7 with the following.

Set
$$m(NROW(j), NCOL(i)) = \frac{a(NROW(j), NCOL(i))}{a(NROW(i), NCOL(i))}.$$

Replace in Step 8:
$$m(NROW(j), i) \quad \text{by} \quad m(NROW(j), NCOL(i))$$

Replace in Step 9:
$$a(NROW(n), n) \quad \text{by} \quad a(NROW(n), NCOL(n))$$

Replace Step 10 with the following.

Set
$$X(NCOL(n)) = \frac{a(NROW(n), n+1)}{a(NROW(n), NCOL(n))}.$$

Replace Step 11 with the following.

Set   X($NCOL$(i))=

$$\frac{a(NROW(i), n+1) - \sum_{j=i+1}^{n} a(NROW(i), NCOL(j)) \cdot X(NCOL(j))}{A(NROW(i), NCOL(i))}.$$

Replace Step 12 with the following.

OUTPUT $('X(', NCOL(i), ',) =', X(NCOL(i))$ for $i = 1, \ldots, n)$.

Exercise Set 6.3, page 390

2. c. Perform the matrix-vector multiplication

$$\begin{bmatrix} 2 & 1 & 0 \\ 1 & -1 & 2 \\ 0 & 2 & 4 \end{bmatrix} \begin{bmatrix} 2 \\ 5 \\ 1 \end{bmatrix}$$

SOLUTION: We have

$$\begin{bmatrix} 2 & 1 & 0 \\ 1 & -1 & 2 \\ 0 & 2 & 4 \end{bmatrix} \begin{bmatrix} 2 \\ 5 \\ 1 \end{bmatrix} = \begin{bmatrix} 2 \cdot 2 + 1 \cdot 5 + 0 \cdot 1 \\ 1 \cdot 2 - 1 \cdot 5 + 2 \cdot 1 \\ 0 \cdot 2 + 2 \cdot 5 + 4 \cdot 1 \end{bmatrix} = \begin{bmatrix} 9 \\ -1 \\ 14 \end{bmatrix}$$

4. b. Perform the matrix-matrix multiplication

$$\begin{bmatrix} -1 & 3 \\ -2 & 4 \end{bmatrix} \begin{bmatrix} 2 & -2 & 3 \\ -3 & 2 & 2 \end{bmatrix}$$

SOLUTION: We have

$$\begin{bmatrix} -1 & 3 \\ -2 & 4 \end{bmatrix} \begin{bmatrix} 2 & -2 & 3 \\ -3 & 2 & 2 \end{bmatrix} = \begin{bmatrix} (-1)(2) + (3)(-3) & (-1)(-2) + (3)(2) & (-1)(3) + (3)(2) \\ (-2)(2) + (4)(-3) & (-2)(-2) + (4)(2) & (-2)(3) + (4)(2) \end{bmatrix}$$

$$= \begin{bmatrix} -11 & 8 & 3 \\ -16 & 12 & 2 \end{bmatrix}$$

5. b. Show that

$$\begin{bmatrix} 1 & 2 & 0 \\ 2 & 1 & -1 \\ 3 & 1 & 1 \end{bmatrix}$$

is nonsingular and compute its inverse.

SOLUTION: The triple augmented matrix used to determine the inverse is

$$\begin{bmatrix} 1 & 2 & 0 & \vdots & 1 & 0 & 0 \\ 2 & 1 & -1 & \vdots & 0 & 1 & 0 \\ 3 & 1 & 1 & \vdots & 0 & 0 & 1 \end{bmatrix}.$$

The inverse exists if and only if the left portion of this matrix can be reduced using row operations to an upper triangular matrix with nonzero diagonal elements.

The pivots for the second and third rows are, respectively,

$$\frac{2}{1} = 2 \quad \text{and} \quad \frac{3}{1} = 3.$$

Performing $(E_2 - 2E_1) \to E_2$ and $(E_3 - 3E_1) \to E_3$ produces the augmented matrix

$$\begin{bmatrix} 1 & 2 & 0 & \vdots & 1 & 0 & 0 \\ 0 & -3 & -1 & \vdots & -2 & 1 & 0 \\ 0 & -5 & 1 & \vdots & -3 & 0 & 1 \end{bmatrix}.$$

The new pivot is

$$\frac{-5}{-3} = \frac{5}{3},$$

and $\left(E_3 - \frac{5}{3}E_2\right) \to E_3$ produces

$$\begin{bmatrix} 1 & 2 & 0 & \vdots & 1 & 0 & 0 \\ 0 & -3 & -1 & \vdots & -2 & 1 & 0 \\ 0 & 0 & \frac{8}{3} & \vdots & \frac{1}{3} & -\frac{5}{3} & 1 \end{bmatrix}.$$

Continuing in this manner until the left side of the matrix is the 3×3 identity matrix produces

$$\begin{bmatrix} 1 & 0 & 0 & \vdots & \frac{1}{4} & \frac{1}{4} \\ 0 & 1 & 0 & \vdots & -\frac{1}{8} & -\frac{1}{8} \\ 0 & 0 & 1 & \vdots & -\frac{5}{8} & \frac{3}{8} \end{bmatrix}.$$

So the inverse matrix is

$$\begin{bmatrix} -\frac{1}{4} & \frac{1}{4} & \frac{1}{4} \\ \frac{5}{8} & -\frac{1}{8} & -\frac{1}{8} \\ \frac{1}{8} & -\frac{5}{8} & \frac{3}{8} \end{bmatrix}.$$

12. Suppose m linear systems
$$A\mathbf{x}^{(p)} = \mathbf{b}^{(p)}, \quad \text{for } p = 1, 2, \ldots, m,$$

are to be solved, each with the $n \times n$ coefficient matrix A.

a. Show that Gaussian elimination with backward substitution applied to the augmented matrix $\left[A : \mathbf{b}^{(1)} \cdots \mathbf{b}^{(m)}\right]$ requires

$\frac{1}{3}n^3 + mn^2 - \frac{1}{3}n$ Multiplications/Divisions   and $\frac{1}{3}n^3 + mn^2 - \frac{1}{2}n^2 - mn + \frac{1}{6}n$ Additions/Subtractions.

b. Show that the Gauss-Jordan method (See Exercise 12, Section 6.1) applied to the augmented matrix $\left[A : \mathbf{b}^{(1)} \cdots \mathbf{b}^{(m)}\right]$ requires

$$\frac{1}{2}n^3 + mn^2 - \frac{1}{2}n \quad \text{Multiplications/Divisions;}$$

and

$$\frac{1}{2}n^3 + (m-1)n^2 + \left(\frac{1}{2} - m\right)n \quad \text{Additions/Subtractions.}$$

d. Construct an algorithm using Gaussian elimination to find A^{-1}, but do not perform multiplications when one of the multipliers is known to be unity, and do not perform additions/subtractions when one of the elements involved is known to be zero. Show that the required computations are reduced to n^3 multiplications/divisions and $n^3 - 2n^2 + n$ additions/subtractions.

SOLUTION: **a.** Following the steps of Algorithm 6.1 with $m - 1$ additional columns in the augmented matrix gives the following:

Reduction Steps 1–6:

Multiplications/Divisions:

$$\sum_{i=1}^{n-1}\sum_{j=i+1}^{n}\{1 + (m + n - i)\} = \sum_{i=1}^{n-1}\{n(m+n+1) - (m+2n+1)i + i^2\}$$

$$= \frac{1}{2}mn^2 - \frac{1}{2}mn + \frac{1}{3}n^3 - \frac{1}{3}n;$$

Additions/Subtractions:

$$\sum_{i=1}^{n-1}\sum_{j=i+1}^{n}\{m + n - i\} = \sum_{i=1}^{n-1}\{n(m+n) - (m+2n)i + i^2\}$$

$$= \frac{1}{2}mn^2 - \frac{1}{2}mn + \frac{1}{3}n^3 - \frac{1}{2}n^2 + \frac{1}{6}n.$$

Backward Substitution Steps 8-9:

Multiplications/Divisions:

$$m\left[1 + \sum_{i=1}^{n-1}(n - i + 1)\right] = m\left[1 + \frac{n(n+1)}{2} - 1\right] = \frac{1}{2}mn^2 + \frac{1}{2}mn;$$

Additions/Subtractions:

$$m\left[\sum_{i=1}^{n-1}(n - i)\right] = \frac{1}{2}mn^2 - \frac{1}{2}mn.$$

Total:

$$\text{Multiplications/Divisions: } \tfrac{1}{3}n^3 + mn^2 - \tfrac{1}{3}n;$$

$$\text{Additions/Subtractions: } \tfrac{1}{3}n^3 + mn^2 - \tfrac{1}{2}n^2 - mn + \tfrac{1}{6}n.$$

b. For the reduction phase:

Multiplications/Divisions:

$$\sum_{i=1}^{n}\sum_{j=1,j\neq i}^{n}\left\{1 + \sum_{k=i+1}^{n+m}1\right\} = \sum_{i=1}^{n}\sum_{j=1,j\neq i}^{n}(m + n + 1 - i) = \sum_{i=1}^{n}\{(n-1)(m+n+1) - (n-1)i\}$$

$$= \frac{1}{2}n^3 + mn^2 - mn - \frac{1}{2}n$$

Additions/Subtractions:

$$\sum_{i=1}^{n}\sum_{j=1,j\neq i}^{n}\sum_{k=i+1}^{n+m}1=\sum_{i=1}^{n}\sum_{j=1,j\neq i}^{n}(n+m-i)=\sum_{i=1}^{n}\{(n-1)(m+n)-(n-1)i\}$$

$$=\frac{1}{2}n^3+mn^2-mn-n^2+\frac{1}{2}n$$

Backward Substitution Steps:
Multiplications/Divisions:

$$\sum_{k=1}^{m}\sum_{i=1}^{n}1=mn;$$

Additions/Subtractions: none.

Totals:

$$\text{Multiplications/Divisions: } \tfrac{1}{2}n^3+mn^2-\tfrac{1}{2}n;$$

$$\text{Additions/Subtractions: } \tfrac{1}{2}n^3+mn^2-n^2-mn+\tfrac{1}{2}n.$$

d. To find the inverse of the $n \times n$ matrix A:

INPUT $n \times n$ matrix $A = (a_{ij})$.
OUTPUT $n \times n$ matrix $B = A^{-1}$.
Step 1   Initialize the $n \times n$ matrix $B = (b_{ij})$ to

$$b_{ij}=\begin{cases} 0 & i\neq j, \\ 1 & i=j \end{cases}$$

Step 2   For $i = 1, \ldots, n - 1$ do Steps 3, 4, and 5.
Step 3   Let p be the smallest integer with $i \leq p \leq n$ and $a_{p,i} \neq 0$.
             If no integer p can be found then
             OUTPUT ('A is singular');
             STOP.
Step 4   If $p \neq i$ then perform $(E_p) \leftrightarrow (E_i)$.
Step 5   For $j = i + 1, \ldots, n$ do Steps 6 through 9.

     Step 6   Set $m_{ji} = a_{ji}/a_{ii}$.
     Step 7   For $k = i + 1, \ldots, n$
                 set $a_{jk} = a_{jk} - m_{ji}a_{ik}$; $a_{ij} = 0$.
     Step 8   For $k = 1, \ldots, i - 1$
                 set $b_{jk} = b_{jk} - m_{ji}b_{ik}$.
     Step 9   Set $b_{ji} = -m_{ji}$.

Step 10   If $a_{nn} = 0$ then OUTPUT ('A is singular');
                          STOP.

Step 11   For $j = 1, \ldots, n$ do Steps 12, 13 and 14.

     Step 12   Set $b_{nj} = b_{nj}/a_{nn}$.
     Step 13   For $i = n - 1, \ldots, j$
                  set $b_{ij} = \left(b_{ij} - \sum_{k=i+1}^{n} a_{ik}b_{kj}\right)/a_{ii}$.
     Step 14   For $i = j - 1, \ldots, 1$
                  set $b_{ij} = -\left[\sum_{k=i+1}^{n} a_{ik}b_{kj}\right]/a_{ii}$.

Step 15 OUTPUT (B);

STOP.

Reduction Steps 2–9:

Multiplications/Divisions:

$$\sum_{i=1}^{n-1}\sum_{j=i+1}^{n}\{1+\sum_{k=i+1}^{n}1+\sum_{k=1}^{i-1}1\} = \sum_{i=1}^{n-1}\sum_{j=i+1}^{n}\{1+n-i+i-1\} = \frac{n^2(n-1)}{2};$$

Additions/Subtractions:

$$\sum_{i=1}^{n-1}\sum_{j=i+1}^{n}\left\{\sum_{k=i+1}^{n}1+\sum_{k=1}^{i-1}1\right\} = \sum_{i=1}^{n-1}\sum_{j=i+1}^{n}\{n-i+i-1\} = \frac{n(n-1)^2}{2}.$$

Backward Substitution Steps 11–14:

Multiplications/Divisions:

$$\sum_{j=1}^{n}\left\{1+\sum_{i=j}^{n-1}\left(1+\sum_{k=i+1}^{n}1\right)+\sum_{i=1}^{j-1}\left(1+\sum_{k=i+1}^{n}1\right)\right\} = \sum_{j=1}^{n}\left\{1+\sum_{i=j}^{n-1}(n+1-i)+\sum_{i=1}^{j-1}(n+1-i)\right\}$$

$$= \sum_{j=1}^{n}\left(1+\sum_{i=1}^{n-1}(n+1-i)\right)$$

$$= \sum_{j=1}^{n}\frac{n(n+1)}{2} = \frac{n^2(n+1)}{2};$$

Additions/Subtractions:

$$\sum_{j=1}^{n}\left\{\sum_{i=j}^{n-1}(1+n-i-1)+\sum_{i=1}^{j-1}(n-i-1)\right\} = \sum_{j=1}^{n}\left\{\left(\sum_{i=1}^{n-1}n-i\right)-j+1\right\}$$

$$= \sum_{j=1}^{n}\left(\frac{n(n-1)}{2}+1-j\right)$$

$$= \frac{n^2(n-1)}{2}+n-\frac{n(n+1)}{2} = \frac{n^3}{2}-n^2+\frac{1}{2}n.$$

Totals:

Multiplications/Divisions:

$$\frac{n^2(n-1)}{2}+\frac{n^2(n+1)}{2} = n^3;$$

Additions/Subtractions:

$$\frac{n(n-1)^2}{2}+\frac{n^3}{2}-n^2+\frac{1}{2}n = n^3-2n^2+n.$$

The remainder of this exercise follows similar lines.

18. Consider the 2 by 2 linear system $(A+iB)(\mathbf{x}+i\mathbf{y}) = \mathbf{c}+i\mathbf{d}$ with complex entries in component form

$$(a_{11}+ib_{11})(x_1+iy_1)+(a_{12}+ib_{12})(x_2+iy_2) = c_1+id_1,$$
$$(a_{21}+ib_{21})(x_1+iy_1)+(a_{22}+ib_{22})(x_2+iy_2) = c_2+id_2.$$

a. Use the properties of complex numbers to convert this system to the equivalent 4 by 4 real linear system

$$A\mathbf{x} - B\mathbf{y} = \mathbf{c},$$
$$B\mathbf{x} + A\mathbf{y} = \mathbf{d}.$$

b. Solve the linear system

$$(1 - 2i)(x_1 + iy_1) + (3 + 2i)(x_2 + iy_2) = 5 + 2i,$$
$$(2 + i)(x_1 + iy_1) + (4 + 3i)(x_2 + iy_2) = 4 - i.$$

SOLUTION: **a.** In component form:

$$(a_{11}x_1 - b_{11}y_1 + a_{12}x_2 - b_{12}y_2) + (b_{11}x_1 + a_{11}y_1 + b_{12}x_2 + a_{12}y_2)i = c_1 + id_1$$

and

$$(a_{21}x_1 - b_{21}y_1 + a_{22}x_2 - b_{22}y_2) + (b_{21}x_1 + a_{21}y_1 + b_{22}x_2 + a_{22}y_2)i = c_2 + id_2,$$

which yields

$$a_{11}x_1 + a_{12}x_2 - b_{11}y_1 - b_{12}y_2 = c_1,$$
$$b_{11}x_1 + b_{12}x_2 + a_{11}y_1 + a_{12}y_2 = d_1,$$
$$a_{21}x_1 + a_{22}x_2 - b_{21}y_1 - b_{22}y_2 = c_2,$$
$$b_{21}x_1 + b_{22}x_2 + a_{21}y_1 + a_{22}y_2 = d_2.$$

b. The system

$$\begin{bmatrix} 1 & 3 & 2 & -2 \\ -2 & 2 & 1 & 3 \\ 2 & 4 & -1 & -3 \\ 1 & 3 & 2 & 4 \end{bmatrix} \begin{bmatrix} x_1 \\ x_2 \\ y_1 \\ y_2 \end{bmatrix} = \begin{bmatrix} 5 \\ 2 \\ 4 \\ -1 \end{bmatrix}$$

has the solution $x_1 = -1.2$, $x_2 = 1$, $y_1 = 0.6$, and $y_2 = -1$.

Exercise Set 6.4, page 399

1. a. Compute the determinant of

$$\begin{bmatrix} 1 & 2 & 0 \\ 2 & 1 & -1 \\ 3 & 1 & 1 \end{bmatrix}.$$

SOLUTION: Expanding along the first row gives

$$\det \begin{bmatrix} 1 & 2 & 0 \\ 2 & 1 & -1 \\ 3 & 1 & 1 \end{bmatrix} = 1 \cdot \det \begin{bmatrix} 1 & -1 \\ 1 & 1 \end{bmatrix} - 2 \cdot \det \begin{bmatrix} 2 & -1 \\ 3 & 1 \end{bmatrix} + 0 \cdot \det \begin{bmatrix} 2 & 1 \\ 3 & 1 \end{bmatrix} = 1(1+1) - 2(2+3) = -8.$$

5. Find all values of α that make the following matrix singular.

$$A = \begin{bmatrix} 1 & -1 & \alpha \\ 2 & 2 & 1 \\ 0 & \alpha & -\frac{3}{2} \end{bmatrix}$$

SOLUTION: We compute

$$\det A = \det \begin{bmatrix} 1 & -1 & \alpha \\ 2 & 2 & 1 \\ 0 & \alpha & -\frac{3}{2} \end{bmatrix} = \det \begin{bmatrix} 2 & 1 \\ \alpha & -\frac{3}{2} \end{bmatrix} + \det \begin{bmatrix} 2 & 1 \\ 0 & -\frac{3}{2} \end{bmatrix} + \alpha \det \begin{bmatrix} 2 & 2 \\ 0 & \alpha \end{bmatrix}$$

$$= -3 - \alpha - 3 + 2\alpha^2 = 2\alpha^2 - \alpha - 6.$$

Setting $\det A = 0$ and solving for α gives $\alpha = 2$ or $\alpha = -\frac{3}{2}$.

7. Find all values of α so that the following linear system has no solutions.

$$\begin{aligned} 2x_1 - x_2 + 3x_3 &= 5, \\ 4x_1 + 2x_2 + 2x_3 &= 6, \\ -2x_1 + \alpha x_2 + 3x_3 &= 4. \end{aligned}$$

SOLUTION: We form the augmented matrix

$$\begin{bmatrix} 2 & -1 & 3 & \vdots & 5 \\ 4 & 2 & 2 & \vdots & 6 \\ -2 & \alpha & 3 & \vdots & 4 \end{bmatrix}.$$

Performing the operations $E_2 - 2E_1 \rightarrow E_2$ and $E_3 + E_1 \rightarrow E_3$ gives

$$\begin{bmatrix} 2 & -1 & 3 & \vdots & 5 \\ 0 & 4 & -4 & \vdots & -4 \\ 0 & \alpha - 1 & 6 & \vdots & 9 \end{bmatrix}.$$

Performing the operation $E_3 - \dfrac{\alpha - 1}{4} E_2 \rightarrow E_3$ gives

$$\begin{bmatrix} 2 & -1 & 3 & \vdots & 5 \\ 0 & 4 & -4 & \vdots & -4 \\ 0 & 0 & 5 + \alpha & \vdots & 8 + \alpha \end{bmatrix}.$$

The linear system has a unique solution unless $\alpha = -5$. If $\alpha = -5$, there are no solutions since the other row would be

$$\begin{bmatrix} 0 & 0 & 0 & \vdots & 3 \end{bmatrix}.$$

9. Show that when $n > 1$, the evaluation of the determinant of an $n \times n$ matrix requires

Multiplications/Divisions:

$$n! \sum_{k=1}^{n-1} \frac{1}{k!};$$

and Additions/Subtractions:

$$n! - 1.$$

SOLUTION: When $n = 2$, $\det A = a_{11}a_{22} - a_{12}a_{21}$ requires 2 multiplications and 1 subtraction. Since

$$2! \sum_{k=1}^{1} \frac{1}{k!} = 2 \quad \text{and} \quad 2! - 1 = 1,$$

the formula holds for $n = 2$. Assume the formula is true for $n = 2, ..., m$, and let A be an $(m + 1) \times (m + 1)$ matrix. Then

$$\det A = \sum_{j=1}^{m+1} a_{ij}A_{ij},$$

for any i, where $1 \le i \le m + 1$. To compute each A_{ij} requires

$$m! \sum_{k=1}^{m-1} \frac{1}{k!} \quad \text{Multiplications} \quad \text{and} \quad m! - 1 \quad \text{Additions/Subtractions.}$$

The number of Multiplications for $\det A$ is

$$(m + 1) \left[m! \sum_{k=1}^{m-1} \frac{1}{k!} \right] + (m + 1) = (m + 1)! \left[\sum_{k=1}^{m-1} \frac{1}{k!} + \frac{1}{m!} \right] = (m + 1)! \sum_{k=1}^{m} \frac{1}{k!},$$

and the number of Additions/Subtractions is

$$(m + 1)[m! - 1] + m = (m + 1)! - 1.$$

By the principle of mathematical induction, the formula is valid for any $n \ge 2$.

11. Prove that AB is nonsingular if and only if both A and B are nonsingular.

SOLUTION: The result can be shown in various ways. One way is to use the fact that $\det AB = \det A \det B$ and apply Theorem 6.16.

The matrix AB is nonsingular if and only if $\det AB \neq 0$. Since $\det AB = \det A \det B$, the matrix AB is nonsingular if and only if $\det A \neq 0$ and $\det B \neq 0$. This is true precisely when both A and B are nonsingular.

13. a. Generalize Cramer's rule to an $n \times n$ linear system.

b. Use the result in Exercise 9 to determine the number of Multiplications/Divisions and Additions/Subtractions required for Cramer's rule on an $n \times n$ system.

SOLUTION: If D_i is the determinant of the matrix formed by replacing the ith column of A with **b** and if $D = \det A$, then

$$x_i = D_i/D, \quad \text{for each } i = 1, \ldots, n.$$

b. In Exercise 9 we found that the determinant of an arbitrary $n \times n$ matrix requires

$$n! \sum_{k=1}^{n-1} \frac{1}{k!} \quad \text{Multiplications} \quad \text{and} \quad n! - 1 \quad \text{Additions/Subtractions.}$$

Cramer's rule to solve an $n \times n$ system requires that determinants of $n + 1$ matrices of size $n \times n$ be found and then n quotients be computed. Hence, Cramer's rule requires

Multiplications/Divisions:

$$(n + 1) \left[n! \left(\sum_{k=1}^{n-1} \frac{1}{k!} \right) \right] + n = (n + 1)! \sum_{k=1}^{n-1} \frac{1}{k!} + n;$$

and Additions/Subtractions:

$$(n + 1)[n! - 1] = (n + 1)! - n - 1.$$

The table below lists the numbers corresponding to these values for some small values of n and the comparative values for $n^3/3$, the approximate number of calculations needed to solve a system of n equations in n unknowns using Gaussian elimination. It should be quite clear from this table that Cramer's rule should not be used except for extremely small values of n.

		n	2	3	4	5	10
Cramer's Rule: M/D	$(n + 1)! \left(\sum_{k=1}^{n-1} \frac{1}{k!} \right) + n$		8	39	204	1235	68,588,310
Cramer's Rule: A/S	$(n + 1)! - n - 1$		3	20	115	714	39,916,789
Gaussian Elimination: M/D or A/S	$\dfrac{n^3}{3}$		3	9	21	42	334

Exercise Set 6.5, page 409

1. a. Solve the following linear system.

$$\begin{bmatrix} 1 & 0 & 0 \\ 2 & 1 & 0 \\ -1 & 0 & 1 \end{bmatrix} \begin{bmatrix} 2 & 3 & -1 \\ 0 & -2 & 1 \\ 0 & 0 & 3 \end{bmatrix} \begin{bmatrix} x_1 \\ x_2 \\ x_3 \end{bmatrix} = \begin{bmatrix} 2 \\ -1 \\ 1 \end{bmatrix}.$$

SOLUTION: We first solve the linear system

$$\begin{bmatrix} 1 & 0 & 0 \\ 2 & 1 & 0 \\ -1 & 0 & 1 \end{bmatrix} \begin{bmatrix} y_1 \\ y_2 \\ y_3 \end{bmatrix} = \begin{bmatrix} 2 \\ -1 \\ 1 \end{bmatrix}$$

using forward substitution. This gives

$$y_1 = 2;$$
$$2y_1 + y_2 = -1, \quad \text{so} \quad y_2 = -1 - 2y_1 = -5;$$
$$-y_1 + y_3 = 1, \quad \text{so} \quad y_3 = 1 + y_1 = 3.$$

We then solve the linear system

$$\begin{bmatrix} 2 & 3 & -1 \\ 0 & -2 & 1 \\ 0 & 0 & 3 \end{bmatrix} \begin{bmatrix} x_1 \\ x_2 \\ x_3 \end{bmatrix} = \begin{bmatrix} y_1 \\ y_2 \\ y_3 \end{bmatrix} = \begin{bmatrix} 2 \\ -5 \\ 3 \end{bmatrix}$$

using backward substitution. This gives

$$3x_3 = 3, \quad \text{so} \quad x_3 = 1;$$
$$-2x_1 + x_3 = -5, \quad \text{so} \quad x_2 = -\frac{1}{2}(-5 - 1) = 3;$$
$$2x_1 + 3x_2 - x_3 = 2, \quad \text{so} \quad x_1 = \frac{1}{2}[2 - 3(3) + 1] = -3.$$

3. b. Find the permutation matrix P so that PA can be factored into the product LU, where L is lower triangular with 1s on its diagonal and U is upper triangular for the following matrix.

$$A = \begin{bmatrix} 0 & 1 & 1 \\ 1 & -2 & -1 \\ 1 & -1 & 1 \end{bmatrix}.$$

SOLUTION: We have

$$A = \begin{bmatrix} 0 & 1 & 1 \\ 1 & -2 & -1 \\ 1 & -1 & 1 \end{bmatrix},$$

and we need to interchange rows 1 and 2 to obtain

$$\begin{bmatrix} 1 & -2 & -1 \\ 0 & 1 & 1 \\ 1 & -1 & 1 \end{bmatrix}.$$

The operation $E_3 - E_1 \rightarrow E_1$ gives

$$\begin{bmatrix} 1 & -2 & -1 \\ 0 & 1 & 1 \\ 0 & 1 & 2 \end{bmatrix}.$$

Thus, the only interchange required is rows 1 and 2 and

$$P = \begin{bmatrix} 0 & 1 & 0 \\ 1 & 0 & 0 \\ 0 & 0 & 1 \end{bmatrix}.$$

5. a. Factor the matrix

$$A = \begin{bmatrix} 2 & -1 & 1 \\ 3 & 3 & 9 \\ 3 & 3 & 5 \end{bmatrix}$$

into LU using the Factorization Algorithm with $l_i = 1$ for all i.

SOLUTION: There are two approaches we can take in this problem. We can first assume that we have a factorization of the form $A = LU$ written as

$$\begin{bmatrix} 2 & -1 & 1 \\ 3 & 3 & 9 \\ 3 & 3 & 5 \end{bmatrix} = \begin{bmatrix} 1 & 0 & 0 \\ l_1 & 1 & 0 \\ l_2 & l_3 & 1 \end{bmatrix} \cdot \begin{bmatrix} u_1 & u_2 & u_3 \\ 0 & u_4 & u_5 \\ 0 & 0 & u_6 \end{bmatrix}.$$

Multiplying the two matrices on the right gives

$$\begin{bmatrix} 2 & -1 & 1 \\ 3 & 3 & 9 \\ 3 & 3 & 5 \end{bmatrix} = \begin{bmatrix} u_1 & u_2 & u_3 \\ l_1 u_1 & l_1 u_2 + u_4 & l_1 u_3 + u_5 \\ l_2 u_1 & l_2 u_2 + l_3 u_4 & l_2 u_3 + l_3 u_5 + u_6 \end{bmatrix}.$$

We can now solve for the entries in L and U by matching the entries in these matrices. Proceeding column by column gives

$$2 = u_1, \quad \text{so} \quad u_1 = 2;$$
$$3 = l_1 u_1 = 2l_1, \quad \text{so} \quad l_1 = \frac{3}{2};$$
$$3 = l_2 u_1 = 2l_2, \quad \text{so} \quad l_2 = \frac{3}{2}.$$

Also

$$-1 = u_2, \quad \text{so} \quad u_2 = -1;$$
$$3 = l_1 u_2 + u_4 = \frac{3}{2}(-1) + u_4, \quad \text{so} \quad u_4 = \frac{9}{2};$$
$$3 = l_2 u_2 + l_3 u_4 = \frac{3}{2}(-1) + \frac{9}{2}l_3, \quad \text{so} \quad l_3 = 1.$$

Finally

$$1 = u_3, \quad \text{so} \quad u_3 = 1;$$
$$9 = l_1 u_3 + u_5 = \frac{3}{2}(1) + u_5, \quad \text{so} \quad u_5 = \frac{15}{2};$$
$$5 = l_2 u_3 + l_3 u_5 + u_6 = \frac{3}{2}(1) + (1)\frac{15}{2} + u_6, \quad \text{so} \quad u_6 = -4.$$

This produces the factorization

$$\begin{bmatrix} 2 & -1 & 1 \\ 3 & 3 & 9 \\ 3 & 3 & 5 \end{bmatrix} = \begin{bmatrix} 1 & 0 & 0 \\ 1.5 & 1 & 0 \\ 1.5 & 1 & 1 \end{bmatrix} \cdot \begin{bmatrix} 2 & -1 & 1 \\ 0 & 4.5 & 7.5 \\ 0 & 0 & -4 \end{bmatrix}.$$

We can also produce this factorization by applying Gaussian elimination to the matrix. The pivots for the first row are

$$m_{21} = \frac{3}{2} = 1.5 \quad \text{and} \quad m_{31} = \frac{3}{2} = 1.5,$$

and the reduced matrix that results from $(E_2 - m_{21}E_1) \to E_2$ and $(E_3 - m_{31}E_1) \to E_3$ is

$$\begin{bmatrix} 2 & -1 & 1 \\ 0 & 4.5 & 7.5 \\ 0 & 4.5 & 3.5 \end{bmatrix}.$$

The final pivot is

$$m_{31} = \frac{4.5}{4.5} = 1,$$

which produces the upper triangular matrix in the factorization

$$U = \begin{bmatrix} 2 & -1 & 1 \\ 0 & 4.5 & 7.5 \\ 0 & 0 & -4 \end{bmatrix}.$$

The lower triangular matrix has as its off-diagonal elements the pivots from the Gaussian elimination, so

$$L = \begin{bmatrix} 1 & 0 & 0 \\ 1.5 & 1 & 0 \\ 1.5 & 1 & 1 \end{bmatrix}.$$

The latter approach is preferable when factoring a matrix A into the form LU, where L has 1s along the diagonal. When this is not the form of L, we will need to resort to the first technique or to Algorithm 6.4.

9. a. Obtain a factorization of the form $P^t L U$ for the matrix

$$A = \begin{bmatrix} 0 & 2 & 3 \\ 1 & 1 & -1 \\ 0 & -1 & 1 \end{bmatrix}.$$

SOLUTION: Before we factor A we multiply by the permutation matrix

$$P = P^t = \begin{bmatrix} 0 & 1 & 0 \\ 1 & 0 & 0 \\ 0 & 0 & 1 \end{bmatrix},$$

which has the effect of interchanging the first two rows of the matrix A and produces

$$P^t A = \begin{bmatrix} 0 & 1 & 0 \\ 1 & 0 & 0 \\ 0 & 0 & 1 \end{bmatrix} \begin{bmatrix} 0 & 2 & 3 \\ 1 & 1 & -1 \\ 0 & -1 & 1 \end{bmatrix} = \begin{bmatrix} 1 & 1 & -1 \\ 0 & 2 & 3 \\ 0 & -1 & 1 \end{bmatrix}.$$

The pivots of $P^t A$ are

$$m_{21} = 0, \quad m_{31} = 0, \quad \text{and} \quad m_{32} = \frac{-1}{2} = -0.5,$$

so the lower triangular matrix for the factorization of $P^t A$ is

$$\begin{bmatrix} 1 & 0 & 0 \\ 0 & 1 & 0 \\ 0 & -\frac{1}{2} & 1 \end{bmatrix}.$$

The upper triangular matrix in the factorization is the result of applying the pivots to $P^t A$, that is,

$$U = \begin{bmatrix} 1 & 1 & -1 \\ 0 & 2 & 3 \\ 0 & 0 & \frac{5}{2} \end{bmatrix}.$$

Hence,

$$P^t L U = \begin{bmatrix} 0 & 1 & 0 \\ 1 & 0 & 0 \\ 0 & 0 & 1 \end{bmatrix} \begin{bmatrix} 1 & 0 & 0 \\ 0 & 1 & 0 \\ 0 & -\frac{1}{2} & 1 \end{bmatrix} \begin{bmatrix} 1 & 1 & -1 \\ 0 & 2 & 3 \\ 0 & 0 & \frac{5}{2} \end{bmatrix}.$$

10. Suppose $A = P^t L U$, where P is a permutation matrix, L is a lower-triangular matrix with 1s on the diagonal, and U is an upper-triangular matrix.

a. Count the number of operations needed to compute $P^t L U$ for a given $n \times n$ matrix A.

b. Show that if P contains k row interchanges, then

$$\det P = \det P^t = (-1)^k.$$

c. Use $\det A = \det P^t \det L \det U = (-1)^k \det U$ to count the number of operations for determining $\det A$ by factoring.

d. Compute $\det A$, and count the number of operations for the matrix

$$A = \begin{bmatrix} 0 & 2 & 1 & 4 & -1 & 3 \\ 1 & 2 & -1 & 3 & 4 & 0 \\ 0 & 1 & 1 & -1 & 2 & -1 \\ 2 & 3 & -4 & 2 & 0 & 5 \\ 1 & 1 & 1 & 3 & 0 & 2 \\ -1 & -1 & 2 & -1 & 2 & 0 \end{bmatrix}.$$

SOLUTION: **a.** To compute $P^t L U$ requires

Multiplications/Divisions: $\frac{1}{3}n^3 - \frac{1}{3}n$; and Additions/Subtractions: $\frac{1}{3}n^3 - \frac{1}{2}n^2 + \frac{1}{6}n$.

b. If $\tilde{P}$ is obtained from P by a simple row interchange, then $\det \tilde{P} = -\det P$. Thus, if $\tilde{P}$ is obtained from P by k interchanges, we have $\det \tilde{P} = (-1)^k \det P$.

c. Only $n-1$ multiplications are needed in addition to the operations in part (a).

d. We have $\det A = -741$. Factoring and computing $\det A$ requires 75 Multiplications/Divisions and 55 Additions/Subtractions.

11. a. Determine the number of computations needed for the Factorization Algorithm, and count the number of operations needed to solve m linear systems of the form $A\mathbf{x}^{(k)} = \mathbf{b}^{(k)}$ by first factoring A.

b. Show that solving $L\mathbf{y} = \mathbf{b}$, where L is a lower-triangular matrix with $l_{ii} = 1$ for all i, requires

$$\text{Multiplications/Divisions: } \frac{1}{2}n^2 - \frac{1}{2}n; \quad \text{and Additions/Subtractions: } \frac{1}{2}n^2 - \frac{1}{2}n.$$

c. Show that solving $A\mathbf{x} = \mathbf{b}$ by factoring A as $A = LU$, solving $L\mathbf{y} = \mathbf{b}$, and $U\mathbf{x} = \mathbf{y}$ requires the same number of operations as Gaussian Elimination.

d. Count the number of operations needed to solve $A\mathbf{x}^{(k)} = \mathbf{b}^{(k)}$ by first factoring A and then using the method in (c) m times.

SOLUTION: **a.** The steps in Algorithm 6.4 give the following:

	Multiplications/Divisions	Additions/Subtractions
Step 2	$n-1$	0
Step 4	$\sum_{i=2}^{n-1} i - 1$	$\sum_{i=2}^{n-1} i - 1$
Step 5	$\sum_{i=2}^{n-1} \sum_{j=i+1}^{n} [2(i-1)+1]$	$\sum_{i=2}^{n-1} \sum_{j=i+1}^{n} 2(i-1)$
Step 6	$n-1$	$n-1$
Totals	$\frac{1}{3}n^3 - \frac{1}{3}n$	$\frac{1}{3}n^3 - \frac{1}{2}n^2 + \frac{1}{6}n$

b. The equations are given by

$$y_1 = \frac{b_1}{l_{11}} \quad \text{and} \quad y_i = b_i - \sum_{j=1}^{i-1} \frac{l_{ij}y_j}{l_{ii}}, \quad \text{for } i = 2, \ldots, n.$$

If we assume that $l_{ii} = 1$, for each $i = 1, 2, \ldots, n$, then the number of Multiplications/Divisions is

$$\sum_{i=2}^{n} (i-1) = \frac{n(n-1)}{2},$$

and the number of Additions/Subtractions is the same.

c.

	Multiplications/Divisions	Additions/Subtractions
Factoring into LU	$\frac{1}{3}n^3 - \frac{1}{3}n$	$\frac{1}{3}n^3 - \frac{1}{2}n^2 + \frac{1}{6}n$
Solving $Ly = b$	$\left(\frac{1}{2}n^2 - \frac{1}{2}n\right)$	$\left(\frac{1}{2}n^2 - \frac{1}{2}n\right)$
Solving $Ux = y$	$\left(\frac{1}{2}n^2 + \frac{1}{2}n\right)$	$\left(\frac{1}{2}n^2 - \frac{1}{2}n\right)$
Total	$\frac{1}{3}n^3 + n^2 - \frac{1}{3}n$	$\frac{1}{3}n^3 + \frac{1}{2}n^2 - \frac{5}{6}n$

d.

	Multiplications/Divisions	Additions/Subtractions
Factoring into LU	$\frac{1}{3}n^3 - \frac{1}{3}n$	$\frac{1}{3}n^3 - \frac{1}{2}n^2 + \frac{1}{6}n$
Solving $Ly^{(k)} = b^{(k)}$	$\left(\frac{1}{2}n^2 - \frac{1}{2}n\right)m$	$\left(\frac{1}{2}n^2 - \frac{1}{2}n\right)m$
Solving $Ux^{(k)} = y^{(k)}$	$\left(\frac{1}{2}n^2 + \frac{1}{2}n\right)m$	$\left(\frac{1}{2}n^2 - \frac{1}{2}n\right)m$
Total	$\frac{1}{3}n^3 + mn^2 - \frac{1}{3}n$	$\frac{1}{3}n^3 + \left(m - \frac{1}{2}\right)n^2 - \left(m - \frac{1}{6}\right)n$

Exercise Set 6.6, page 425

3. a. Use the LDL^t Factorization Algorithm to find a factorization of

$$A = \begin{bmatrix} 2 & -1 & 0 \\ -1 & 2 & -1 \\ 0 & -1 & 2 \end{bmatrix}.$$

SOLUTION: This factorization has the form

$$\begin{bmatrix} 2 & -1 & 0 \\ -1 & 2 & -1 \\ 0 & -1 & 2 \end{bmatrix} = \begin{bmatrix} 1 & 0 & 0 \\ l_{21} & 1 & 0 \\ l_{31} & l_{32} & 1 \end{bmatrix} \cdot \begin{bmatrix} d_1 & 0 & 0 \\ 0 & d_2 & 0 \\ 0 & 0 & d_3 \end{bmatrix} \cdot \begin{bmatrix} 1 & l_{21} & l_{31} \\ 0 & 1 & l_{32} \\ 0 & 0 & 1 \end{bmatrix}$$

$$= \begin{bmatrix} d_1 & d_1 l_{21} & d_1 l_{31} \\ d_1 l_{21} & d_2 + d_1 l_{21}^2 & d_1 l_{21} l_{31} + d_2 l_{32} \\ d_1 l_{31} & d_1 l_{21} l_{31} + d_2 l_{32} & d_1 l_{31}^2 + d_2 l_{32}^2 + d_3 \end{bmatrix}.$$

Matching in order the entries (1,1), (2,1), (3,1), (2,2), (3,2), and (3,3), gives

$$d_1 = 2, \quad l_{21} = \frac{-1}{d_1} = -\frac{1}{2}, \quad l_{31} = \frac{0}{d_1} = 0,$$

$$d_2 = 2 - 2\left[-\frac{1}{2}\right]^2 = \frac{3}{2}, \quad l_{32} = \frac{1}{3/2}[-1-0] = -\frac{2}{3},$$

and

$$d_3 = 2 - 0 - \frac{3}{2}\left[-\frac{2}{3}\right]^2 = \frac{4}{3}.$$

Thus, we have

$$L = \begin{bmatrix} 1 & 0 & 0 \\ -\frac{1}{2} & 1 & 0 \\ 0 & -\frac{2}{3} & 1 \end{bmatrix} \quad \text{and} \quad D = \begin{bmatrix} 2 & 0 & 0 \\ 0 & \frac{3}{2} & 0 \\ 0 & 0 & \frac{4}{3} \end{bmatrix}.$$

5. a. Use Cholesky's Algorithm to find a factorization of the form $A = LL^t$ for the matrix in Exercise 3(a).

SOLUTION: This factorization has the form

$$\begin{bmatrix} 2 & -1 & 0 \\ -1 & 2 & -1 \\ 0 & -1 & 2 \end{bmatrix} = \begin{bmatrix} l_{11} & 0 & 0 \\ l_{21} & l_{22} & 0 \\ l_{31} & l_{32} & l_{33} \end{bmatrix} \cdot \begin{bmatrix} l_{11} & l_{21} & l_{31} \\ 0 & l_{22} & l_{32} \\ 0 & 0 & l_{33} \end{bmatrix}$$

$$= \begin{bmatrix} l_{11}^2 & l_{11}l_{21} & l_{11}l_{31} \\ l_{11}l_{21} & l_{21}^2 + l_{22}^2 & l_{21}l_{31} + l_{22}l_{32} \\ l_{11}l_{31} & l_{21}l_{31} + l_{22}l_{32} & l_{31}^2 + l_{32}^2 + l_{33}^2 \end{bmatrix}.$$

Matching in order the entries (1,1), (2,1), (3,1), (2,2), (3,2), and (3,3), gives

$$l_{11} = \sqrt{2}, \quad l_{21} = \frac{-1}{l_{11}} = -\frac{\sqrt{2}}{2}, \quad l_{31} = \frac{0}{l_{11}} = 0,$$

$$l_{22} = \sqrt{2 - \left[-\frac{\sqrt{2}}{2}\right]^2} = \frac{\sqrt{6}}{2}, \quad l_{32} = \frac{1}{\sqrt{6}/2}[-1-0] = -\frac{\sqrt{6}}{3},$$

and

$$l_{33} = \sqrt{2 - 0 - \left[-\frac{\sqrt{6}}{3}\right]^2} = \frac{2\sqrt{3}}{3}.$$

Thus, we have

$$L = \begin{bmatrix} \sqrt{2} & 0 & 0 \\ -\frac{\sqrt{2}}{2} & \frac{\sqrt{6}}{2} & 0 \\ 0 & -\frac{\sqrt{6}}{3} & \frac{2\sqrt{3}}{3} \end{bmatrix}.$$

11. a. Use the Crout Factorization Algorithm to solve

$$\begin{aligned} x_1 - \quad\; x_2 &= 0, \\ -2x_1 + 4x_2 - 2x_3 &= -1, \\ -\; x_2 + 2x_3 &= 1.5. \end{aligned}$$

SOLUTION: This factorization has the form

$$
\begin{bmatrix} 1 & -1 & 0 \\ -2 & 4 & -2 \\ 0 & -1 & 2 \end{bmatrix} = \begin{bmatrix} l_{11} & 0 & 0 \\ l_{21} & l_{22} & 0 \\ 0 & l_{32} & l_{33} \end{bmatrix} \cdot \begin{bmatrix} 1 & u_{12} & 0 \\ 0 & 1 & u_{23} \\ 0 & 0 & 1 \end{bmatrix} = \begin{bmatrix} l_{11} & l_{11}u_{12} & 0 \\ l_{21} & l_{21}u_{12} + l_{22} & l_{22}u_{23} \\ 0 & l_{32} & l_{32}u_{23} + l_{33} \end{bmatrix}.
$$

Matching in order the entries (1,1), (2,1), (3,1), (2,2), (3,2), and (3,3), gives

$$
l_{11} = 1, \quad l_{21} = -2, \quad l_{32} = -1,
$$

$$
u_{12} = \frac{-1}{1} = -1, \quad l_{22} = 4 - (-2)(-1) = 2, \quad u_{23} = \frac{-2}{2} = -1,
$$

and

$$
l_{33} = 2 - (-1)(-1) = 1.
$$

Thus,

$$
\begin{bmatrix} 1 & -1 & 0 \\ -2 & 4 & -2 \\ 0 & -1 & 2 \end{bmatrix} = \begin{bmatrix} 1 & 0 & 0 \\ -2 & 2 & 0 \\ 0 & -1 & 1 \end{bmatrix} \cdot \begin{bmatrix} 1 & -1 & 0 \\ 0 & 1 & -1 \\ 0 & 0 & 1 \end{bmatrix}.
$$

To solve the system

$$
\begin{bmatrix} 1 & 0 & 0 \\ -2 & 2 & 0 \\ 0 & -1 & 1 \end{bmatrix} \cdot \begin{bmatrix} 1 & -1 & 0 \\ 0 & 1 & -1 \\ 0 & 0 & 1 \end{bmatrix} \begin{bmatrix} x_1 \\ x_2 \\ x_3 \end{bmatrix} = \begin{bmatrix} 1 & -1 & 0 \\ -2 & 4 & -2 \\ 0 & -1 & 2 \end{bmatrix} \begin{bmatrix} x_1 \\ x_2 \\ x_3 \end{bmatrix} = \begin{bmatrix} 0 \\ -1 \\ \frac{3}{2} \end{bmatrix},
$$

we first let

$$
\begin{bmatrix} y_1 \\ y_2 \\ y_3 \end{bmatrix} = \begin{bmatrix} 1 & -1 & 0 \\ 0 & 1 & -1 \\ 0 & 0 & 1 \end{bmatrix} \begin{bmatrix} x_1 \\ x_2 \\ x_3 \end{bmatrix}
$$

and solve for $\mathbf{y} = (y_1, y_2, y_3)^t$ in

$$
\begin{bmatrix} 1 & 0 & 0 \\ -2 & 2 & 0 \\ 0 & -1 & 1 \end{bmatrix} \begin{bmatrix} y_1 \\ y_2 \\ y_3 \end{bmatrix} = \begin{bmatrix} 0 \\ -1 \\ \frac{3}{2} \end{bmatrix}.
$$

This gives

$$
y_1 = 0, \quad y_2 = \frac{1}{2}[-1 - 0] = -\frac{1}{2}, \quad \text{and} \quad y_3 = \frac{3}{2} + \left(-\frac{1}{2}\right) = 1.
$$

Now we solve for $\mathbf{x} = (x_1, x_2, x_3)^t$ in

$$
\begin{bmatrix} 0 \\ -\frac{1}{2} \\ 1 \end{bmatrix} = \begin{bmatrix} 1 & -1 & 0 \\ 0 & 1 & -1 \\ 0 & 0 & 1 \end{bmatrix} \begin{bmatrix} x_1 \\ x_2 \\ x_3 \end{bmatrix},
$$

to obtain the solution to the original system

$$
x_3 = 1, \quad x_2 = -\frac{1}{2} + 1 = \frac{1}{2}, \quad \text{and} \quad x_1 = 0 + \frac{1}{2} = \frac{1}{2}.
$$

17. Find α so that

$$A = \begin{bmatrix} 2 & \alpha & -1 \\ \alpha & 2 & 1 \\ -1 & 1 & 4 \end{bmatrix}$$

is positive definite.

SOLUTION: The matrix

$$A = \begin{bmatrix} 2 & \alpha & -1 \\ \alpha & 2 & 1 \\ -1 & 1 & 4 \end{bmatrix}$$

is positive definite if and only if

$$\det \begin{bmatrix} 2 & \alpha \\ \alpha & 2 \end{bmatrix} = 4 - \alpha^2 > 0$$

and

$$\det \begin{bmatrix} 2 & \alpha & -1 \\ \alpha & 2 & 1 \\ -1 & 1 & 4 \end{bmatrix} = 2(8-1) - \alpha(4\alpha + 1) - (\alpha + 2) = 12 - 2\alpha - 4\alpha^2 > 0.$$

Thus, A is positive definite if and only if

$$\alpha^2 < 4 \quad \text{and} \quad -(\alpha - 3/2)(\alpha + 2) > 0.$$

This implies that

$$-2 < \alpha < 2 \quad \text{and} \quad -2 < \alpha < \frac{3}{2}.$$

Consequently, the matrix A is positive definite if and only if $-2 < \alpha < \frac{3}{2}$.

19. Find $\alpha > 0$ and $\beta > 0$ so that the matrix

$$A = \begin{bmatrix} 3 & 2 & \beta \\ \alpha & 5 & \beta \\ 2 & 1 & \alpha \end{bmatrix}$$

is strictly diagonally dominant.

SOLUTION: For

$$A = \begin{bmatrix} 3 & 2 & \beta \\ \alpha & 5 & \beta \\ 2 & 1 & \alpha \end{bmatrix}$$

to be strictly diagonally dominant, with $\alpha > 0$ and $\beta > 0$, we need

$$3 > 2 + \beta, \quad \alpha + \beta < 5, \quad \text{and} \quad 3 < \alpha.$$

So we need $\beta < 1$, $\alpha < 5 - \beta$, and $\alpha > 3$. Thus,

$$0 < \beta < 1 \quad \text{and} \quad 3 < \alpha < 5 - \beta.$$

20. Suppose that A and B are strictly diagonally dominant $n \times n$ matrices.

  a. Is $-A$ strictly diagonally dominant?

  b. Is A^t strictly diagonally dominant?

  c. Is $A + B$ strictly diagonally dominant?

  d. Is A^2 strictly diagonally dominant?

  e. Is $A - B$ strictly diagonally dominant?

SOLUTION: **a.** Yes, since the definition of strict diagonal dominance considers the absolute values of the entries.

b. Not necessarily. Consider, for example,

$$A = \begin{bmatrix} 2 & -1 \\ 3 & 4 \end{bmatrix}.$$

c. Not necessarily. Consider, for example,

$$A = \begin{bmatrix} 2 & 1 \\ 1 & 2 \end{bmatrix} \quad \text{and} \quad B = \begin{bmatrix} -2 & 1 \\ 1 & -2 \end{bmatrix}.$$

d. Not necessarily. Consider, for example,

$$A = \begin{bmatrix} 2 & -1 \\ 3 & 4 \end{bmatrix}.$$

e. Not necessarily. Consider, for example,

$$A = \begin{bmatrix} 2 & 1 \\ 1 & 2 \end{bmatrix} \quad \text{and} \quad B = \begin{bmatrix} 2 & -1 \\ -1 & 2 \end{bmatrix}.$$

22. Let

$$A = \begin{bmatrix} 1 & 0 & -1 \\ 0 & 1 & 1 \\ -1 & 1 & \alpha \end{bmatrix}.$$

Find all values of α for which

  a. A is singular.

  b. A is strictly diagonally dominant.

  c. A is symmetric.

  d. A is positive definite.

SOLUTION: **a.** The determinant of A is $\alpha - 2$, so A is singular if and only if $\alpha = 2$.

b. A cannot be strictly diagonally dominant regardless of α since the condition fails for the first and second rows.

c. Symmetry of a matrix has no dependence on the diagonal elements. The matrix is symmetric for all values of α.

d. Consider the product

$$\mathbf{x} A \mathbf{x}^t = \begin{bmatrix} x_1 & x_2 & x_3 \end{bmatrix} \begin{bmatrix} 1 & 0 & -1 \\ 0 & 1 & 1 \\ -1 & 1 & \alpha \end{bmatrix} \begin{bmatrix} x_1 \\ x_2 \\ x_3 \end{bmatrix} = x_1^2 - 2x_1 x_3 + x_2^2 + 2x_2 x_3 + \alpha x_3^2.$$

If we rewrite this result as

$$\mathbf{x} A \mathbf{x}^t = (x_1 - x_3)^2 + (x_2 + x_3)^2 + (\alpha - 2)x_3^2,$$

we see that this term will be positive for all nonzero vectors $\mathbf{x}$ if and only if $\alpha > 2$. In Theorem 9.18, we will see an alternative characterization of positive definite matrices that gives an easier solution to this problem.

27. Tridiagonal matrices are usually labeled by using notation that involves only terms with a single index. Rewrite the Crout Factorization Algorithm, using this notation, and change the notation of the l_{ij} and u_{ij} in a similar manner.

   SOLUTION: Using the notation in the exercise, the Crout Factorization Algorithm can be rewritten as follows:

   Step 1 Set $l_1 = a_1$; $u_1 = c_1/l_1$.
   Step 2 For $i = 2, \ldots, n-1$ set $l_i = a_i - b_i u_{i-1}$; $u_i = c_i/l_i$.
   Step 3 Set $l_n = a_n - b_n u_{n-1}$.
   Step 4 Set $z_1 = d_1/l_1$.
   Step 5 For $i = 2, \ldots, n$ set $z_i = (d_i - b_i z_{i-1})/l_i$.
   Step 6 Set $x_n = z_n$.
   Step 7 For $i = n-1, \ldots, 1$ set $x_i = z_i - u_i x_{i+1}$.
   Step 8 OUTPUT $(x_1, \ldots, x_n)$;
            STOP.

28. Prove Theorem 6.31.

   SOLUTION: First, $|l_{11}| = |a_{11}| > 0$ and $|u_{12}| = \frac{|a_{12}|}{|l_{11}|} < 1$.

   In general, assume $|l_{jj}| > 0$ and $|u_{j,j+1}| < 1$ for $j = 1, \ldots, i-1$. Then

   $$\begin{aligned} |l_{ii}| &= |a_{ii} - l_{i,i-1} u_{i-1,i}| = |a_{ii} - a_{i,i-1} u_{i-1,i}| \\ &\geq |a_{ii}| - |a_{i,i-1} u_{i-1,i}| > |a_{ii}| - |a_{i,i-1}| > 0, \end{aligned}$$

   and

   $$|u_{i,i+1}| = \frac{|a_{i,i+1}|}{|l_{ii}|} < \frac{|a_{i,i+1}|}{|a_{ii}| - |a_{i,i-1}|} \leq 1,$$

   for $i = 2, \ldots, n-1$. Further,

   $$|l_{nn}| = |a_{nn} - l_{n,n-1} u_{n-1,n}| = |a_{nn} - a_{n,n-1} u_{n-1,n}| \geq |a_{nn}| - |a_{n,n-1}| > 0.$$

   So

   $$\det A = \det L \cdot \det U = l_{11} \cdot l_{22} \ldots l_{nn} \cdot 1 > 0.$$

32. Suppose that the positive definite matrix A has the Cholesky factorization $A = LL^t$ and also the factorization $A = \hat{L}D\hat{L}^t$, where D is the diagonal matrix with positive diagonal entries $d_{11}, d_{22}, \ldots, d_{nn}$. Let $D^{1/2}$ be the diagonal matrix with diagonal entries $\sqrt{d_{11}}, \sqrt{d_{22}}, \ldots, \sqrt{d_{nn}}$.

a. Show that $D = D^{1/2}D^{1/2}$.

b. Show that $L = \hat{L}D^{1/2}$.

SOLUTION: **a.**

$$D^{1/2}D^{1/2} = \begin{bmatrix} \sqrt{d_{11}} & 0 & \cdots & 0 \\ 0 & \sqrt{d_{22}} & \ddots & \vdots \\ \vdots & \ddots & \ddots & 0 \\ 0 & \cdots & 0 & \sqrt{d_{nn}} \end{bmatrix} \cdot \begin{bmatrix} \sqrt{d_{11}} & 0 & \cdots & 0 \\ 0 & \sqrt{d_{22}} & \ddots & \vdots \\ \vdots & \ddots & \ddots & 0 \\ 0 & \cdots & 0 & \sqrt{d_{nn}} \end{bmatrix}$$

$$= \begin{bmatrix} d_{11} & 0 & \cdots & 0 \\ 0 & d_{22} & \ddots & \vdots \\ \vdots & \ddots & \ddots & 0 \\ 0 & \cdots & 0 & d_{nn} \end{bmatrix} = D.$$

b.

$$\left(\hat{L}D^{1/2}\right)\left(\hat{L}D^{1/2}\right)^t = \hat{L}D^{1/2}\left(D^{1/2}\right)^t \hat{L}^t = \hat{L}D^{1/2}D^{1/2}\hat{L}^t = \hat{L}D\hat{L}^t = A.$$

Since $LL^t = A$, and the factorization is unique, we have $\hat{L}D^{1/2} = L$.

Iterative Techniques in Matrix Algebra

Exercise Set 7.1, page 441

2. a. Show that $\|\mathbf{x}\|_1 = \sum_{i=1}^{n} |x_i|$ is a norm on $\mathbb{R}^n$.

SOLUTION: Since $\|\mathbf{x}\|_1 = \sum_{i=1}^{n} |x_i| \geq 0$ with equality only if $x_i = 0$ for all i, properties (i) and (ii) in Definition 7.1 hold.

Also,

$$\|\alpha \mathbf{x}\|_1 = \sum_{i=1}^{n} |\alpha x_i| = \sum_{i=1}^{n} |\alpha||x_i| = |\alpha| \sum_{i=1}^{n} |x_i| = |\alpha|\|\mathbf{x}\|_1,$$

so property (iii) holds.

Finally,

$$\|\mathbf{x} + \mathbf{y}\|_1 = \sum_{i=1}^{n} |x_i + y_i| \leq \sum_{i=1}^{n} (|x_i| + |y_i|) = \sum_{i=1}^{n} |x_i| + \sum_{i=1}^{n} |y_i| = \|\mathbf{x}\|_1 + \|\mathbf{y}\|_1,$$

so property (iv) also holds.

6. d. The matrix norm $\|\cdot\|_1$ defined by $\|A\|_1 = \max \|A\mathbf{x}\|_1$ can be computed using the formula

$$\|A\|_1 = \max_{1 \leq j \leq n} \sum_{i=1}^{n} |a_{ij}|,$$

where the vector norm $\|\cdot\|_1$ is defined in Exercise 2. Find $\|\cdot\|_1$ for the matrix in Exercise 4 part (d).

SOLUTION: This result can be proved by modifying the proof of Theorem 7.11. Applying the result to the exercise gives

$$\|A\|_1 = \max\{|4| + |-1| + |-7|, |-1| + |4| + |0|, |7| + |0| + |4|\} = 12.$$

7. Show by example that $\|\cdot\|_\infty$, defined by $\|A\|_\infty = \max_{1 \leq i,j \leq n} |a_{ij}|$, does not define a matrix norm.

SOLUTION: Let $A = \begin{bmatrix} 1 & 1 \\ 0 & 1 \end{bmatrix}$ and $B = \begin{bmatrix} 1 & 0 \\ 1 & 1 \end{bmatrix}$. Then $\|AB\|_\infty = 2 > 1 = \|A\|_\infty \cdot \|B\|_\infty$.

151

11. Let S be a positive definite $n \times n$ matrix. For any $\mathbf{x}$ in $\mathbb{R}^n$ define $\|\mathbf{x}\| = (\mathbf{x}^t S \mathbf{x})^{\frac{1}{2}}$. Show that this defines a norm on $\mathbb{R}^n$.

SOLUTION: That $\|\mathbf{x}\| \geq 0$ follows easily. That $\|\mathbf{x}\| = 0$ if and only if $\mathbf{x} = \mathbf{0}$ follows from the definition of positive definite. In addition,

$$\|\alpha\mathbf{x}\| = \left[(\alpha\mathbf{x}^t)\, S(\alpha\mathbf{x})\right]^{\frac{1}{2}} = \left[\alpha^2 \mathbf{x}^t S \mathbf{x}\right]^{\frac{1}{2}} = |\alpha|\, (\mathbf{x}^t S \mathbf{x})^{\frac{1}{2}} = |\alpha|\|\mathbf{x}\|.$$

From Cholesky's factorization, let $S = LL^t$. Then

$$\mathbf{x}^t S \mathbf{y} = \mathbf{x}^t L L^t \mathbf{y} = \left(L^t\mathbf{x}\right)^t \left(L^t\mathbf{y}\right) \leq \left[\left(L^t\mathbf{x}\right)^t \left(L^t\mathbf{x}\right)\right]^{1/2} \left[\left(L^t\mathbf{y}\right)^t \left(L^t\mathbf{y}\right)\right]^{1/2}$$
$$= \left(\mathbf{x}^t L L^t \mathbf{x}\right)^{1/2} \left(\mathbf{y}^t L L^t \mathbf{y}\right)^{1/2} = \left(\mathbf{x}^t S \mathbf{x}\right)^{1/2} \left(\mathbf{y}^t S \mathbf{y}\right)^{1/2}.$$

Thus

$$\|\mathbf{x} + \mathbf{y}\|^2 = \left[(\mathbf{x} + \mathbf{y})^t S(\mathbf{x} + \mathbf{y})\right] = \left[\mathbf{x}^t S \mathbf{x} + \mathbf{y}^t S \mathbf{x} + \mathbf{x}^t S \mathbf{y} + \mathbf{y}^t S \mathbf{y}\right]$$
$$\leq \mathbf{x}^t S \mathbf{x} + 2\left(\mathbf{x}^t S \mathbf{x}\right)^{1/2} \left(\mathbf{y}^t S \mathbf{y}\right)^{1/2} + \left(\mathbf{y}^t S \mathbf{y}\right)^{1/2}$$
$$= \mathbf{x}^t S \mathbf{x} + 2\|\mathbf{x}\|\|\mathbf{y}\| + \mathbf{y}^t S \mathbf{y} = (\|\mathbf{x}\| + \|\mathbf{y}\|)^2.$$

This demonstrates properties **(i)** – **(iv)** of Definition 7.1.

13. Prove that if $\| \cdot \|$ is a vector norm on $\mathbb{R}$, then $\|A\| = \max_{\|\mathbf{x}\|=1} \|A\mathbf{x}\|$ is a matrix norm.

SOLUTION: It is not difficult to show that (*i*) holds. If $\|A\| = 0$, then $\|A\mathbf{x}\| = 0$ for all vectors $\mathbf{x}$ with $\|\mathbf{x}\| = 1$. Using

$$\mathbf{x} = (1, 0, \ldots, 0)^t, \quad \mathbf{x} = (0, 1, 0, \ldots, 0)^t, \ldots, \quad \text{and} \quad \mathbf{x} = (0, \ldots, 0, 1)^t$$

successively implies that each column of A is zero. Thus, $\|A\| = 0$ if and only if $A = O$. Moreover,

$$\|\alpha A\| = \max_{\|\mathbf{x}\|=1} \|\alpha A\mathbf{x}\| = |\alpha| \max_{\|\mathbf{x}\|=1} \|A\mathbf{x}\| = |\alpha| \cdot \|A\|,$$
$$\|A + B\| = \max_{\|\mathbf{x}\|=1} \|(A + B)\mathbf{x}\| \leq \max_{\|\mathbf{x}\|=1} (\|A\mathbf{x}\| + \|B\mathbf{x}\|),$$

so

$$\|A + B\| \leq \max_{\|\mathbf{x}\|=1} \|A\mathbf{x}\| + \max_{\|\mathbf{x}\|=1} \|B\mathbf{x}\| = \|A\| + \|B\|$$

and

$$\|AB\| = \max_{\|\mathbf{x}\|=1} \|(AB)\mathbf{x}\| = \max_{\|\mathbf{x}\|=1} \|A(B\mathbf{x})\|.$$

Thus,

$$\|AB\| \leq \max_{\|\mathbf{x}\|=1} \|A\| \|B\mathbf{x}\| = \|A\| \max_{\|\mathbf{x}\|=1} \|B\mathbf{x}\| = \|A\| \|B\|.$$

Exercise Set 7.2, page 449

1. d. Compute the eigenvalues and associated eigenvectors of

$$\begin{bmatrix} 2 & 1 & 0 \\ 1 & 2 & 0 \\ 0 & 0 & 3 \end{bmatrix}.$$

SOLUTION: The eigenvalues are determined by solving

$$0 = \det \begin{bmatrix} 2 - \lambda & 1 & 0 \\ 1 & 2 - \lambda & 0 \\ 0 & 0 & 3 - \lambda \end{bmatrix} = (2 - \lambda) \det \begin{bmatrix} 2 - \lambda & 0 \\ 0 & 3 - \lambda \end{bmatrix} - \det \begin{bmatrix} 1 & 0 \\ 0 & 3 - \lambda \end{bmatrix}.$$

So

$$0 = (2 - \lambda)(2 - \lambda)(3 - \lambda) - (3 - \lambda) = (3 - \lambda)(\lambda^2 - 4\lambda + 3) = -(\lambda - 3)^2(\lambda - 1),$$

and the eigenvalues are $\lambda_1 = \lambda_2 = 3$ and $\lambda_3 = 1$. To determine eigenvectors associated with the eigenvalue $\lambda = 3$, we solve the system

$$\begin{bmatrix} 0 \\ 0 \\ 0 \end{bmatrix} = \begin{bmatrix} 2 - 3 & 1 & 0 \\ 1 & 2 - 3 & 0 \\ 0 & 0 & 3 - 3 \end{bmatrix} \cdot \begin{bmatrix} x_1 \\ x_2 \\ x_3 \end{bmatrix} = \begin{bmatrix} -x_1 + x_2 \\ x_1 - x_2 \\ 0 \end{bmatrix}.$$

This implies that we must have $x_1 = x_2$, but that x_3 is arbitrary. Two linearly independent choices for the eigenvectors associated with the double eigenvalue $\lambda = 3$ are

$$\mathbf{x}_1 = (1, 1, 0)^t \quad \text{and} \quad \mathbf{x}_2 = (1, 1, 1)^t.$$

The eigenvector associated with the eigenvalue $\lambda = 1$ must satisfy

$$\begin{bmatrix} 0 \\ 0 \\ 0 \end{bmatrix} = \begin{bmatrix} 2 - 1 & 1 & 0 \\ 1 & 2 - 1 & 0 \\ 0 & 0 & 3 - 1 \end{bmatrix} \cdot \begin{bmatrix} x_1 \\ x_2 \\ x_3 \end{bmatrix} = \begin{bmatrix} x_1 + x_2 \\ x_1 + x_2 \\ 2x_3 \end{bmatrix}.$$

This implies that we must have $x_1 = -x_2$, and that $x_3 = 0$. One choice for the eigenvector associated with the eigenvalue $\lambda = 1$ is

$$\mathbf{x}_3 = (1, -1, 0)^t.$$

3. b. Find the complex eigenvalues and associated eigenvectors for the matrix

$$\begin{bmatrix} 1 & 2 \\ -1 & 2 \end{bmatrix}$$

SOLUTION: We have

$$\det \begin{bmatrix} 1 - \lambda & 2 \\ -1 & 2 - \lambda \end{bmatrix} = (1 - \lambda)(2 - \lambda) + 2 = \lambda^2 - 3\lambda + 4.$$

So the eigenvalues are the solutions to $\lambda^2 - 3\lambda + 4 = 0$, that is,

$$\lambda_1 = \frac{3 + \sqrt{7}i}{2} \quad \text{and} \quad \lambda_2 = \frac{3 - \sqrt{7}i}{2}$$

When $\lambda_1 = \dfrac{3 + \sqrt{7}i}{2}$, the singular matrix

$$A - \lambda_1 I = \begin{bmatrix} 1 - \dfrac{3 + \sqrt{7}i}{2} & 2 \\ -1 & 2 - \dfrac{3 + \sqrt{7}i}{2} \end{bmatrix}$$

has the nonzero solution $((1 - \sqrt{7}\,i)/2, 1)^t$, which is an eigenvector corresponding to λ_1.
In a similar manner, $((1 + \sqrt{7}\,i)/2, 1)^t$ is an eigenvector corresponding to λ_2.

5. d. Find the spectral radius for the matrix A in part (d) of Exercise 1.

SOLUTION: The spectral radius is

$$\rho(A) = \max\{3, 3, 1\} = 3.$$

7. Which of the matrices in Exercise 1 are convergent?

SOLUTION: For a matrix to be convergent it is necessary and sufficient that all its eigenvalues have magnitude strictly less than one. In Exercise 1 we found that the matrices have eigenvalues:

$$\text{Part(a): } \lambda_1 = 3, \quad \lambda_2 = 1;$$
$$\text{Part(b): } \lambda_1 = \frac{1 + \sqrt{5}}{2}, \quad \lambda_2 = \frac{1 - \sqrt{5}}{2};$$
$$\text{Part(c): } \lambda_1 = \frac{1}{2}, \quad \lambda_2 = -\frac{1}{2};$$
$$\text{Part(d): } \lambda_1 = \lambda_2 = 3, \quad \lambda_3 = 1;$$
$$\text{Part(e): } \lambda_1 = 7, \quad \lambda_2 = 3, \quad \lambda_3 = -1;$$
$$\text{Part(f): } \lambda_1 = 5, \quad \lambda_2 = \lambda_3 = 1.$$

So only the matrix in part (c) is convergent.

12. An $n \times n$ matrix A is called *nilpotent* if an integer m exists with $A^m = O_n$. Show that if λ is an eigenvalue of a nilpotent matrix, then $\lambda = 0$.

SOLUTION: Suppose the λ is an eigenvalue of A with corresponding eigenvector $\mathbf{x}$. Then $A\mathbf{x} = \lambda\mathbf{x}$, $A^2\mathbf{x} = A(\lambda\mathbf{x}) = \lambda A\mathbf{x} = \lambda^2\mathbf{x}$, and in general, $A^m\mathbf{x} = \lambda^m\mathbf{x}$, for any positive integer m. As a consequence,

$$\mathbf{0} = A^m\mathbf{x} = \lambda^m\mathbf{x} \quad \text{which implies that } \lambda^m = 0.$$

Hence $\lambda = 0$.

14. a. Show that if A is an $n \times n$ matrix with eigenvalues $\lambda_1, \ldots, \lambda_n$, then

$$\det A = \lambda_1 \cdot \lambda_2 \cdots \lambda_n = \prod_{i=1}^{n} \lambda_i.$$

SOLUTION: The characteristic polynomial of A is

$$P(\lambda) = (\lambda_1 - \lambda)(\lambda_2 - \lambda) \ldots (\lambda_n - \lambda) = \det(A - \lambda I), \quad \text{so} \quad P(0) = \lambda_1 \cdot \lambda_2 \cdots \lambda_n = \det A.$$

b. Show that A is singular if and only if $\lambda = 0$ is an eigenvalue of A.

SOLUTION: The matrix A singular if and only if $\det A = 0$. Part (a) implies that this is true if and only if at least one of the λ_i is 0.

Exercise Set 7.3, page 459

1. a. Use the Jacobi method and $\mathbf{x}^{(0)} = \mathbf{0}$ to approximate the solution to

$$\begin{aligned}
3x_1 - x_2 + x_3 &= 1, \\
3x_1 + 6x_2 + 2x_3 &= 0, \\
3x_1 + 3x_2 + 7x_3 &= 4.
\end{aligned}$$

SOLUTION: The initial approximation $\mathbf{x}^{(0)} = (0, 0, 0)^t$ gives

$$\begin{aligned}
x_1^{(1)} &= \frac{1}{3}\left(1 + x_2^{(0)} - x_3^{(0)}\right) = \frac{1}{3}(1 + 0 - 0) = 0.3333333, \\
x_2^{(1)} &= \frac{1}{6}\left(0 - 3x_1^{(0)} - 2x_3^{(0)}\right) = \frac{1}{6}(0 - 0 - 0) = 0, \\
x_3^{(1)} &= \frac{1}{7}\left(4 - 3x_1^{(0)} - 3x_2^{(0)}\right) = \frac{1}{7}(4 - 0 - 0) = 0.5714286.
\end{aligned}$$

The second approximation is

$$\begin{aligned}
x_1^{(2)} &= \frac{1}{3}(1 + 0 - 0.5714286) = 0.1428571, \\
x_2^{(2)} &= \frac{1}{6}(0 - 3(0.3333333) - 2(0.5714286)) = -0.3571429, \\
x_3^{(2)} &= \frac{1}{7}(4 - 3(0.3333333) - 0) = 0.4285714.
\end{aligned}$$

3. a. Use the Gauss-Seidel method and $\mathbf{x}^{(0)} = \mathbf{0}$ to approximate the solution to the system in part (a) of Exercise 1.

SOLUTION: The initial approximation $\mathbf{x}^{(0)} = (0, 0, 0)^t$ gives

$$\begin{aligned}
x_1^{(1)} &= \frac{1}{3}\left(1 + x_2^{(0)} - x_3^{(0)}\right) = \frac{1}{3}(1 + 0 - 0) = 0.3333333, \\
x_2^{(1)} &= \frac{1}{6}\left(0 - 3x_1^{(1)} - 2x_3^{(0)}\right) = \frac{1}{6}(0 - 3(0.3333333) - 0) = -0.1666667, \\
x_3^{(1)} &= \frac{1}{7}\left(4 - 3x_1^{(1)} - 3x_2^{(1)}\right) = \frac{1}{7}(4 - 3(0.3333333) - 3(-0.1666667)) = 0.5000000.
\end{aligned}$$

The second approximation is

$$\begin{aligned}
x_1^{(2)} &= \frac{1}{3}(1 - 0.1666667 - 0.5000000) = 0.1111111, \\
x_2^{(2)} &= \frac{1}{6}(0 - 3(0.1111111) - 2(0.5000000)) = -0.2222222, \\
x_3^{(2)} &= \frac{1}{7}(4 - 3(0.1111111) - 3(-0.2222222)) = 0.6190476.
\end{aligned}$$

5. a. Use the Jacobi method to solve the linear systems in Exercise 1(a), with $TOL = 10^{-3}$ in the l_∞ norm.

SOLUTION: The first two iterations were computed in Exercise 1 a). The entire set of approximations is given in table along with an approximation to the error.

i	$x_1^{(i)}$	$x_2^{(i)}$	$x_3^{(i)}$	$\|\mathbf{x}^{(i)} - \mathbf{x}^{(i-1)}\|_\infty$
1	0.33333333	0.00000000	0.57142857	0.57142857
2	0.14285714	−0.35714286	0.42857143	0.35714286
3	0.07142857	−0.21428571	0.66326531	0.23469388
4	0.04081633	−0.25680272	0.63265306	0.04251701
5	0.03684807	−0.23129252	0.66399417	0.03134111
6	0.03490444	−0.23975543	0.65476190	0.00923226
7	0.03516089	−0.23570619	0.65922185	0.00445995
8	0.03502399	−0.23732106	0.65737656	0.00184530
9	0.03510079	−0.23663751	0.65812732	0.00075076

7. a. Use the Gauss-Seidel method to solve the linear systems in Exercise 1(a), with $TOL = 10^{-3}$ in the l_∞ norm.

SOLUTION: The first two iterations were computed in Exercise 3(a). The entire set of approximations is given in table along with an approximation to the error.

i	$x_1^{(i)}$	$x_2^{(i)}$	$x_3^{(i)}$	$\|\mathbf{x}^{(i)} - \mathbf{x}^{(i-1)}\|_\infty$
1	0.33333333	−0.16666667	0.50000000	0.50000000
2	0.11111111	−0.22222222	0.61904762	0.22222222
3	0.05291005	−0.23280423	0.64852608	0.05820106
4	0.03955656	−0.23595364	0.65559875	0.01335349
5	0.03614920	−0.23660752	0.65733928	0.00340736
6	0.03535107	−0.23678863	0.65775895	0.00079814

14. Show that if A is strictly diagonally dominant, then $\|T_j\|_\infty < 1$.

SOLUTION: The matrix $T_j = (t_{ik})$ has entries given by

$$t_{ik} = \begin{cases} 0, & i = k, \text{ for } 1 \le i, k \le n \\ -\frac{a_{ik}}{a_{ii}}, & i \ne k, \text{ for } 1 \le i, k \le n. \end{cases}$$

Thus

$$\|T_j\|_\infty = \max_{1 \le i \le n} \sum_{\substack{k=1 \\ k \ne i}}^{n} \left| \frac{a_{ik}}{a_{ii}} \right| < 1,$$

since A is strictly diagonally dominant.

Exercise Set 7.4, page 467

3. a. Use the SOR method with $w = 1.3$ to approximate the solution to the system in Exercise 1 part (a).

SOLUTION: The initial approximation $\mathbf{x}^{(0)} = (0, 0, 0)^t$ gives

$$\mathbf{x}_1^{(1)} = (1 - 1.3)x_1^{(0)} + \frac{1.3}{3}(1 + x_2^{(0)} - x_3^{(0)}) = 0.4333333,$$

$$\mathbf{x}_2^{(1)} = (1 - 1.3)x_2^{(0)} + \frac{1.3}{6}(0 - 3x_1^{(1)} - 2x_3^{(0)}) = -\frac{1.3}{6}(3)(0.4333333) = -1.126667,$$

$$\mathbf{x}_3^{(1)} = (1 - 1.3)x_3^{(0)} + \frac{1.3}{7}(4 - 3x_1^{(1)} - 3x_2^{(1)}) = \frac{1.3}{7}(4 - 3(0.4333333) - 3(1.126667)) = -0.1136667.$$

Repeating the process gives

$$\mathbf{x}_1^{(2)} = -0.1040103, \quad \mathbf{x}_2^{(2)} = -0.1331814, \quad \text{and} \quad \mathbf{x}_3^{(2)} = 0.6774997.$$

5. a. Use the SOR method with $\omega = 1.2$ to solve the linear system in Exercise 1(a) with a tolerance $TOL = 10^{-3}$ in the l_∞ norm.

SOLUTION: The first iteration is

$$x_1^{(1)} = (1 - 1.2)x_1^{(0)} + \frac{1.2}{3}\left(1 - (-x_2^{(0)} + x_3^{(0)})\right) = 0.4$$

$$x_2^{(1)} = (1 - 1.2)x_2^{(0)} + \frac{1.2}{6}\left(0 - (3x_1^{(1)} - 2x_3^{(0)})\right) = -0.24$$

$$x_3^{(1)} = (1 - 1.2)x_3^{(0)} + \frac{1.2}{7}\left(4 - (3x_1^{(1)} + 3x_2^{(1)})\right) = 0.60342857$$

The entire set of approximations is given in table along with an approximation to the error.

i	$x_1^{(i)}$	$x_2^{(i)}$	$x_3^{(i)}$	$\|\mathbf{x}^{(i)} - \mathbf{x}^{(i-1)}\|_\infty$
1	0.40000000	−0.24000000	0.60342857	0.60342857
2	−0.01737143	−0.18294857	0.66805029	0.41737143
3	0.06307474	−0.26847525	0.65773877	0.08552667
4	0.01689944	−0.21954013	0.65838174	0.04893512
5	0.04545136	−0.24671549	0.65754520	0.02855192
6	0.02920545	−0.23119825	0.65808726	0.01624591
7	0.03844471	−0.24006208	0.65778577	0.00923925
8	0.03317192	−0.23500504	0.65795702	0.00527278
9	0.03618079	−0.23789027	0.65785919	0.00300887
10	0.03446406	−0.23624406	0.65791502	0.00171673
11	0.03544356	−0.23718333	0.65788317	0.00097950

7. Show that the matrix in part (b) of Exercise 1 is tridiagonal and positive definite, and find the first two iterations of the SOR method with using the optimal choice of ω.

SOLUTION: The matrix is symmetric and the leading principal submatrices have the positive determinants 10, 99, and 950, so the matrix is positive definite, as well as being tridiagonal. The matrix

$$T_j = \begin{bmatrix} 0 & \frac{1}{10} & 0 \\ \frac{1}{10} & 0 & \frac{1}{5} \\ 0 & \frac{1}{5} & 0 \end{bmatrix}$$

has characteristic polynomial $\det(T_j - \lambda I) = -\lambda(\lambda^2 - 1/20)$ so its eigenvalues are 0, $\sqrt{5}/10$, and $-\sqrt{5}/10$. So the spectral radius of T_j is $\rho(T_j) = \sqrt{5}/10$ and the optimal choice of ω is

$$\omega = \frac{2}{1 - \sqrt{1 - \rho(T_j)^2}} = \frac{2}{1 - \sqrt{1 - \frac{1}{20}}} \approx 1.01282262.$$

With this choice of ω and $\mathbf{x}^{(0)} = (0,0,0)^t$, the first iteration of the SOR method gives

$$x_1^{(1)} = (1 - 1.01282262)x_1^{(0)} + \frac{1.01282262}{10}(9 + x_2^{(0)}) = 0.9115403580$$

$$x_2^{(1)} = (1 - 1.01282262)x_2^{(0)} + \frac{1.01282262}{10}(7 + x_1^{(1)} + 2x_3^{(0)}) = 0.8012987034$$

$$x_3^{(1)} = (1 - 1.01282262)x_3^{(0)} + \frac{1.01282262}{10}(6 + 2x_2^{(1)}) = 0.7700082624$$

The second iteration gives

$$x_1^{(2)} = (1 - 1.01282262)x_1^{(1)} + \frac{1.01282262}{10}(9 + x_2^{(1)}) = 0.9810093676$$

$$x_2^{(2)} = (1 - 1.01282262)x_2^{(1)} + \frac{1.01282262}{10}(7 + x_1^{(2)} + 2x_3^{(1)}) = 0.9540362902$$

$$x_3^{(2)} = (1 - 1.01282262)x_3^{(1)} + \frac{1.01282262}{10}(6 + 2x_2^{(2)}) = 0.7910739557$$

Exercise Set 7.5, page 482

2. d. Compute the condition number of the matrix

$$A = \begin{bmatrix} 0.04 & 0.01 & -0.01 \\ 0.2 & 0.5 & -0.2 \\ 1 & 2 & 4 \end{bmatrix}.$$

SOLUTION: We have

$$A^{-1} = \begin{bmatrix} 27.586207 & -0.68965517 & 0.034482759 \\ -11.494253 & 1.9540230 & 0.068965517 \\ -1.1494253 & -0.80459770 & 0.20689655 \end{bmatrix}.$$

Thus,

$$\|A\|_\infty = 7 \quad \text{and} \quad \|A^{-1}\|_\infty = 28.310345,$$

so

$$K_\infty(A) = \|A\|_\infty \|A^{-1}\|_\infty = 7(28.310345) = 198.17.$$

4. d. The linear system

$$0.04x_1 + 0.01x_2 - 0.01x_3 = 0.06,$$
$$0.2x_1 + 0.5x_2 - 0.2x_3 = 0.3,$$
$$x_1 + 2x_2 + 4x_3 = 11$$

has the exact solution

$$\mathbf{x} = (1.827586, 0.6551724, 1.965517)^t$$

and the approximate solution

$$\tilde{\mathbf{x}} = (1.8, 0.64, 1.9)^t.$$

Use the results of part (d) of Exercise 2 to compute

$$\|\mathbf{x} - \tilde{\mathbf{x}}\|_\infty \quad \text{and} \quad K_\infty(A)\frac{\|\mathbf{b} - A\tilde{\mathbf{x}}\|_\infty}{\|A\|_\infty}.$$

SOLUTION: We have $\tilde{\mathbf{x}} = (1.827586, 0.6551724, 1.965517)^t$ and $\mathbf{x} = (1.8, 0.64, 1.9)^t$, so

$$\|\tilde{\mathbf{x}} - \mathbf{x}\|_\infty = \max\{|1.827586 - 1.8|, |0.6551724 - 0.64|, |1.965517 - 1.9|\}$$
$$= \max\{0.027586, 0.01551724, 0.065517\} = 0.065517.$$

Further,

$$\mathbf{b} - A\tilde{\mathbf{x}} = \begin{bmatrix} 0.06 \\ 0.3 \\ 11 \end{bmatrix} - \begin{bmatrix} 0.04 & 0.01 & -0.01 \\ 0.2 & 0.5 & -0.2 \\ 1 & 2 & 4 \end{bmatrix} \begin{bmatrix} 1.8 \\ 0.64 \\ 1.9 \end{bmatrix}$$

$$= \begin{bmatrix} 0.06 \\ 0.3 \\ 11 \end{bmatrix} - \begin{bmatrix} 0.0594 \\ 0.3 \\ 10.68 \end{bmatrix} = \begin{bmatrix} 0.0006 \\ 0 \\ 0.32 \end{bmatrix}$$

and

$$K_\infty(A)\frac{\|\mathbf{b} - A\tilde{\mathbf{x}}\|_\infty}{\|A\|_\infty} = (198.17)\left(\frac{0.32}{7}\right) = 9.0592.$$

9. Show that if B is singular, then

$$\frac{1}{K(A)} \leq \frac{\|A - B\|}{\|A\|}.$$

SOLUTION: Since

$$\|\mathbf{x}\| = \|A^{-1}A\mathbf{x}\| \leq \|A^{-1}\|\,\|A\mathbf{x}\|, \quad \text{we have} \quad \|A\mathbf{x}\| \geq \frac{\|\mathbf{x}\|}{\|A^{-1}\|}.$$

With $\mathbf{x} \neq \mathbf{0}$ such that $\|\mathbf{x}\| = 1$ and $B\mathbf{x} = \mathbf{0}$, we have

$$\|(A - B)\mathbf{x}\| = \|A\mathbf{x}\| \geq \frac{\|\mathbf{x}\|}{\|A^{-1}\|}.$$

Thus

$$\frac{\|(A - B)\mathbf{x}\|}{\|A\|} \geq \frac{1}{\|A^{-1}\|\|A\|} = \frac{1}{K(A)}.$$

But

$$\|(A - B)\mathbf{x}\| \leq \|A - B\|\|\mathbf{x}\| = \|A - B\|,$$

giving the result.

10. a. Using Exercise 9, estimate the condition number for

$$\begin{bmatrix} 1 & 2 \\ 1.0001 & 2 \end{bmatrix}.$$

SOLUTION: Exercise 9 implies that

$$\frac{1}{K_\infty(A)} \leq \frac{\|A - B\|_\infty}{\|A\|_\infty} = \frac{\left\|\begin{bmatrix} 1 & 2 \\ 1.0001 & 2 \end{bmatrix} - \begin{bmatrix} 1 & 2 \\ 1 & 2 \end{bmatrix}\right\|_\infty}{\left\|\begin{bmatrix} 1 & 2 \\ 1.0001 & 2 \end{bmatrix}\right\|_\infty} = \frac{\left\|\begin{bmatrix} 0 & 0 \\ 0.0001 & 0 \end{bmatrix}\right\|_\infty}{\left\|\begin{bmatrix} 1 & 2 \\ 1.0001 & 2 \end{bmatrix}\right\|_\infty} = \frac{0.0001}{3.0001}.$$

Hence, we have

$$K_\infty(A) \geq \frac{3.0001}{0.0001} = 30,001.$$

Exercise Set 7.6, page 492

5. b. Perform only two steps of the conjugate gradient method with $C = C^{-1} = I$ on the linear system

$$\begin{aligned}
10x_1 - x_2 \qquad\quad &= 9, \\
-x_1 + 10x_2 - 2x_3 &= 7, \\
- 2x_2 + 10x_3 &= 6.
\end{aligned}$$

SOLUTION: Let $\mathbf{x}_0 = (0, 0, 0)^t$. Then

$$\mathbf{r}_0 = \mathbf{b} - A\mathbf{x}_0 = \mathbf{b} = (9, 7, 4)^t, \quad \text{so} \quad \mathbf{w} = \mathbf{r}_0 = (9, 7, 4)^t \quad \text{and} \quad \mathbf{v}_1 = \mathbf{w} = (9, 7, 4)^t.$$

Hence

$$\alpha = \mathbf{w}^t\mathbf{w} = 166, \quad \mathbf{u} = A\mathbf{v}_1 = (83, 49, 46)^t, \quad t_1 = \frac{\alpha}{\mathbf{v}_1^t\mathbf{u}} = 0.121523,$$

and

$$\mathbf{x}_1 = \mathbf{x}_0 + t_1\mathbf{v}1 = (1.093704, 0.850659, 0.729136)^t.$$

The residual for this first iteration is

$$\mathbf{r}_1 = \mathbf{r}_0 - t_1\mathbf{u} = (-1.0863841, 1.045388, 0.409956)^t.$$

For the second iteration we have

$$\mathbf{w} = \mathbf{r}_1, \quad \beta = \mathbf{w}^t\mathbf{w} = 2.441129 \quad s_1 = \frac{\beta}{\alpha} = 0.014706$$

and

$$\mathbf{v}_2 = \mathbf{w} + s_1\mathbf{v}_1 = (-0.954033, 1.148327, 0.498190)^t.$$

This gives the new value of $\alpha = \beta = 2.441129$.

$$\mathbf{v}_2 = \mathbf{w} + s_1\mathbf{v}_1 = (-0.954033, 1.148327, 0.498190)^t,$$

$$\mathbf{u} = A\mathbf{v}_2 = (-10.688659, 11.4409257, 2.68524231)^t, \quad t_2 = \frac{\alpha}{\mathbf{v}_2^t\mathbf{u}} = 0.098939,$$

and

$$\mathbf{x}_2 = \mathbf{x}_1 + t_2\mathbf{v}_2 = (0.999313, 0.964273, 0.778427)^t.$$

The residual of this second iteration is

$$\mathbf{r}_2 = \mathbf{r}_1 - t_2\mathbf{u} = (-0.028856, -0.086568, 0.144280)^t.$$

6. b. Repeat Exercise 5(b) using $C^{-1} = D^{-1/2}$.

SOLUTION: Let

$$C = \begin{bmatrix} \frac{\sqrt{10}}{10} & 0 & 0 \\ 0 & \frac{\sqrt{10}}{10} & 0 \\ 0 & 0 & \frac{\sqrt{10}}{10} \end{bmatrix}.$$

With $\mathbf{x}_0 = (0, 0, 0,)^t$ we have

$$\mathbf{r}_0 = \mathbf{b} - A\mathbf{x}_0 = \mathbf{b} = (9, 7, 4)^t, \quad \mathbf{w} = C^{-1}\mathbf{r}_0 = (2.846050, 2.213594, 1.897367)^t$$

and

$$\mathbf{v}_1 = C^{-1}\mathbf{w} = (0.9, 0.7, 0.6)^t.$$

So

$$\alpha = \mathbf{w}^t\mathbf{w} = 166, \quad \mathbf{u} = A\mathbf{v}_1 = (8.3, 4.9, 4.6)^t, \quad t_1 = \frac{\alpha}{\mathbf{v}_1^t\mathbf{u}} = 1.215227,$$

and

$$\mathbf{x}_1 = \mathbf{x}_0 + t_1\mathbf{v}_1 = (1.093704, 0.850659, 0.729136)^t$$

with residual

$$\mathbf{r}_1 = (-1.086384, 1.045388, 0.409956)^t.$$

The second iteration employs these steps in the same manner, and the result is

$$\mathbf{x}_2 = (0.999313, 0.964273, 0.778427)^t$$

with residual

$$\mathbf{r}_2 = (-0.028856, -0.086568, 0.144280)^t.$$

7. b. Repeat Exercise 5(b) with $TOL = 10^{-3}$ in the l_∞ norm.

SOLUTION: Exercise 5(b) found 2 iterations but the tolerance was not met. Because we have a 3×3 system, the exact solution will theoretically be obtained in 3 iterations. In fact, the third iteration gives

$$\mathbf{x}_3 = (0.995789, 0.957895, 0.791579)^t,$$

which is accurate to the places listed.

8. b. Repeat Exercise 7(b) using $C^{-1} = D^{-1/2}$.

SOLUTION: Exercise 6(b) found 2 iterations but the tolerance was not met. Because we have a 3×3 system, the exact solution will theoretically be obtained in 3 iterations. In fact, the third iteration gives, as in 7(b),

$$\mathbf{x}_3 = (0.995789, 0.957895, 0.791579)^t,$$

which is accurate to the places listed.

12. Use the transpose properties given in Theorem 6.14 on page 390 to prove Theorem 7.30.

SOLUTION:

a. We have

$$\langle \mathbf{x}, \mathbf{y} \rangle = \mathbf{x}^t \cdot \mathbf{y} = \sum_1^n x_i y_i = \mathbf{y}^t \cdot \mathbf{x} = \langle \mathbf{y}, \mathbf{x} \rangle.$$

b. For every real number α we have both

$$\alpha \langle \mathbf{x}, \mathbf{y} \rangle = \alpha \mathbf{x}^t \cdot \mathbf{y} = \sum_1^n (\alpha x_i) y_i = \langle \alpha \mathbf{x}, \mathbf{y} \rangle,$$

and

$$\langle \alpha \mathbf{x}, \mathbf{y} \rangle = \alpha \mathbf{x}^t \cdot \mathbf{y} = \sum_1^n (\alpha x_i) y_i = \sum_1^n x_i (\alpha y_i) = \mathbf{x}^t \cdot \alpha \mathbf{y} = \langle \mathbf{x}, \alpha \mathbf{y} \rangle.$$

c. We have

$$\langle \mathbf{x} + \mathbf{z}, \mathbf{y} \rangle = (\mathbf{x} + \mathbf{z})^t \cdot \mathbf{y} = (\mathbf{x}^t + \mathbf{z}^t) \cdot \mathbf{y} = \mathbf{x}^t \cdot \mathbf{y} + \mathbf{z}^t \cdot \mathbf{y} = \langle \mathbf{x}, \mathbf{y} \rangle + \langle \mathbf{z}, \mathbf{y} \rangle.$$

d. For all $\mathbf{x}$,

$$\langle \mathbf{x}, \mathbf{x} \rangle = \mathbf{x}^t \cdot \mathbf{x} = \sum_1^n x_i^2 \geq 0.$$

e. The equation in part (d) implies that $\langle \mathbf{x}, \mathbf{x} \rangle = 0$ if and only if $x_i = 0$ for each $i = 1, 2, \ldots, n$. But this is true if and only if $\mathbf{x} = \mathbf{0}$.

14. Prove Theorem 7.33 using mathematical induction as follows:

a. Show that $\langle \mathbf{r}^{(1)}, \mathbf{v}^{(1)} \rangle = 0$.

SOLUTION: We have

$$\mathbf{x}^{(1)} = \mathbf{x}^{(0)} + t_1 \mathbf{v}^{(1)} = \mathbf{x}^{(0)} + \frac{\langle \mathbf{v}^{(1)}, \mathbf{r}^{(0)} \rangle}{\langle \mathbf{v}^{(1)}, A\mathbf{v}^{(1)} \rangle} \mathbf{v}^{(1)}.$$

Thus,

$$A\mathbf{x}^{(1)} = A\mathbf{x}^{(0)} + \frac{\langle \mathbf{v}^{(1)}, \mathbf{r}^{(0)} \rangle}{\langle \mathbf{v}^{(1)}, A\mathbf{v}^{(1)} \rangle} A\mathbf{v}^{(1)}$$

and

$$\mathbf{b} - A\mathbf{x}^{(1)} = \mathbf{b} - A\mathbf{x}^{(0)} - \frac{\langle \mathbf{v}^{(1)}, \mathbf{r}^{(0)} \rangle}{\langle \mathbf{v}^{(1)}, A\mathbf{v}^{(1)} \rangle} A\mathbf{v}^{(1)}.$$

Hence,

$$\mathbf{r}^{(1)} = \mathbf{r}^{(0)} - \frac{\left\langle \mathbf{v}^{(1)}, \mathbf{r}^{(0)} \right\rangle}{\left\langle \mathbf{v}^{(1)}, A\mathbf{v}^{(1)} \right\rangle} A\mathbf{v}^{(1)}.$$

Taking the inner product with $\mathbf{v}^{(1)}$ gives

$$\left\langle \mathbf{r}^{(1)}, \mathbf{v}^{(1)} \right\rangle = \left\langle \mathbf{r}^{(0)}, \mathbf{v}^{(1)} \right\rangle - \frac{\left\langle \mathbf{v}^{(1)}, \mathbf{r}^{(0)} \right\rangle}{\left\langle \mathbf{v}^{(1)}, A\mathbf{v}^{(1)} \right\rangle} \left\langle A\mathbf{v}^{(1)}, \mathbf{v}^{(1)} \right\rangle = \left\langle \mathbf{r}^{(0)}, \mathbf{v}^{(1)} \right\rangle - \left\langle \mathbf{v}^{(1)}, \mathbf{r}^{(0)} \right\rangle = 0.$$

This establishes the base step.

b. Assume that $\left\langle \mathbf{r}^{(k)}, \mathbf{v}^{(j)} \right\rangle = 0$, for each $k \leq l$ and $j = 1, 2, \ldots, k$, and show that this implies that $\left\langle \mathbf{r}^{(l+1)}, \mathbf{v}^{(j)} \right\rangle = 0$, for each $j = 1, 2, \ldots, l$.

SOLUTION: For the inductive hypothesis we assume that $\left\langle \mathbf{r}^{(k)}, \mathbf{v}^{(j)} \right\rangle = 0$, for all $k \leq l$ and each $j = 1, 2, \ldots, k$. We must then show that

$$\left\langle \mathbf{r}^{(l+1)}, \mathbf{v}^{(j)} \right\rangle = 0, \quad \text{for} \quad j = 1, 2, \ldots, l+1.$$

We do this in two parts. First, for each $j = 1, 2, \ldots, l$, we will show that $\left\langle \mathbf{r}^{(l+1)}, \mathbf{v}^{(j)} \right\rangle = 0$. We have

$$\mathbf{x}^{(l+1)} = \mathbf{x}^{(l)} + t_{l+1} \mathbf{v}^{(l+1)} = \mathbf{x}^{(l)} + \frac{\left\langle \mathbf{v}^{(l+1)}, \mathbf{r}^{(l)} \right\rangle}{\left\langle \mathbf{v}^{(l+1)}, A\mathbf{v}^{(l+1)} \right\rangle} \mathbf{v}^{(l+1)},$$

so

$$A\mathbf{x}^{(l+1)} = A\mathbf{x}^{(l)} + \frac{\left\langle \mathbf{v}^{(l+1)}, \mathbf{r}^{(l)} \right\rangle}{\left\langle \mathbf{v}^{(l+1)}, A\mathbf{v}^{(l+1)} \right\rangle} A\mathbf{v}^{(l+1)}.$$

Subtracting **b** from both sides gives

$$-\mathbf{r}^{(l+1)} = -\mathbf{r}^{(l)} + \frac{\left\langle \mathbf{v}^{(l+1)}, \mathbf{r}^{(l)} \right\rangle}{\left\langle \mathbf{v}^{(l+1)}, A\mathbf{v}^{(l+1)} \right\rangle} A\mathbf{v}^{(l+1)}.$$

Taking the inner product of both sides of $-\mathbf{r}^{(l+1)}$ with $\mathbf{v}^{(j)}$ gives

$$-\left\langle \mathbf{r}^{(l+1)}, \mathbf{v}^{(j)} \right\rangle = -\left\langle \mathbf{r}^{(l)}, \mathbf{v}^{(j)} \right\rangle + \frac{\left\langle \mathbf{v}^{(l+1)}, \mathbf{r}^{(l)} \right\rangle}{\left\langle \mathbf{v}^{(l+1)}, A\mathbf{v}^{(l+1)} \right\rangle} \left\langle A\mathbf{v}^{(l+1)}, \mathbf{v}^{(j)} \right\rangle.$$

The first term on the right-hand side of this equation is 0 by the inductive hypothesis, and the factor $\left\langle A\mathbf{v}^{(l+1)}, \mathbf{v}^{(j)} \right\rangle$ is 0 because of A-orthogonality. Thus, $\left\langle \mathbf{r}^{(l+1)}, \mathbf{v}^{(j)} \right\rangle = 0$, for each $j = 1, 2, \ldots, l$.

c. Show that $\left\langle \mathbf{r}^{(l+1)}, \mathbf{v}^{(l+1)} \right\rangle = 0$.

SOLUTION: For the second part we take the inner product of both sides of

$$-\mathbf{r}^{(l+1)} = -\mathbf{r}^{(l)} + \frac{\left\langle \mathbf{v}^{(l+1)}, \mathbf{r}^{(l)} \right\rangle}{\left\langle \mathbf{v}^{(l+1)}, A\mathbf{v}^{(l+1)} \right\rangle} A\mathbf{v}^{(l+1)}$$

with $\mathbf{v}^{(l+1)}$ to get

$$-\left\langle \mathbf{r}^{(l+1)}, \mathbf{v}^{(l+1)} \right\rangle = -\left\langle \mathbf{r}^{(l)}, \mathbf{v}^{(l+1)} \right\rangle + \frac{\left\langle \mathbf{v}^{(l+1)}, \mathbf{r}^{(l)} \right\rangle}{\left\langle \mathbf{v}^{(l+1)}, A\mathbf{v}^{(l+1)} \right\rangle} \left\langle A\mathbf{v}^{(l+1)}, \mathbf{v}^{(l+1)} \right\rangle.$$

Thus,

$$-\left\langle \mathbf{r}^{(l+1)}, \mathbf{v}^{(l+1)} \right\rangle = -\left\langle \mathbf{r}^{(l)}, \mathbf{v}^{(l+1)} \right\rangle + \left\langle \mathbf{v}^{(l+1)}, \mathbf{r}^{(l)} \right\rangle = 0.$$

This completes the proof by induction.

Approximation Theory

Exercise Set 8.1, page 506

1. Compute the linear least-squares polynomial for the data of Example 2.

SOLUTION: The data needed for the solution is listed in the following table.

i	1	2	3	4	5	Totals
x_i	0	0.25	0.5	0.75	1	2.5
y_i	1.0000	1.2840	1.6487	2.1170	2.7183	8.7680
x_i^2	0	0.0625	0.25	0.5625	1	1.875
$x_i y_i$	0	0.3211	0.8244	1.5878	2.7813	5.4514

This gives

$$a = \frac{5(5.4516) - 2.5(8.7685)}{5(1.8750) - (2.5)^2} = 1.70784$$

and

$$b = \frac{1.8750(8.7685) - 5.4516(2.5)}{5(1.8750) - (2.5)^2} = 0.89968.$$

So the linear least-squares polynomial is

$$P_1(x) = 1.70784x + 0.89968.$$

The error is

$$E = \sum_{i=1}^{s}(y_i - (1.70784x_i + 0.89968))^2 = 0.39198.$$

2. Compute the least-squares polynomial of degree two for the data given in Example 1.

SOLUTION: The linear system describing the solution has the form

$$\begin{bmatrix} 10 & \sum_{i=1}^{10} x_i & \sum_{i=1}^{10} x_i^2 \\ \sum_{i=1}^{10} x_i & \sum_{i=1}^{10} x_i^2 & \sum_{i=1}^{10} x_i^3 \\ \sum_{i=1}^{10} x_i^2 & \sum_{i=1}^{10} x_i^3 & \sum_{i=1}^{10} x_i^4 \end{bmatrix} \begin{bmatrix} a_0 \\ a_1 \\ a_2 \end{bmatrix} = \begin{bmatrix} \sum_{i=1}^{10} y_i \\ \sum_{i=1}^{10} x_i y_i \\ \sum_{i=1}^{10} x_i^2 y_i \end{bmatrix},$$

165

which, for the data in Example 1, is

$$\begin{bmatrix} 10 & 55 & 385 \\ 55 & 385 & 3025 \\ 385 & 3025 & 25333 \end{bmatrix} \begin{bmatrix} a_0 \\ a_1 \\ a_2 \end{bmatrix} = \begin{bmatrix} 81 \\ 572.4 \\ 4532.8 \end{bmatrix}.$$

Solving this system gives

$$a_0 = 0.40667, \quad a_1 = 1.1548, \quad \text{and} \quad a_2 = 0.034848,$$

so the least-squares polynomial of degree two is

$$P_2(x) = 0.40667 + 1.1548x + 0.034848x^2.$$

8. A collection of data is given in the text for 30 students taking a numerical analysis course. For each student, a numerical score is given for total homework and one for the final examination. The question to answer is how well the scores correlate. Specifically, does the least squares line for this data do a good job of predicting a 60% passing grade and a 90% minimal A grade?

SOLUTION: The least-square line for the data has equation

$$P(x) = 0.22335x - 0.80283.$$

This predicts a homework grade of 272 for a minimal D, which seems somewhat in line with the data. However the prediction of over 400 for a 90% A is certainly not reasonable.

14. Show that the normal equations yield a symmetric and nonsingular matrix and hence have a unique solution.

SOLUTION: For each $i = 1, \ldots, n+1$ and $j = 1, \ldots, n+1$, we have

$$a_{ij} = a_{ji} = \sum_{k=1}^{m} x_k^{i+j-2},$$

so $A = (a_{ij})$ is symmetric.
Suppose A is singular and $\mathbf{c} \neq \mathbf{0}$ satisfies $\mathbf{c}^t A \mathbf{c} = \mathbf{0}$. Then,

$$0 = \sum_{i=1}^{n+1} \sum_{j=1}^{n+1} a_{ij} c_i c_j = \sum_{i=1}^{n+1} \sum_{j=1}^{n+1} \left(\sum_{k=1}^{m} x_k^{i+j-2} \right) c_i c_j = \sum_{k=1}^{m} \left[\sum_{i=1}^{n+1} \sum_{j=1}^{n+1} c_i c_j x_k^{i+j-2} \right],$$

so

$$\sum_{k=1}^{m} \left(\sum_{i=1}^{n+1} c_i x_k^{i-1} \right)^2 = 0.$$

Define $P(x) = c_1 + c_2 x + \ldots + c_{n+1} x^n$. Then

$$\sum_{k=1}^{m} [P(x_k)]^2 = 0$$

and $P(x) = 0$ has roots $x_1, \ldots, x_m$. Since the roots are distinct and $m > n$, $P(x)$ must be the zero polynomial. Thus,

$$c_1 = c_2 = \ldots = c_{n+1} = 0,$$

and A must be nonsingular.

Exercise Set 8.2, page 518

1. b. Find the linear-least squares polynomial approximation to $f(x) = x^3$ on $[0, 2]$.

SOLUTION: Because we have

$$\int_0^2 1 \, dx = 2, \quad \int_0^2 x \, dx = 2, \quad \int_0^2 x^2 \, dx = \frac{8}{3},$$

$$\int_0^2 x^3 \, dx = 4, \quad \text{and} \quad \int_0^2 x^4 \, dx = \frac{32}{5},$$

and the linear system has the form

$$a_0 \int_0^2 1 \, dx + a_1 \int_0^2 x \, dx = \int_0^2 x^3 \, dx$$

$$a_0 \int_0^2 x \, dx + a_1 \int_0^2 x^2 \, dx = \int_0^2 x^4 \, dx,$$

the system reduces to

$$2a_0 + 2a_1 = 4 \quad \text{and} \quad 2a_0 + \frac{8}{3}a_1 = \frac{32}{5}.$$

The solution $a_0 = -\frac{8}{5}$ and $a_1 = \frac{18}{5}$, and the linear least-square approximation is

$$P_1(x) = \frac{2}{5}(9x - 4).$$

3. b. Find the least-squares polynomial approximation of degree two to the function and interval in Exercise 1(b).

SOLUTION: The 3×3 linear system in this case has the form

$$a_0 \int_0^2 1 \, dx + a_1 \int_0^2 x \, dx + a_2 \int_0^2 x^2 \, dx = \int_0^2 x^3 \, dx,$$

$$a_0 \int_0^2 x \, dx + a_1 \int_0^2 x^2 \, dx + a_2 \int_0^2 x^3 \, dx = \int_0^2 x^4 \, dx,$$

$$a_0 \int_0^2 x^2 \, dx + a_1 \int_0^2 x^3 \, dx + a_2 \int_0^2 x^4 \, dx = \int_0^2 x^5 \, dx,$$

so in addition to the integrals evaluated in Exercise 1(b) we need $\int_0^2 x^5 \, dx = \frac{32}{3}$.

The linear system is consequently

$$2a_0 + 2a_1 + \frac{8}{3}a_2 = 4,$$

$$2a_0 + \frac{8}{3}a_1 + 4a_2 = \frac{32}{5},$$

$$\frac{8}{3}a_0 + 4a_1 + \frac{32}{5}a_2 = \frac{32}{3}.$$

This system has the solution $a_0 = \frac{2}{5}$, $a_1 = -\frac{12}{5}$, and $a_2 = 3$, so

$$P_2(x) = \frac{1}{5}(15x^2 - 12x + 2).$$

5. b. Compute the error E for the approximation in Exercise 3(b).

SOLUTION: The approximation error is

$$\int_0^2 \left(x^3 - \frac{1}{5}(15x^2 - 12x + 2)\right)^2 dx = \int_0^2 x^6 - 6x^5 + \frac{69}{5}x^4 - \frac{76}{5}x^3 + \frac{204}{25}x^2 - \frac{48}{25}x + \frac{4}{25}\, dx = \frac{8}{175}.$$

7. b. Use the Gram-Schmidt process to construct $\phi_0(x)$, $\phi_1(x)$, $\phi_2(x)$, and $\phi_3(x)$ for the interval $[0, 2]$

SOLUTION: First we have

$$\Phi_0(x) = 1 \quad \text{and} \quad \Phi_1 = x - B_1, \quad \text{where} \quad B_1 = \frac{\int_0^2 x\Phi_0(x)^2\, dx}{\int_0^2 \Phi_0(x)^2\, dx} = 1.$$

so $\Phi_1(x) = x - 1$.

Then

$$\Phi_2(x) = (x - B_2)\Phi_1(x) - C_2\Phi_0(x), \quad \text{where}$$

$$B_2 = \frac{\int_0^2 x\Phi_1(x)^2\, dx}{\int_0^2 \Phi_1(x)^2\, dx} = \frac{\int_0^2 x(x-1)^2\, dx}{\int_0^2 (x-1)^2\, dx} = 1 \quad \text{and}$$

$$C_2 = \frac{\int_0^2 x\Phi_0(x)\Phi_1(x)\, dx}{\int_0^2 \Phi_0(x)^2\, dx} = \frac{\int_0^2 x(x-1)\, dx}{\int_0^2 1\, dx} = \frac{1}{3}.$$

This gives

$$\phi_2(x) = x^2 - 2x + \frac{2}{3}.$$

Finally,

$$\Phi_3(x) = (x - B_3)\Phi_2(x) - C_3\Phi_1(x), \quad \text{where}$$

$$B_3 = \frac{\int_0^2 x\Phi_2(x)^2\, dx}{\int_0^2 \Phi_2(x)^2\, dx} = \frac{\int_0^2 x\left(x^2 - 2x + \frac{2}{3}\right)^2 dx}{\int_0^2 \left(x^2 - 2x + \frac{2}{3}\right)^2 dx} = 1 \quad \text{and}$$

$$C_3 = \frac{\int_0^2 x\Phi_1(x)\Phi_2(x)\, dx}{\int_0^2 \Phi_1(x)^2\, dx} = \frac{\int_0^2 x(x-1)(x^2 - 2x + \frac{2}{3})\, dx}{\int_0^2 (x-1)^2\, dx} = \frac{4}{15}.$$

This gives

$$\Phi_3(x) = x^3 - 3x^2 + \frac{12}{5}x - \frac{2}{5}.$$

8. b. Repeat Exercise 1(b) using the results of Exercise 7(b).

SOLUTION: The orthogonality of the set $\{\Phi_0(x), \Phi_1(x)\}$ means that the calculations are reduced to

$$a_0 = \frac{\int_0^2 x^3\phi_0\, dx}{\int_0^2 \phi_0^2\, dx} = \frac{\int_0^2 x^3\, dx}{\int_0^2 1\, dx} = 2 \quad \text{and} \quad a_1 = \frac{\int_0^2 x^3\phi_1\, dx}{\int_0^2 \phi_1^2\, dx} = \frac{\int_0^2 x^3(x-1)\, dx}{\int_0^2 (x-1)^2\, dx} = \frac{18}{5}.$$

So

$$P_1(x) = 2\Phi_0(x) + \frac{18}{5}\Phi_1(x) = 2 + \frac{18}{5}(x - 1) = \frac{2}{5}(9x - 4).$$

10. b. Repeat Exercise 3(b) using the results of Exercise 7(a).

SOLUTION: In Exercise 8(b) we found that for the orthogonal set $\{\Phi_0(x), \Phi_1(x), \Phi_2(x)\}$ we have

$$a_0 = 2 \quad \text{and} \quad a_1 = \frac{18}{5}.$$

So we only need to determine a_2, and this is given by

$$a_2 = \frac{\int_0^2 x^3 \phi_2 \, dx}{\int_0^2 \phi_2^2 \, dx} = \frac{\int_0^2 x^3 \left(x^2 - 2x + \frac{2}{3}\right) dx}{\int_0^2 \left(x^2 - 2x + \frac{2}{3}\right)^2 dx} = 3.$$

So

$$P_2(x) = 2\Phi_0(x) + \frac{18}{5}\Phi_1(x) + 3\Phi_2(x) = 2 + \frac{18}{5}(x-1) + 3\left(x^2 - 2x + \frac{2}{3}\right) = \frac{1}{5}(15x^2 - 12x + 2).$$

13. Suppose $\{\phi_0, \phi_1, \ldots, \phi_n\}$ is any linearly independent set in $\prod_n$. Show that for any element $Q \in \prod_n$, there exist unique constants $c_0, c_1, \ldots, c_n$, such that

$$Q(x) = \sum_{k=0}^n c_k \phi_k(x).$$

SOLUTION: Let $\{\phi_0(x), \phi_1(x), \ldots, \phi_n(x)\}$ be a linearly independent set of polynomials in $\prod_n$. For each $i = 0, 1, \ldots, n$, let $\phi_i(x) = \sum_{k=0}^n b_{ki} x^k$. Let $Q(x) = \sum_{k=0}^n a_k x^k \in \prod_n$. We want to find constants $c_0, \ldots, c_n$ so that

$$Q(x) = \sum_{i=0}^n c_i \phi_i(x).$$

This equation becomes

$$\sum_{k=0}^n a_k x^k = \sum_{i=0}^n c_i \left(\sum_{k=0}^n b_{ki} x^k\right) = \sum_{k=0}^n \left(\sum_{i=0}^n c_i b_{ki}\right) x^k = \sum_{k=0}^n \left(\sum_{i=0}^n b_{ki} c_i\right) x^k.$$

But $\{1, x, \ldots, x^n\}$ is linearly independent, so, for each $k = 0, \ldots, n$, we have

$$\sum_{i=0}^n b_{ki} c_i = a_k,$$

which expands to the linear system

$$\begin{bmatrix} b_{01} & b_{02} & \cdots & b_{0n} \\ b_{11} & b_{12} & \cdots & b_{1n} \\ \vdots & \vdots & & \vdots \\ b_{n1} & b_{n2} & \cdots & b_{nn} \end{bmatrix} \begin{bmatrix} c_0 \\ c_1 \\ \vdots \\ c_n \end{bmatrix} = \begin{bmatrix} a_0 \\ a_1 \\ \vdots \\ a_n \end{bmatrix}.$$

This linear system must have a unique solution $\{c_0, c_1, \ldots, c_n\}$, or else there is a nontrivial set of constants $\{c'_0, c'_1, \ldots, c'_n\}$, for which

$$\begin{bmatrix} b_{01} & \cdots & b_{0n} \\ \vdots & & \vdots \\ b_{n1} & \cdots & b_{nn} \end{bmatrix} \begin{bmatrix} c'_0 \\ \vdots \\ c'_n \end{bmatrix} = \begin{bmatrix} 0 \\ \vdots \\ 0 \end{bmatrix}.$$

Thus,

$$c_0'\phi_0(x) + c_1'\phi_1(x) + \cdots + c_n'\phi_n(x) = \sum_{k=0}^{n} 0 \cdot x^k = 0,$$

which contradicts the linear independence of the set $\{\phi_0, \ldots, \phi_n\}$. Thus, there is a unique set of constants $\{c_0, \ldots, c_n\}$, for which

$$Q(x) = c_0\phi_0(x) + c_1\phi_1(x) + \cdots + c_n\phi_n(x).$$

14. Show that if $\{\phi_0, \phi_1, \ldots, \phi_n\}$ is an orthogonal set of functions on $[a, b]$ with respect to the weight function w, then $\{\phi_0, \phi_1, \ldots, \phi_n\}$ is a linearly independent set.

SOLUTION: Suppose that $c_0, c_1, \ldots, c_n$ are constants such that for all x on $[a, b]$,

$$0 = c_0\phi_0(x) + c_1\phi_1(x) + \cdots + c_i\phi_i(x) + \cdots + c_n\phi_n(x).$$

Then for each $j \leq n$ with $j \neq i$ we have

$$\int_a^b 0 \cdot w(x)\phi_j(x)\,dx = \sum_{i=0}^{n} c_i \int_a^b w(x)\phi_i(x)\phi_j(x)\,dx.$$

But the orthogonality of the functions with respect to $w(x)$ implies that all but one of the integrals on the right side of the equation are 0, and the integral on the left side of the equation is also 0. As a consequence, for each $j = 0, 1, 2, \ldots, n$ we have

$$0 = c_j \int_a^b w(x)(\phi_j(x))^2\,dx.$$

Since the integral on the right side is not zero, we must have $c_j = 0$ for each $j = 0, 1, 2, \ldots, n$. Hence the set of functions is linear independent.

Exercise Set 8.3, page 527

1. a. Use the zeros of $\tilde{T}_3$ to construct an interpolating polynomial of degree two for $f(x) = e^x$.

SOLUTION: The zeros of $\tilde{T}_3$ are

$$\bar{x}_1 = \cos\frac{1}{6}\pi = \frac{\sqrt{3}}{2}, \quad \bar{x}_2 = \cos\frac{3}{6}\pi = 0, \quad \text{and} \quad \bar{x}_3 = \cos\frac{5}{6}\pi = -\frac{\sqrt{3}}{2}.$$

These nodes are used in the following divided-difference table.

$\bar{x}_1 = \sqrt{3}/2 \approx 0.8660254$	$e^{\bar{x}_1} = 2.377443$		
$\bar{x}_2 = 0$	$e^0 = 1$	1.590534	
$\bar{x}_3 = -\sqrt{3}/2 \approx -0.8660254$	$e^{\bar{x}_3} = 0.4206200$	0.6690104	0.5320418

The Newton form of the interpolation polynomial is

$$P_2(x) = 2.377443 + 1.590534(x - 0.8660254) + 0.5320418(x - 0.8660254)x.$$

3. a. Find a bound for the maximum error of the approximation in Exercise 1.

SOLUTION: By Corollary 8.11, the approximation error for the Lagrange polynomial in Exercise 1 has the following bound:

$$\max_{x \in [-1,1]} |e^x - P_2(x)| \leq \frac{1}{2^2(3!)} \max_{x \in [-1,1]} \left| f^{(3)}(x) \right| \leq \frac{e}{24} = 0.1132617.$$

8. Show that for any positive integers i and j with $i > j$, we have $T_i(x)T_j(x) = \frac{1}{2}[T_{i+j}(x) + T_{i-j}(x)]$.

SOLUTION: Since $i > j$, we have

$$\frac{1}{2}(T_{i+j}(x) + T_{i-j}(x)) = \frac{1}{2}(\cos(i+j)\theta + \cos(i-j)\theta) = \cos i\theta \cos j\theta = T_i(x)T_j(x).$$

9. Show that for each Chebyshev polynomial $T_n(x)$, we have

$$\int_{-1}^{1} \frac{(T_n(x))^2}{\sqrt{1 - x^2}} \, dx = \frac{\pi}{2}.$$

SOLUTION: The change of variable $x = \cos\theta$ produces

$$\int_{-1}^{1} \frac{(T_n(x))^2}{\sqrt{1 - x^2}} \, dx = \int_{-1}^{1} \frac{[\cos(n \arccos x)]^2}{\sqrt{1 - x^2}} \, dx = \int_{0}^{\pi} (\cos(n\theta))^2 \, d\theta = \frac{1}{2} \int_{0}^{\pi} (1 + \cos(2n\theta)) \, d\theta = \frac{\pi}{2}.$$

Exercise Set 8.4, page 537

1. Determine all Padé approximations for $f(x) = e^{2x}$ of degree two.

SOLUTION: The Taylor series for $f(x) = e^{2x}$ has the form

$$e^{2x} = 1 + 2x + 2x^2 + \frac{8}{3!}x^3 + \cdots.$$

When $n = 2$ and $m = 0$, $r_{2,0}(x)$ is simply the first three terms of this series; that is,

$$r_{2,0}(x) = 1 + 2x + 2x^2.$$

When $n = 1$ and $m = 1$, we need to have the constant, linear, and quadratic terms be zero in the expansion

$$\left(1 + 2x + 2x^2 + \cdots\right)(1 + q_1 x) - (p_0 + p_1 x) = (1 - p_0) + (q_1 + 2 - p_1)x + (2q_1 + 2)x^2 + \cdots.$$

This implies that $p_0 = 1$, $q_1 = -1$, and $p_1 = 1$, so

$$r_{1,1}(x) = \frac{1 + x}{1 - x}.$$

Similarly, when $n = 0$ and $m = 2$, we need to have the constant, linear, and quadratic terms be zero in the expansion

$$\left(1 + 2x + 2x^2 + \dots\right)\left(1 + q_1 x + q_2 x^2\right) - p_0 = (1 - p_0) + (q_1 + 2)x + (q_2 + 2q_1 + 2)x^2 + \cdots.$$

This implies that $p_0 = 1$, $q_1 = -2$, and $q_2 = 2$, so

$$r_{0,2}(x) = \frac{1}{1 - 2x + 2x^2}.$$

The following table compares the results from these Padé approximations to the exact results.

i	x_i	$f(x_i)$	$r_{2,0}(x_i)$	$r_{1,1}(x_i)$	$r_{0,2}(x_i)$
1	0.2	1.4918	1.4800	1.5000	1.4706
2	0.4	2.2255	2.1200	2.3333	1.9231
3	0.6	3.3201	2.9200	4.0000	1.9231
4	0.8	4.9530	3.8800	9.0000	1.4706
5	1.0	7.3891	5.0000	undefined	1.0000

8. a. Express

$$\frac{x^2 + 3x + 2}{x^2 - x + 1}$$

in continued-fraction form.

SOLUTION: First we divide the polynomial $x^2 - x + 1$ into the polynomial $x^2 + 3x + 2$ to produce

$$\frac{x^2 + 3x + 2}{x^2 - x + 1} = 1 + \frac{4x + 1}{x^2 - x + 1}.$$

Then we write this as

$$\frac{x^2 + 3x + 2}{x^2 - x + 1} = 1 + \frac{4}{\frac{x^2 - x + 1}{x + 1/4}}.$$

Finally, we divide $x + \frac{1}{4}$ into $x^2 - x + 1$ to obtain

$$\frac{x^2 + 3x + 2}{x^2 - x + 1} = 1 + \frac{4}{x - \frac{5}{4} + \frac{21/16}{x + 1/4}}.$$

14. To accurately approximate $\sin x$ and $\cos x$ for inclusion in a mathematical library, we first restrict their domains. Given a real number x, divide by π to obtain the relation

$$|x| = M\pi + s, \quad \text{where } M \text{ is an integer and } |s| \leq \frac{\pi}{2}.$$

a. Show that $\sin x = \sin x \cdot (-1)^M \cdot \sin s$.

b. Construct a rational function approximation to $\sin s$ using $n = m = 4$. Estimate the error when $0 \leq |s| \leq \frac{\pi}{2}$.

c. Design an implementation of $\sin x$ using parts (a) and (b).

SOLUTION: **a.** Since

$$\sin |x| = \sin(M\pi + s) = \sin M\pi \cos s + \cos M\pi \sin s = (-1)^M \sin s,$$

we have

$$\sin x = \operatorname{sgn} x \sin |x| = \operatorname{sgn}(x)(-1)^M \sin s.$$

b. We have

$$\sin x \approx \frac{s - \frac{31}{294}s^3}{1 + \frac{3}{49}s^2 + \frac{11}{5880}s^4},$$

with $|error| \le 2.84 \times 10^{-4}$.

c. Set

$$M = \operatorname{Round}\frac{|x|}{\pi}, \quad s = |x| - M\pi, \quad \text{and} \quad f_1 = \frac{s - \frac{31}{294}s^3}{1 + \frac{3}{49}s^2 + \frac{11}{5880}s^4}.$$

Then $f = (-1)^M f_1 x/|x|$ is the approximation.

Exercise Set 8.5, page 546

1. Find the continuous least-squares trigonometric polynomial S_2 for $f(x) = x^2$ on $[-\pi, \pi]$.

SOLUTION: We have

$$S_2(x) = \frac{a_0}{2} + a_1 \cos x + a_2 \cos 2x + b_1 \sin x,$$

where

$$a_0 = \frac{1}{\pi} \int_{-\pi}^{\pi} x^2 \, dx = \frac{1}{3\pi} \left(\pi^3 - (-\pi)^3\right) = \frac{2\pi^2}{3},$$

$$a_1 = \frac{1}{\pi} \int_{-\pi}^{\pi} x^2 \cos x \, dx = \frac{1}{\pi} \left[x^2 \sin x + 2x \cos x - 2 \sin x\right]_{-\pi}^{\pi} = -4,$$

$$a_2 = \frac{1}{\pi} \int_{-\pi}^{\pi} x^2 \cos 2x \, dx = \frac{1}{\pi} \left[\frac{x^2}{2} \sin 2x + \frac{x}{2} \cos 2x - \frac{1}{4} \sin 2x\right]_{-\pi}^{\pi} = 1,$$

$$b_1 = \frac{1}{\pi} \int_{-\pi}^{\pi} x^2 \sin x \, dx = 0.$$

Hence, we have

$$S_2(x) = \frac{\pi^2}{3} - 4 \cos x + \cos 2x.$$

7. a. Determine the discrete least-squares trigonometric polynomial $S_n(x)$ on the interval $[-\pi, \pi]$ for $f(x) = \cos 2x$, using $m = 4$ and $n = 2$.

SOLUTION: We have $x_0 = -\pi$, $x_1 = -\frac{3\pi}{4}$, $x_2 = -\frac{\pi}{2}$, $x_3 = -\frac{\pi}{4}$, $x_4 = 0$, $x_5 = \frac{\pi}{4}$, $x_6 = \frac{\pi}{2}$, and $x_7 = \frac{3\pi}{4}$. Further, $y_0 = \cos 2x_0 = 1$, $y_1 = \cos 2x_1 = 0$, $y_2 = \cos 2x_2 = -1$, $y_3 = \cos 2x_3 = 0$, $y_4 = \cos 2x_4 = 1$, $y_5 = \cos 2x_5 = 0$, $y_6 = \cos 2x_6 = -1$, and $y_7 = \cos 2x_7 = 0$.

Thus

$$a_0 = \frac{1}{4}[y_0 + y_1 + y_2 + y_3 + y_4 + y_5 + y_6 + y_7] = 0,$$

$$a_1 = \frac{1}{4}\left[\sum_{j=0}^{7} y_j \cos x_j\right] = \frac{1}{4}[\cos x_0 - \cos x_2 + \cos x_4 - \cos x_6] = \frac{1}{4}[-1 - 0 + 1 + 0] = 0,$$

$$a_2 = \frac{1}{4}\left[\sum_{j=0}^{7} y_j \cos 2x_j\right] = \frac{1}{4}[\cos 2x_0 - \cos 2x_2 + \cos 2x_4 - \cos 2x_6] = \frac{1}{4}[1 + 1 + 1 + 1] = 1,$$

$$b_1 = \frac{1}{4}\left[\sum_{j=0}^{7} y_j \sin x_j\right] = \frac{1}{4}[\sin x_0 - \sin x_2 + \sin x_4 - \sin x_6] = \frac{1}{4}[0 + 1 + 0 - 1] = 0.$$

So

$$S_2(x) = \frac{a_0}{2} + a_1 \cos x + a_2 \cos 2x + b_1 \sin x = \cos 2x.$$

8. a. Compute the error $E(S_n)$ for the function in part (a) of Exercise 7.

SOLUTION: We have

$$E(S_2) = \sum_{j=0}^{7}[y_j - S_2(x_j)]^2 = \sum_{j=0}^{7}[y_j - \cos 2x_j]^2 = \sum_{j=0}^{7}[\cos 2x_j - \cos 2x_j]^2 = 0.$$

10. Repeat Exercise 9 using $m = 8$. Compare the values of the approximating polynomials with the values of f at the points $\xi_j = -\pi + 0.2j\pi$, for $0 \le j \le 10$. Which approximation is better?

SOLUTION: The trigonometric least-squares polynomial is

$$S_3(x) = 0.06201467 - 0.8600803 \cos x + 2.549330 \cos 2x - 0.6409933 \cos 3x$$
$$- 0.8321197 \sin x - 0.6695062 \sin 2x,$$

with least-squares error 107.913.

The approximation in Exercise 10 is better because, in this case,

$$\sum_{j=0}^{10}(f(\xi_j) - S_3(\xi_j))^2 = 397.3678,$$

whereas the approximation in Exercise 9 has

$$\sum_{j=0}^{10}(f(\xi_j) - S_3(\xi_j))^2 = 569.3589.$$

15. Show that the functions $\phi_0(x) = \frac{1}{2}$, $\phi_1(x) = \cos x, \ldots, \phi_n(x) = \cos nx$, $\phi_{n+1}(x) = \sin x, \ldots,$ $\phi_{2n-1}(x) = \sin(n-1)x$ are orthogonal on $[-\pi, \pi]$ with respect to $w(x) \equiv 1$.

SOLUTION: The following integrations establish the orthogonality.

$$\int_{-\pi}^{\pi} [\phi_0(x)]^2 \, dx = \frac{1}{2} \int_{-\pi}^{\pi} dx = \pi,$$

$$\int_{-\pi}^{\pi} [\phi_k(x)]^2 \, dx = \int_{-\pi}^{\pi} (\cos kx)^2 \, dx = \int_{-\pi}^{\pi} \left[\frac{1}{2} + \frac{1}{2} \cos 2kx\right] dx = \pi + \left[\frac{1}{4k} \sin 2kx\right]_{-\pi}^{\pi} = \pi,$$

$$\int_{-\pi}^{\pi} [\phi_{n+k}(x)]^2 \, dx = \int_{-\pi}^{\pi} (\sin kx)^2 \, dx = \int_{-\pi}^{\pi} \left[\frac{1}{2} - \frac{1}{2} \cos 2kx\right] dx = \pi - \left[\frac{1}{4k} \sin 2kx\right]_{-\pi}^{\pi} = \pi,$$

$$\int_{-\pi}^{\pi} \phi_k(x)\phi_0(x) \, dx = \frac{1}{2} \int_{-\pi}^{\pi} \cos kx \, dx = \frac{1}{2k} \sin kx \Big]_{-\pi}^{\pi} = 0,$$

$$\int_{-\pi}^{\pi} \phi_{n+k}(x)\phi_0(x) \, dx = \frac{1}{2} \int_{-\pi}^{\pi} \sin kx \, dx = \frac{-1}{2k} \cos kx \Big|_{-\pi}^{\pi} = \frac{-1}{2k} [\cos k\pi - \cos(-k\pi)] = 0,$$

$$\int_{-\pi}^{\pi} \phi_k(x)\phi_j(x) \, dx = \int_{-\pi}^{\pi} \cos kx \cos jx \, dx = \frac{1}{2} \int_{-\pi}^{\pi} [\cos(k+j)x + \cos(k-j)x] \, dx = 0,$$

$$\int_{-\pi}^{\pi} \phi_{n+k}(x)\phi_{n+j}(x) \, dx = \int_{-\pi}^{\pi} \sin kx \sin jx \, dx = \frac{1}{2} \int_{-\pi}^{\pi} [\cos(k-j)x - \cos(k+j)x] \, dx = 0,$$

and

$$\int_{-\pi}^{\pi} \phi_k(x)\phi_{n+j}(x) \, dx = \int_{-\pi}^{\pi} \cos kx \sin jx \, dx = \frac{1}{2} \int_{-\pi}^{\pi} [\sin(k+j)x - \sin(k-j)x] \, dx = 0.$$

Exercise Set 8.6, page 557

1. **a.** Determine the trigonometric interpolating polynomial S_2 of degree two on $[-\pi, \pi]$ for $f(x) = \pi(x - \pi)$.

SOLUTION: The polynomial has the form

$$S_2(x) = \frac{a_0 + a_2 \cos 2x}{2} + a_1 \cos x + b_1 \sin x,$$

where

$$a_0 = \frac{1}{2} \sum_{j=0}^{3} y_j, \quad a_1 = \frac{1}{2} \sum_{j=0}^{3} y_j \cos x_j, \quad a_2 = \frac{1}{2} \sum_{j=0}^{3} y_j \cos 2x_j, \quad \text{and} \quad b_1 = \frac{1}{2} \sum_{j=0}^{3} y_j \sin x_j.$$

The values of x_j and y_j are listed in the following table.

j	0	1	2	3
x_j	$-\pi$	$-\frac{1}{2}\pi$	0	$\frac{1}{2}\pi$
y_j	$-2\pi^2$	$-\frac{3}{2}\pi^2$	$-\pi^2$	$-\frac{1}{2}\pi^2$

Thus

$$a_0 = \frac{1}{2}\left[-2\pi^2 - \frac{3}{2}\pi^2 - \pi^2 - \frac{1}{2}\pi^2\right] = -\frac{5}{2}\pi^2,$$

$$a_1 = \frac{1}{2}\left[-2\pi^2\cos(-\pi) - \frac{3}{2}\pi^2\cos\left(-\frac{\pi}{2}\right) - \pi^2\cos 0 - \frac{1}{2}\pi^2\cos\left(-\frac{\pi}{2}\right)\right] = \frac{1}{2}(2\pi^2 - \pi^2) = \frac{\pi^2}{2},$$

$$a_2 = \frac{1}{2}\left[-2\pi^2\cos(-2\pi) - \frac{3}{2}\pi^2\cos(-\pi) - \pi^2\cos 0 - \frac{1}{2}\pi^2\cos(\pi)\right] = \frac{1}{2}\left[-2\pi^2 + \frac{3}{2}\pi^2 - \pi^2 + \frac{1}{2}\pi^2\right] = -\frac{\pi^2}{2},$$

and

$$b_1 = \frac{1}{2}\left[-2\pi^2\sin(-\pi) - \frac{3}{2}\pi^2\sin\left(-\frac{\pi}{2}\right) - \pi^2\sin 0 - \frac{1}{2}\pi^2\sin\left(-\frac{\pi}{2}\right)\right] = \frac{1}{2}\left[\frac{3}{2}\pi^2 - \frac{1}{2}\pi^2\right] = \frac{\pi^2}{2}.$$

This gives

$$S_2(x) = -\frac{5\pi^2}{4} + \frac{\pi^2}{2}\cos x - \frac{\pi^2}{4}\cos 2x + \frac{\pi^2}{2}\sin x.$$

4. a. Determine the trigonometric interpolating polynomial $S_4(x)$ of degree 4 for $f(x) = x^2 \sin x$ on the interval $[0, 1]$.

b. Compute $\int_0^1 S_4(x)\,dx$.

c. Compare the integral in part (b) to $\int_0^1 x^2 \sin x\,dx$.

SOLUTION: **a.** First transform the interval $[0, 1]$ to $[-\pi, \pi]$ with $z = \pi(2x - 1)$ so that for each $j = 0, 1, \ldots, 7$ we have

$$z_j = \pi\left(\frac{j}{4} - 1\right) \quad \text{and} \quad y_j = f\left(\frac{z_j}{2\pi} + \frac{1}{2}\right).$$

Then

$$S_4(z) = \frac{1}{2}\left(a_0 + a_4\cos(4z)\right) + \sum_{j=1}^{3}\left(a_j\cos(jz) + b_j\sin(jz)\right)$$

where

$$a_0 = \frac{1}{4}\sum_{j=0}^{7} y_j\cos(0 \cdot z_j) = 0.34710, \qquad a_1 = \frac{1}{4}\sum_{j=0}^{7} y_j\cos(1 \cdot z_j) = -0.02475$$

$$a_2 = \frac{1}{4}\sum_{j=0}^{7} y_j\cos(2 \cdot z_j) = -0.06975, \qquad a_3 = \frac{1}{4}\sum_{j=0}^{7} y_j\cos(3 \cdot z_j) = 0.08468$$

$$a_4 = \frac{1}{4}\sum_{j=0}^{7} y_j\cos(4 \cdot z_j) = -0.08773, \qquad b_1 = \frac{1}{4}\sum_{j=0}^{7} y_j\sin(1 \cdot z_j) = 0.22683$$

$$b_2 = \frac{1}{4}\sum_{j=0}^{7} y_j\sin(2 \cdot z_j) = -0.10216, \qquad b_3 = \frac{1}{4}\sum_{j=0}^{7} y_j\sin(3 \cdot z_j) = 0.04285$$

Converting back to x gives

$$S_4(x) = \frac{1}{2}\left(0.34710 - 0.08773\cos(4\pi(2x - 1))\right) - 0.02475\cos(\pi(2x - 1)) - 0.06975\cos(2\pi(2x - 1))$$

$$+ 0.08468\cos(2\pi(2x - 1)) + 0.22683\sin(\pi(2x - 1)) - 0.10216\sin(2\pi(2x - 1))$$

$$+ 0.04285\sin(2\pi(2x - 1)).$$

b. Integrating the expression for $S_4(x)$ in part (a) gives

$$\int_0^1 S_4(x)\,dx = 0.17355.$$

c. Integration by parts gives

$$\int_0^1 x^2 \sin x\,dx = 0.22324, \quad \text{so} \quad \left| \int_0^1 x^2 \sin x\,dx - \int_0^1 S_4(x)\,dx \right| = 0.04970.$$

8. Show that

$$\sum_{j=0}^{2m-1} (\cos mx_j)^2 = 2m.$$

SOLUTION: Using a basic trigonometric identity gives

$$\sum_{j=0}^{2m-1} (\cos mx_j)^2 = \sum_{j=0}^{2m-1} \frac{1 + \cos 2mx_j}{2} = \frac{1}{2}\sum_{j=0}^{2m-1} 1 + \frac{1}{2}\sum_{j=0}^{2m-1} \cos 2mx_j$$

$$= \frac{1}{2}(2m) + \frac{1}{2}\left(\sum_{j=0}^{2m-1} \cos 2m\left(\frac{j}{m}\pi - \pi \right) \right),$$

so

$$\sum_{j=0}^{2m-1} (\cos mx_j)^2 = m + \sum_{j=0}^{2m-1} [\cos 2j\pi \, \cos 2m\pi + \sin 2j\pi \, \sin 2m\pi]$$

$$= m + \frac{1}{2}\sum_{j=0}^{2m-1} [(1)(1) + (0)(0)] = m + \frac{1}{2}(2m) = 2m.$$

9. Show that $c_0, \ldots, c_{2m-1}$ in Algorithm 8.3 are given by

$$\begin{bmatrix} c_0 \\ c_1 \\ \vdots \\ c_{2m-1} \end{bmatrix} =: \begin{bmatrix} 1 & 1 & \cdots & 1 \\ 1 & \zeta & \cdots & \zeta^{2m-1} \\ \vdots & \vdots & & \vdots \\ 1 & \zeta^{2m-1} & \cdots & \zeta^{(2m-1)^2} \end{bmatrix} \begin{bmatrix} y_0 \\ y_1 \\ \vdots \\ y_{2m-1} \end{bmatrix},$$

where $\zeta = e^{\pi i/m}$.

SOLUTION: From Eq. (8.28), we have

$$c_k = \sum_{j=0}^{2m-1} y_j e^{\frac{\pi i j k}{m}} = \sum_{j=0}^{2m-1} y_j (\zeta)^{jk} = \sum_{j=0}^{2m-1} y_j (\zeta^k)^j.$$

Thus

$$c_k = \left(1, \zeta^k, \zeta^{2k}, \ldots, \zeta^{(2m-1)k}\right)^t \begin{bmatrix} y_0 \\ y_1 \\ \vdots \\ y_{2m-1} \end{bmatrix},$$

and the result follows.

Approximating Eigenvalues

Exercise Set 9.1, page 568

1. a. Find the eigenvalues and associated eigenvectors of

$$A = \begin{bmatrix} 2 & -3 & 6 \\ 0 & 3 & -4 \\ 0 & 2 & -3 \end{bmatrix}.$$

SOLUTION: Note that the Geršgorin Circle Theorem implies that the eigenvalues are contained within the circle centered at the real number 2 with radius 9, since the circle for the first row contains the circles from the other two rows.

To determine the eigenvalues, consider

$$0 = \det(A - \lambda I) = \det \begin{bmatrix} 2 - \lambda & -3 & 6 \\ 0 & 3 - \lambda & -4 \\ 0 & 2 & -3 - \lambda \end{bmatrix} = (2 - \lambda) \det \begin{bmatrix} 3 - \lambda & -4 \\ 2 & -3 - \lambda \end{bmatrix}$$

$$= (2 - \lambda)[(3 - \lambda)(-3 - \lambda) + 8] = -(\lambda - 2)(\lambda - 1)(\lambda + 1).$$

Hence, the eigenvalues are $\lambda = 2$, $\lambda = 1$, and $\lambda = -1$. To determine an eigenvector associated with $\lambda = 2$, we need $\mathbf{x} = (x_1, x_2, x_3)^t$ with

$$\begin{bmatrix} 0 \\ 0 \\ 0 \end{bmatrix} = \begin{bmatrix} 0 & -3 & 6 \\ 0 & 1 & -4 \\ 0 & 2 & -5 \end{bmatrix} \begin{bmatrix} x_1 \\ x_2 \\ x_3 \end{bmatrix} = \begin{bmatrix} -3x_2 + 6x_3 \\ x_2 - 4x_3 \\ 2x_2 - 5x_3 \end{bmatrix}.$$

A nonzero solution to this system is $\mathbf{x} = (1, 0, 0)^t$. To determine an eigenvector associated with $\lambda = 1$, we need $\mathbf{x} = (x_1, x_2, x_3)^t$ with

$$\begin{bmatrix} 0 \\ 0 \\ 0 \end{bmatrix} = \begin{bmatrix} 1 & -3 & 6 \\ 0 & 2 & -4 \\ 0 & 2 & -4 \end{bmatrix} \begin{bmatrix} x_1 \\ x_2 \\ x_3 \end{bmatrix} = \begin{bmatrix} x_1 - 3x_2 + 6x_3 \\ 2x_2 - 4x_3 \\ 2x_2 - 4x_3 \end{bmatrix}.$$

A nonzero solution to this system is $\mathbf{x} = (0, 2, 1)^t$. To determine an eigenvector associated with $\lambda = -1$, we need $\mathbf{x} = (x_1, x_2, x_3)^t$ with

$$\begin{bmatrix} 0 \\ 0 \\ 0 \end{bmatrix} = \begin{bmatrix} 3 & -3 & 6 \\ 0 & 4 & -4 \\ 0 & 2 & -2 \end{bmatrix} \begin{bmatrix} x_1 \\ x_2 \\ x_3 \end{bmatrix} = \begin{bmatrix} 3x_1 - 3x_2 + 6x_3 \\ 4x_2 - 4x_3 \\ 2x_2 - 2x_3 \end{bmatrix}.$$

179

A nonzero solution to this system is $\mathbf{x} = (-1, 1, 1)^t$.

To show linear independence, suppose that

$$(0, 0, 0)^t = \mathbf{0} = c_1\mathbf{x}_1 + c_2\mathbf{x}_2 + c_3\mathbf{x}_3 = c_1(1, 0, 0)^t + c_2(0, 2, 1)^t + c_3(-1, 1, 1)^t.$$

Then

$$0 = c_1(1) + c_2(0) + c_3(-1),$$
$$0 = c_1(0) + c_2(2) + c_3(1),$$
$$0 = c_1(0) + c_2(1) + c_3(1).$$

Subtracting the third equation from the second implies that $c_2 = 0$. Substituting this value into one of these equations produces $c_3 = 0$. The first equation now implies that $c_1 = c_3 = 0$. This ensures the linear independence of the eigenvectors.

4. b. Use the Geršgorin Circle Theorem to determine bounds for the eigenvalues and the spectral radius of the matrix

$$\begin{bmatrix} 1 & 0 & -1 & 1 \\ 2 & 2 & -1 & 1 \\ 0 & 1 & 3 & -2 \\ 1 & 0 & 1 & 4 \end{bmatrix}$$

SOLUTION: The Geršgorin circles are

$$R_1 = \{z \in \mathcal{C} \mid |z - 1| \le 2\}, \quad R_2 = \{z \in \mathcal{C} \mid |z - 2| \le 4\}, \quad R_3 = \{z \in \mathcal{C} \mid |z - 3| \le 3\},$$

and $\quad R_4 = \{z \in \mathcal{C} \mid |z - 4| \le 2\}$.

These are shown in the figure, which also shows that the circle centered at the origin of radius 6 contains all the Geršgorin circle. Hence 6 is a bound for the spectral radius.

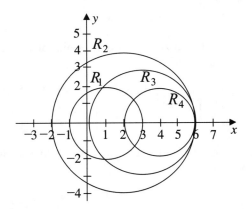

5. a. Find the factorization $A = PDP^{-1}$ for the matrix in Exercise 1(a).

SOLUTION: The matrix A has the three distinct eigenvalues $\lambda_1 = -1$, $\lambda_2 = 1$, and $\lambda_3 = 2$ with eigenvectors $\mathbf{v}_1 = (-1, 1, 1)^t$, $\mathbf{v}_2 = (0, 2, 1)^t$, and $\mathbf{v}_3 = (1, 0, 0)^t$, respectively. So $A = PDP^{-1}$ with

$$P = \begin{bmatrix} -1 & 0 & 1 \\ 1 & 2 & 0 \\ 1 & 1 & 0 \end{bmatrix}, \quad D = \begin{bmatrix} -1 & 0 & 0 \\ 0 & 1 & 0 \\ 0 & 0 & 2 \end{bmatrix}, \quad \text{and} \quad P^{-1} = \begin{bmatrix} 0 & -1 & 2 \\ 0 & 1 & -1 \\ 1 & -1 & 2 \end{bmatrix}.$$

8. Show that the three eigenvectors in Example 3 are linearly independent.

   SOLUTION: The vectors are linearly independent if and only if the matrix formed by having these vectors as columns (or rows) is nonsingular, which is true if and only if the determinant of this matrix is nonzero. Since

$$\det \begin{bmatrix} 0 & 0 & -2 \\ 1 & 2 & 0 \\ 1 & 1 & 1 \end{bmatrix} = 2,$$

   the vectors are linearly independent.

10. Show that if A is a matrix and $\lambda_1, \lambda_2, \ldots, \lambda_k$ are distinct eigenvalues with associated eigenvectors $\mathbf{x}_1, \mathbf{x}_2, \ldots, \mathbf{x}_k$, then $\{\mathbf{x}_1, \mathbf{x}_2, \ldots, \mathbf{x}_k\}$ is a linearly independent set.

   SOLUTION: There must be a largest subset $\{\mathbf{x}_1, \mathbf{x}_2, \ldots, \mathbf{x}_j\}$, with $j \leq k$ that is linearly independent because the set with the nonzero vector $\{\mathbf{x}_1\}$ is linearly independent. Suppose that we have this largest linearly independent set and that $j < k$. Then $\{\mathbf{x}_1, \mathbf{x}_2, \ldots, \mathbf{x}_j, \mathbf{x}_{j+1}\}$ is linearly dependent and there is a set of constants $\{c_1, c_2, \ldots, c_j\}$, not all zero, with

$$\mathbf{x}_{j+1} = c_1 \mathbf{x}_1 + c_2 \mathbf{x}_2 + \cdots + c_j \mathbf{x}_j.$$

   Because these are all eigenvectors, $A\mathbf{x}_i = \lambda_i \mathbf{x}_i$ for each $i = 1, 2, \ldots j + 1$, and

$$A\mathbf{x}_{j+1} = \lambda_{j+1}\mathbf{x}_{j+1} = c_1\lambda_1\mathbf{x}_1 + c_2\lambda_2\mathbf{x}_2 + \cdots + c_j\lambda_j\mathbf{x}_j.$$

   But we also have

$$\lambda_{j+1}\mathbf{x}_{j+1} = c_1\lambda_{j+1}\mathbf{x}_1 + c_2\lambda_{j+1}\mathbf{x}_2 + \cdots + c_j\lambda_{j+1}\mathbf{x}_j,$$

   and subtracting these equations gives

$$\mathbf{0} = c_1(\lambda_1 - \lambda_{j+1})\mathbf{x}_1 + c_2(\lambda_2 - \lambda_{j+1})\mathbf{x}_2 + \cdots c_j(\lambda_j - \lambda_{j+1})\mathbf{x}_j.$$

   But the set $\{\mathbf{x}_1, \mathbf{x}_2, \ldots, \mathbf{x}_j\}$ was assumed to be linearly independent, so we must have

$$0 = c_1(\lambda_1 - \lambda_{j+1}) = c_2(\lambda_2 - \lambda_{j+1}) = \cdots = c_j(\lambda_j - \lambda_{j+1}).$$

   Since eigenvalues are all distinct, this implies that $0 = c_1 = c_2 = \cdots = c_j$. This contradicts the statement that $\mathbf{x}_{j+1}$ could be written as a combination of the vectors $\mathbf{x}_1, \mathbf{x}_2, \ldots, \mathbf{x}_j$. The only assumption made was that $j < k$, so this statement must be false. As a consequence, the entire set $\{\mathbf{x}_1, \mathbf{x}_2, \ldots, \mathbf{x}_k\}$ is linearly independent.

15. Use the Geršgorin Circle Theorem to show that a strictly diagonally dominant matrix must be nonsingular.

   SOLUTION: If A is a strictly diagonally dominant matrix, then in each row, the sum of the magnitudes of the off-diagonal entries in the row are less than the magnitude of the diagonal entry in that row. By Geršgorin Circle Theorem this implies that for each row the magnitude of the center of the Geršgorin circle for that row exceeds the radius, so the circle does not contain the origin. Hence 0 cannot be in any Geršgorin circle and consequently cannot be an eigenvalue of A. This implies that A in nonsingular.

16. Prove that the set of vectors $(V)_k = \{\mathbf{v}_1, \mathbf{v}_2, \ldots, \mathbf{v}_k\}$ described in the Gram-Schmidt Theorem is orthogonal.

SOLUTION: Let $(X)_k = \{\mathbf{x}_1, \mathbf{x}_2, \ldots, \mathbf{x}_k\}$ and define the set $(V)_k = \{\mathbf{v}_1, \mathbf{v}_2, \ldots, \mathbf{v}_k\}$ in the Gram-Schmidt manner as

$$\mathbf{v}_1 = \mathbf{x}_1, \quad \text{and} \quad \mathbf{v}_k = \mathbf{x}_k - \sum_{i=1}^{k-1} \left(\frac{\mathbf{v}_i^t \mathbf{x}_k}{\mathbf{v}_i^t \mathbf{v}_i} \right) \mathbf{v}_i.$$

for each $k > 1$. We will use Mathematical Induction to show that $(V)_k$ is orthogonal for every integer k.

First note that since

$$\mathbf{v}_1 = \mathbf{x}_1, \quad \text{and} \quad \mathbf{v}_2 = \mathbf{x}_2 - \left(\frac{\mathbf{v}_1^t \mathbf{x}_2}{\mathbf{v}_1^t \mathbf{v}_1} \right) \mathbf{v}_1,$$

we have

$$\mathbf{v}_1^t \cdot \mathbf{v}_2 = \mathbf{v}_1^t \cdot \mathbf{x}_2 - \left(\frac{\mathbf{v}_1^t \mathbf{x}_2}{\mathbf{v}_1^t \mathbf{v}_1} \right) \mathbf{v}_1^t \cdot \mathbf{v}_1 = \mathbf{v}_1^t \cdot \mathbf{x}_2 - \mathbf{v}_1^t \cdot \mathbf{x}_2 = 0,$$

so $(V)_2$ is an orthogonal set.

Now assume that $(V)_j$ is orthogonal for some positive integer j, and consider $(V)_{j+1}$. Since $(V)_j$ is orthogonal the set $(V)_{j+1}$ will be orthogonal if and only if $\mathbf{v}_s^t \cdot \mathbf{v}_{j+1} = 0$ for each $s = 1, 2, \ldots j$.

For each $s = 1, 2, \ldots j$ we have

$$\mathbf{v}_s^t \cdot \mathbf{v}_{j+1}^t = \mathbf{v}_s^t \cdot \mathbf{x}_{j+1} - \sum_{i=1}^{j} \left(\frac{\mathbf{v}_i^t \mathbf{x}_{j+1}}{\mathbf{v}_i^t \mathbf{v}_i} \right) (\mathbf{v}_s^t \cdot \mathbf{v}_i)$$

$$= \mathbf{v}_s^t \cdot \mathbf{x}_{j+1} - \left(\frac{\mathbf{v}_s^t \mathbf{x}_{j+1}}{\mathbf{v}_s^t \mathbf{v}_s} \right) (\mathbf{v}_s^t \cdot \mathbf{v}_s) = \mathbf{v}_s^t \cdot \mathbf{x}_{j+1} - \mathbf{v}_s^t \cdot \mathbf{x}_{j+1}^t = 0.$$

So $(V)_j$ being orthogonal implies that $(V)_{j+1}$ is also orthogonal. Mathematical Induction implies that $(V)_j$ is true for all positive integers j.

Exercise Set 9.2, page 573

1. c. Show that the following matrices are not similar.

$$A = \begin{bmatrix} 1 & 2 & 1 \\ 0 & 1 & 2 \\ 0 & 0 & 2 \end{bmatrix} \quad \text{and} \quad B = \begin{bmatrix} 1 & 2 & 0 \\ 0 & 1 & 2 \\ 1 & 0 & 2 \end{bmatrix}$$

SOLUTION: One way to see this is to notice that $\det(A) = 2$ and $\det(B) = 6$. Similar matrices must have the same determinants.

3. b. Find A^3 if $A = PDP^{-1}$, where

$$P = \begin{bmatrix} -1 & 2 \\ 1 & 0 \end{bmatrix} \quad \text{and} \quad D = \begin{bmatrix} -2 & 0 \\ 0 & 1 \end{bmatrix}$$

SOLUTION: Because

$$A^3 = \left(PDP^{-1}\right)^3 = PD\left(P^{-1}P\right)D\left(P^{-1}P\right)DP^{-1} = PDIDIDP^{-1} = PD^3P^{-1}$$

We have

$$A^3 = \begin{bmatrix} -1 & 2 \\ 1 & 0 \end{bmatrix} \begin{bmatrix} (-2)^3 & 0 \\ 0 & 1^3 \end{bmatrix} \begin{bmatrix} 0 & 1 \\ \frac{1}{2} & \frac{1}{2} \end{bmatrix} = \begin{bmatrix} 1 & 9 \\ 0 & -8 \end{bmatrix}.$$

4. b. Determine A^4 for the matrix in Exercise 3(b).

SOLUTION: It would be easy to simply use A^3 from Exercise 3(b) and multiply by A. However, we don't explicitly have A. Alternatively,

$$A^4 = PD^4P^{-1} = \begin{bmatrix} -1 & 2 \\ 1 & 0 \end{bmatrix} \begin{bmatrix} 16 & 0 \\ 0 & 1 \end{bmatrix} \begin{bmatrix} 0 & 1 \\ \frac{1}{2} & \frac{1}{2} \end{bmatrix} = \begin{bmatrix} 1 & -15 \\ 0 & 16 \end{bmatrix}.$$

7. c. (i) Determine if the matrix

$$A = \begin{bmatrix} 2 & 0 & 1 \\ 0 & 2 & 0 \\ 1 & 0 & 2 \end{bmatrix}$$

is positive definite.

(ii) If the matrix is positive definite, construct an orthogonal matrix Q for which $Q^t AQ = D$, where D is a diagonal matrix.

SOLUTION: **(i)** The matrix A is symmetric and has eigenvalues $\lambda_1 = 1$, $\lambda_2 = 2$, and $\lambda_3 = 3$ are all positive. By Theorem 9.18, the matrix A is positive definite.

(ii) Eigenvectors corresponding to these eigenvalues are $\mathbf{x}_1 = (1, 0, -1)^t$, $\mathbf{x}_1 = (0, 1, 0)^t$, and $\mathbf{x}_3 = (1, 0, 1)^t$, respectively. Normalizing these vectors and using these for the columns of Q gives the orthogonal matrix

$$Q = \begin{bmatrix} \frac{\sqrt{2}}{2} & 0 & \frac{\sqrt{2}}{2} \\ 0 & 1 & 0 \\ -\frac{\sqrt{2}}{2} & 0 & \frac{\sqrt{2}}{2} \end{bmatrix}$$

and

$$D = Q^t AQ = \begin{bmatrix} \frac{\sqrt{2}}{2} & 0 & -\frac{\sqrt{2}}{2} \\ 0 & 1 & 0 \\ \frac{\sqrt{2}}{2} & 0 & \frac{\sqrt{2}}{2} \end{bmatrix} \begin{bmatrix} 2 & 0 & 1 \\ 0 & 2 & 0 \\ 1 & 0 & 2 \end{bmatrix} \begin{bmatrix} \frac{\sqrt{2}}{2} & 0 & \frac{\sqrt{2}}{2} \\ 0 & 1 & 0 \\ -\frac{\sqrt{2}}{2} & 0 & \frac{\sqrt{2}}{2} \end{bmatrix} = \begin{bmatrix} 1 & 0 & 0 \\ 0 & 2 & 0 \\ 0 & 0 & 3 \end{bmatrix}$$

9. c. Show that the matrix

$$A = \begin{bmatrix} 2 & 1 & -1 \\ 0 & 2 & 1 \\ 0 & 0 & 3 \end{bmatrix}$$

is nonsingular but not diagonalizable.

SOLUTION: Because A is an upper triangular matrix, the eigenvalues of A are the diagonal entries, that is, $\lambda_1 = \lambda_2 = 2$ and $\lambda_3 = 3$. The multiple eigenvalue at 2 has only one linearly independent eigenvector, so by Theorem 9.13 on page 571, A is not similar to a diagonal matrix.

10. a. Show that the matrix

$$A = \begin{bmatrix} 2 & -1 & 0 \\ -1 & 2 & 0 \\ 0 & 0 & 0 \end{bmatrix}$$

is singular but diagonalizable.

SOLUTION: The matrix is clearly singular because it has a row (and a column) all of whose entries are 0. However, the eigenvalues of A, which are $\lambda_1 = 0$, $\lambda_2 = 1$, and $\lambda_3 = 3$ are distinct so the corresponding eigenvectors, $\mathbf{x}_1 = (0, 0, 1)^t$, $\mathbf{x}_1 = (1, 1, 0)^t$, and $\mathbf{x}_3 = (1, -1, 0)^t$, can be used to form the columns of the matrix P with $A = PDP^{-1}$. The matrix D is the diagonal matrix with diagonal entries in the order 0, 1, and 3.

16. Prove Theorem 9.10.

SOLUTION: **(i)** Let the columns of Q be denoted by the vectors $\mathbf{q}_1, \mathbf{q}_2, \ldots, \mathbf{q}_n$, which are also the rows of Q^t. Because Q is orthogonal, $(\mathbf{q}_i)^t \cdot \mathbf{q}_j$ is zero when $i \neq j$ and 1 when $i = j$. But the ij-entry of Q^tQ is $(\mathbf{q}_i)^t \cdot \mathbf{q}_j$ for each i and j so $Q^tQ = I$. Hence $Q^t = Q^{-1}$.

(ii) From part (i) we have $Q^tQ = I$, so

$$(Q\mathbf{x})^t(Q\mathbf{y}) = (\mathbf{x}^tQ^t)(Q\mathbf{y}) = \mathbf{x}^t(Q^tQ)\mathbf{y} = \mathbf{x}^t(I)\mathbf{y} = \mathbf{x}^t\mathbf{y}.$$

(iii) This follows from part (ii) with $\mathbf{x}$ replacing $\mathbf{y}$, since then

$$\|Q\mathbf{x}\|_2^2 = (Q\mathbf{x})^t(Q\mathbf{x}) = \mathbf{x}^t\mathbf{x} = \|\mathbf{x}\|_2^2.$$

18. Prove that if Q is nonsingular matrix with $Q^t = Q^{-1}$, then Q is orthogonal.

Let the columns of Q be denoted by the vectors $\mathbf{q}_1, \mathbf{q}_2, \ldots, \mathbf{q}_n$, which are the rows of $Q^t = Q^{-1}$. Then $Q^tQ = I$ is equivalent to having $(\mathbf{q}_i)^t \cdot \mathbf{q}_j = 0$ when $i \neq j$ and $(\mathbf{q}_i)^t \cdot \mathbf{q}_i = 1$, for each $i = 1, 2, \ldots, n$ and $j = 1, 2, \ldots, n$. But this is precisely what is required for the columns of Q to form an orthonormal set. Hence Q is an orthogonal matrix.

Exercise Set 9.3, page 590

1. a. Perform the first three iterations of the Power method on

$$\begin{bmatrix} 2 & 1 & 1 \\ 1 & 2 & 1 \\ 1 & 1 & 2 \end{bmatrix},$$

using $\mathbf{x}^{(0)} = (1, -1, 2)^t$.

SOLUTION: The maximum entry of $\mathbf{x}^{(0)}$ is in the third position, so $p = 3$. The unit vector in the same direction of $\mathbf{x}^{(0)}$ is $(1/2, -1/2, 1)^t$ and

$$\mathbf{y} = \begin{bmatrix} 2 & 1 & 1 \\ 1 & 2 & 1 \\ 1 & 1 & 2 \end{bmatrix} \begin{bmatrix} \frac{1}{2} \\ -\frac{1}{2} \\ 1 \end{bmatrix} = \begin{bmatrix} 1 - \frac{1}{2} + 1 \\ \frac{1}{2} - 1 + 1 \\ \frac{1}{2} - \frac{1}{2} + 2 \end{bmatrix} = \begin{bmatrix} \frac{3}{2} \\ \frac{1}{2} \\ 2 \end{bmatrix}.$$

The first approximation to the eigenvalue is the $p = 3$ entry of $\mathbf{y}$; that is, $\mu^{(1)} = 2$. The first approximation to the eigenvector is

$$\mathbf{x}^{(1)} = \frac{\mathbf{y}}{\|\mathbf{y}\|_\infty} = \left(\frac{3}{4}, \frac{1}{4}, 1\right)^t.$$

Continuing to the second iteration, we have

$$\mathbf{y} = \begin{bmatrix} 2 & 1 & 1 \\ 1 & 2 & 1 \\ 1 & 1 & 2 \end{bmatrix} \begin{bmatrix} \frac{3}{4} \\ \frac{1}{4} \\ 1 \end{bmatrix} = \begin{bmatrix} \frac{11}{4} \\ \frac{9}{4} \\ 3 \end{bmatrix},$$

so the second approximation to the eigenvalue is $\mu^{(2)} = 3$, and the approximation to the eigenvector is

$$\mathbf{x}^{(2)} = \frac{\mathbf{y}}{\|\mathbf{y}\|_\infty} = \left(\frac{11}{12}, \frac{3}{4}, 1\right)^t.$$

Finally, we have

$$\mathbf{y} = \begin{bmatrix} 2 & 1 & 1 \\ 1 & 2 & 1 \\ 1 & 1 & 2 \end{bmatrix} \begin{bmatrix} \frac{11}{12} \\ \frac{3}{4} \\ 1 \end{bmatrix} = \begin{bmatrix} \frac{43}{12} \\ \frac{41}{12} \\ \frac{44}{12} \end{bmatrix},$$

which gives the approximate eigenvalue $\mu^{(3)} = 11/3$ with approximate unit eigenvector

$$\mathbf{x}^{(1)} = \frac{\mathbf{y}}{\|\mathbf{y}\|_\infty} = \left(\frac{43}{44}, \frac{41}{44}, 1\right)^t.$$

3. **a.** Perform the first three iterations of the Inverse Power method on the matrix in part (a) of Exercise 1.

   SOLUTION: We first find

$$q = \frac{\begin{bmatrix} 1 & -1 & 2 \end{bmatrix} \begin{bmatrix} 2 & 1 & 1 \\ 1 & 2 & 1 \\ 1 & 1 & 2 \end{bmatrix} \begin{bmatrix} 1 \\ -1 \\ 2 \end{bmatrix}}{\begin{bmatrix} 1 & -1 & 2 \end{bmatrix} \begin{bmatrix} 1 \\ -1 \\ 2 \end{bmatrix}} = \frac{\begin{bmatrix} 1 & -1 & 2 \end{bmatrix} \begin{bmatrix} 3 \\ 1 \\ 4 \end{bmatrix}}{6} = \frac{5}{3}.$$

Then we normalize $\mathbf{x}^{(0)}$ to $(1/2, -1/2, 1)^t$ and solve the linear system

$$\left(A - \frac{5}{3}I\right)\mathbf{y} = \begin{bmatrix} \frac{1}{3} & 1 & 1 \\ 1 & \frac{1}{3} & 1 \\ 1 & 1 & \frac{1}{3} \end{bmatrix} \begin{bmatrix} y_1 \\ y_2 \\ y_3 \end{bmatrix} = \begin{bmatrix} \frac{1}{2} \\ -\frac{1}{2} \\ 1 \end{bmatrix}.$$

This produces $\mathbf{y} = (-3/28, 39/28, -6/7)^t$, and the first approximation to the eigenvalue is

$$\mu^{(1)} = \frac{1}{-6/7} + \frac{5}{3} = \frac{1}{2},$$

with normalized eigenvector

$$\mathbf{x}^{(1)} = (-1/13, 1, -8/13)^t.$$

Proceeding in the same manner for the next iteration produces an approximate eigenvalue $\mu^{(2)} = 0.89873$, with approximate eigenvector $\mathbf{x}^{(2)} = (-0.24051, 1, -0.86076)^t$.

A third and final iteration produces $\mu^{(3)} = 1.02773$, with approximate eigenvector $\mathbf{x}^{(3)} = (-0.18891, 1, -0.78336)^t$.

5. a. Perform the first three iterations of the Symmetric Power method on the matrix in part (a) of Exercise 1.

SOLUTION: The Symmetric Power uses the two norm instead of the infinity norm, so we apply the method to the initial approximation

$$\frac{(1, -1, 2)^t}{\sqrt{1^2 + (-1)^2 + 2^2}} = \frac{1}{\sqrt{6}}(1, -1, 2)^t, \quad \text{which gives} \quad \mathbf{y} = \frac{1}{\sqrt{6}}\begin{bmatrix} 2 & 1 & 1 \\ 1 & 2 & 1 \\ 1 & 1 & 2 \end{bmatrix}\begin{bmatrix} 1 \\ -1 \\ 2 \end{bmatrix} = \frac{1}{\sqrt{6}}\begin{bmatrix} 3 \\ 1 \\ 4 \end{bmatrix}.$$

The first eigenvalue approximation is

$$\mu^{(1)} = \frac{1}{(\sqrt{6})^2}\begin{bmatrix} 1 & -1 & 2 \end{bmatrix}\begin{bmatrix} 3 \\ 1 \\ 4 \end{bmatrix} = \frac{5}{3},$$

with approximate l_2 normalized eigenvector

$$\mathbf{x}^{(1)} = \frac{1}{\sqrt{26}}\begin{bmatrix} 3 \\ 1 \\ 4 \end{bmatrix}.$$

Applying the method again gives

$$\mathbf{y} = \frac{1}{\sqrt{26}}\begin{bmatrix} 11 \\ 9 \\ 12 \end{bmatrix}, \quad \mathbf{x}^{(2)} = \frac{1}{\sqrt{346}}\begin{bmatrix} 11 \\ 9 \\ 12 \end{bmatrix}, \quad \text{and} \quad \mu^{(2)} = \frac{45}{13}.$$

The final iteration produces

$$\mathbf{y} = \frac{1}{\sqrt{346}}\begin{bmatrix} 43 \\ 41 \\ 44 \end{bmatrix}, \quad \mathbf{x}^{(3)} = \frac{1}{\sqrt{5466}}\begin{bmatrix} 43 \\ 41 \\ 44 \end{bmatrix}, \quad \text{and} \quad \mu^{(3)} = \frac{1370}{346}.$$

20. Show that the ith row of $B = A - \lambda_1\mathbf{v}^{(1)}\mathbf{x}^t$ is zero, where λ_1 is the largest eigenvalue of A in absolute value, $\mathbf{v}^{(1)}$ is the associated eigenvector of A, and $\mathbf{x}$ is the vector defined in Eq. (9.7).

SOLUTION: Since

$$\mathbf{x}^t = \frac{1}{\lambda_1 v_i^{(1)}}(a_{i1}, a_{i2}, \ldots, a_{in}),$$

the ith row of B is

$$\left(a_{i1}, a_{i2}, \ldots, a_{in}\right) - \frac{\lambda_1}{\lambda_1 v_i^{(1)}} \left(v_i^{(1)} a_{i1}, v_i^{(1)} a_{i2}, \ldots, v_i^{(1)} a_{in}\right) = \mathbf{0}.$$

Exercise Set 9.4, page 600

1. a. Use Householder's method to place in tridiagonal form the matrix

$$A = \begin{bmatrix} 12 & 10 & 4 \\ 10 & 8 & -5 \\ 4 & -5 & 3 \end{bmatrix}.$$

SOLUTION: Since A is a 3×3 matrix, only one Householder transformation, $P^{(1)}$, is required to produce the tridiagonal matrix

$$A^{(2)} = P^{(1)} A P^{(1)}.$$

To construct $P^{(1)}$, first compute

$$q = \sum_{j=2}^{3} a_{j1}^2 = a_{21}^2 + a_{31}^2 = 116,$$

$$\alpha = -\sqrt{q}\, \frac{a_{21}}{|a_{21}|} = -\sqrt{116} = -10.77033,$$

$$r = \left(\frac{1}{2}\alpha^2 - \frac{1}{2}a_{21}\alpha\right)^{1/2} = 10.57599,$$

and

$$\mathbf{w} = \begin{bmatrix} w_1 \\ w_2 \\ w_3 \end{bmatrix} = \begin{bmatrix} 0 \\ \dfrac{a_{21} - \alpha}{2r} \\ \dfrac{a_{31}}{2r} \end{bmatrix} = \begin{bmatrix} 0 \\ 0.9819564 \\ 0.1891075 \end{bmatrix}.$$

This permits us to determine

$$P^{(1)} = I - 2\mathbf{w}\mathbf{w}^t = \begin{bmatrix} 1 & 0 & 0 \\ 0 & 1 & 0 \\ 0 & 0 & 1 \end{bmatrix} - 2 \begin{bmatrix} 0 \\ 0.9819564 \\ 0.1891075 \end{bmatrix} \begin{bmatrix} 0 & 0.9819564 & 0.1891075 \end{bmatrix}$$

$$= \begin{bmatrix} 1 & 0 & 0 \\ 0 & -0.9284768 & -0.3713907 \\ 0 & -0.3713907 & 0.9284767 \end{bmatrix}$$

and the tridiagonal matrix

$$A^{(2)} = P^{(1)} A P^{(1)} = \begin{bmatrix} 12 & -10.77033 & 0 \\ -10.77033 & 3.862069 & 5.344828 \\ 0 & 5.344828 & 7.137931 \end{bmatrix}.$$

Exercise Set 9.5, page 611

1. a. Apply two iterations of the QR method without shifting to

$$\begin{bmatrix} 2 & -1 & 0 \\ -1 & 2 & -1 \\ 0 & -1 & 2 \end{bmatrix} = \begin{bmatrix} a_1^{(1)} & b_2^{(1)} & 0 \\ b_2^{(1)} & a_2^{(1)} & b_3^{(1)} \\ 0 & b_3^{(1)} & a_3^{(1)} \end{bmatrix}.$$

SOLUTION: We have

$$x_1 = a_1 = 2, \quad y_1 = b_2 = -1, \quad \text{and} \quad z_1 = \sqrt{x_1^2 + b_2^2} = \sqrt{5},$$

so

$$c_2 = x_1/z_1 = 2\sqrt{5}/5, \quad \text{and} \quad s_2 = b_2/z_1 = -\sqrt{5}/5.$$

The rotation matrix is

$$P_2 = \begin{bmatrix} \frac{2\sqrt{5}}{5} & -\frac{\sqrt{5}}{5} & 0 \\ \frac{\sqrt{5}}{5} & \frac{2\sqrt{5}}{5} & 0 \\ 0 & 0 & 1 \end{bmatrix}.$$

and

$$A_2^{(1)} = P_2 A^{(1)} = \begin{bmatrix} \frac{2\sqrt{5}}{5} & -\frac{\sqrt{5}}{5} & 0 \\ \frac{\sqrt{5}}{5} & \frac{2\sqrt{5}}{5} & 0 \\ 0 & 0 & 1 \end{bmatrix} \begin{bmatrix} 2 & -1 & 0 \\ -1 & 2 & -1 \\ 0 & -1 & 2 \end{bmatrix} = \begin{bmatrix} \sqrt{5} & -\frac{4\sqrt{5}}{5} & \frac{\sqrt{5}}{5} \\ 0 & \frac{3\sqrt{5}}{5} & -\frac{2\sqrt{5}}{5} \\ 0 & -1 & 2 \end{bmatrix} = \begin{bmatrix} z_1 & q_1 & r_1 \\ 0 & x_2 & y_2 \\ 0 & b_3 & a_3 \end{bmatrix}.$$

Continuing for the second iteration we have

$$z_2 = \sqrt{x_2^2 + b_3^2} = \sqrt{\left(\frac{3\sqrt{5}}{5}\right)^2 + (-1)^2} = \frac{\sqrt{70}}{5}, \quad c_3 = \frac{x_2}{z_2} = \frac{3\sqrt{14}}{14}, \quad \text{and} \quad s_3 = \frac{b_3}{z_2} = -\frac{\sqrt{70}}{14}.$$

So

$$P_3 = \begin{bmatrix} 1 & 0 & 0 \\ 0 & c_3 & s_3 \\ 0 & -s_3 & c_3 \end{bmatrix} = \begin{bmatrix} 1 & 0 & 0 \\ 0 & \frac{2\sqrt{14}}{14} & \frac{\sqrt{70}}{14} \\ 0 & -\frac{\sqrt{70}}{14} & \frac{2\sqrt{14}}{14} \end{bmatrix} \approx \begin{bmatrix} 1 & 0 & 0 \\ 0 & 0.801784 & -0.597614 \\ 0 & 0.597614 & 0.801784 \end{bmatrix}.$$

Then

$$R^{(1)} = A_3^{(1)} = P_3 A_2^{(1)} = \begin{bmatrix} 1 & 0 & 0 \\ 0 & 0.801784 & -0.597614 \\ 0 & 0.597614 & 0.801784 \end{bmatrix} \begin{bmatrix} 2.236068 & -1.788854 & 0.447214 \\ 0 & 1.34164 & -0.894427 \\ 0 & -1 & 2 \end{bmatrix}$$

$$= \begin{bmatrix} 2.236068 & -1.788854 & 0.447214 \\ 0 & 1.673320 & -1.912366 \\ 0 & 0 & 1.069045 \end{bmatrix}.$$

The orthogonal matrix is

$$Q^{(1)} = P_2^t P_3^t = (P_3 P_2)^t = \left(\begin{bmatrix} 1 & 0 & 0 \\ 0 & 0.801784 & -0.597614 \\ 0 & 0.597614 & 0.801784 \end{bmatrix} \begin{bmatrix} 0.894427 & -0.447214 & 0 \\ 0.447214 & 0.894427 & 0 \\ 0 & 0 & 1 \end{bmatrix} \right)^t$$

$$= \begin{bmatrix} 0.894427 & 0.358569 & 0.267261 \\ -0.447214 & 0.717137 & 0.534522 \\ 0 & -0.597614 & 0.801784 \end{bmatrix} .$$

After the first iteration we have

$$A^{(2)} = R^{(1)} Q^{(1)} = \begin{bmatrix} 2.800000 & -0.748331 & 0 \\ -0.748331 & 2.342857 & -0.638877 \\ 0 & -0.638877 & 0.857143 \end{bmatrix} .$$

Repeating these steps for the second iteration gives

$$A^{(3)} = \begin{bmatrix} 3.142857 & -0.559397 & 0 \\ -0.559397 & 2.248447 & -0.187848 \\ 0 & -0.187848 & 0.608696 \end{bmatrix} .$$

3. **a.** Use the QR Algorithm with shifting to determine, to within 10^{-5}, all the eigenvalues for the matrix given in Exercise 1(a).

SOLUTION: When applying the QR Algorithm, we shift so that the iterations converge faster. So the first two iterations in this exercise will not be the same as in Exercise 1(a) which did not involve shifting.

To determine the acceleration parameter, we need the eigenvalues of

$$\begin{bmatrix} a_2^{(1)} & b_3^{(1)} \\ b_3^{(1)} & a_3^{(1)} \end{bmatrix} = \begin{bmatrix} 2 & -1 \\ -1 & 2 \end{bmatrix},$$

which are $\mu_1 = 3$ and $\mu_2 = 1$. We choose $\mu_1 = 3$ as the closest eigenvalue to $a_3^{(1)} = 2$, so we have *SHIFT* = 3. (Although we could also have chosen $\mu_2 = 1$.) This gives

$$\begin{bmatrix} d_1 & b_2^{(1)} & 0 \\ b_2^{(1)} & d_2 & b_3^{(1)} \\ 0 & b_3^{(1)} & d_3 \end{bmatrix} = \begin{bmatrix} -1 & -1 & 0 \\ -1 & -1 & -1 \\ 0 & -1 & -1 \end{bmatrix},$$

so $x_1 = -1$, $y_1 = -1$, $z_1 = \sqrt{2}$, $c_2 = -\sqrt{2}/2$, $s_2 = -\sqrt{2}/2$, $q_1 = \sqrt{2}$, $x_2 = 0$, $r_1 = \sqrt{2}/2$, and $y_2 = \sqrt{2}/2$. Thus,

$$P_2 = \begin{bmatrix} -\frac{\sqrt{2}}{2} & -\frac{\sqrt{2}}{2} & 0 \\ \frac{\sqrt{2}}{2} & -\frac{\sqrt{2}}{2} & 0 \\ 0 & 0 & 1 \end{bmatrix}$$

and

$$P_2 \begin{bmatrix} -1 & -1 & 0 \\ -1 & -1 & -1 \\ 0 & -1 & -1 \end{bmatrix} = \begin{bmatrix} \sqrt{2} & \sqrt{2} & \frac{\sqrt{2}}{2} \\ 0 & 0 & \frac{\sqrt{2}}{2} \\ 0 & -1 & -1 \end{bmatrix}.$$

Further, $z_2 = 1$, $c_3 = 0$, $s_3 = -1$, $q_2 = 1$, and $x_3 = \sqrt{2}/2$, so

$$P_3 = \begin{bmatrix} 1 & 0 & 0 \\ 0 & 0 & -1 \\ 0 & 1 & 0 \end{bmatrix}$$

and

$$R^{(1)} = P_3 P_2 \begin{bmatrix} -1 & -1 & 0 \\ -1 & -1 & -1 \\ 0 & -1 & -1 \end{bmatrix} = \begin{bmatrix} \sqrt{2} & \sqrt{2} & \frac{\sqrt{2}}{2} \\ 0 & 1 & 1 \\ 0 & 0 & \frac{\sqrt{2}}{2} \end{bmatrix}.$$

Since

$$Q^{(1)} = P_2^t P_3^t = \begin{bmatrix} -\frac{\sqrt{2}}{2} & 0 & \frac{\sqrt{2}}{2} \\ -\frac{\sqrt{2}}{2} & 0 & -\frac{\sqrt{2}}{2} \\ 0 & -1 & 0 \end{bmatrix},$$

we have

$$A^{(1)} = R^{(1)} Q^{(1)} = \begin{bmatrix} -2 & -\frac{\sqrt{2}}{2} & 0 \\ -\frac{\sqrt{2}}{2} & -1 & -\frac{\sqrt{2}}{2} \\ 0 & -\frac{\sqrt{2}}{2} & 0 \end{bmatrix}.$$

For the next iteration, we find the eigenvalues of

$$\begin{bmatrix} -1 & -\frac{\sqrt{2}}{2} \\ -\frac{\sqrt{2}}{2} & 0 \end{bmatrix},$$

which are $\mu_1 = 0.3660254$ and $\mu_2 = -1.366025$. The closest to 0 is μ_1, so the shift is modified to $SHIFT = 3 + 0.366025 = 3.366025$. Completing the steps gives

$$A^{(2)} = \begin{bmatrix} -2.672028 & -0.375975 & 0 \\ -0.375975 & -1.473608 & -0.030397 \\ 0 & -0.030397 & 0.047560 \end{bmatrix}.$$

Repeating the process gives the accumulated shift is now 3.414192 and

$$A^{(3)} = \begin{bmatrix} -2.799713 & -0.199386 & 0 \\ -0.199386 & -1.442885 & -4.5 \times 10^{-6} \\ 0 & -4.5 \times 10^{-6} & 0.000021 \end{bmatrix}.$$

The lower off diagonal elements are less than 10^{-5}, so an approximate eigenvalue is

$$\lambda_1 = 3.414192 + 0.000021 = 3.414213.$$

The remaining two eigenvalues $\lambda_2 = 0.585786$ and $\lambda_2 = 2.000000$ are obtained by finding the eigenvalues of the submatrix of $A^{(3)}$:

$$\begin{bmatrix} -2.799713 & -0.199386 \\ -0.199386 & -1.442885 \end{bmatrix}$$

and then adding the shift.

7. a. Show that the rotation matrix

$$\begin{bmatrix} \cos\theta & -\sin\theta \\ \sin\theta & \cos\theta \end{bmatrix}$$

applied to the vector $\mathbf{x} = (x_1, x_2)^t$ has the geometric effect of rotating $\mathbf{x}$ through the angle θ without changing its magnitude with respect to the l_2 norm.

b. Show that the magnitude of $\mathbf{x}$ with respect to the l_∞ norm can be changed by a rotation matrix.

SOLUTION: **a.** First note that for any vector $\mathbf{x} = (x_1, x_2)^t$ we have

$$\begin{bmatrix} \cos\theta & -\sin\theta \\ \sin\theta & \cos\theta \end{bmatrix}\begin{bmatrix} x_1 \\ x_2 \end{bmatrix} = \begin{bmatrix} x_1\cos\theta - x_2\sin\theta \\ x_1\sin\theta + x_2\cos\theta \end{bmatrix},$$

and that

$$(x_1\cos\theta - x_2\sin\theta)^2 + (x_1\sin\theta + x_2\cos\theta)^2 = x_1^2((\cos\theta)^2 + (\sin\theta)^2) + x_2^2((-\sin\theta)^2 + (\cos\theta)^2)$$
$$= x_1^2 + x_2^2.$$

So the l_2 norms are the same.

Now let $\mathbf{z} = (z_1, z_2)^t$ represent the vector that has the same magnitude as $\mathbf{x}$ but has been rotated by the angle θ. Let ϕ be the angle from the x-axis to the point (x_1, x_2) and let $r = \sqrt{x_1^2 + x_2^2} = \|\mathbf{x}\|_2$. Then

$$x_1 = r\cos\phi \quad \text{and} \quad x_2 = r\sin\phi.$$

In a similar manner,

$$z_1 = r\cos(\phi + \theta) = r\cos\phi\cos\theta - r\sin\phi\sin\theta = x_1\cos\theta - x_2\sin\theta$$
$$z_2 = r\sin(\phi + \theta) = r\cos\phi\sin\theta + r\sin\phi\cos\theta = x_1\sin\theta + x_2\cos\theta$$

So the unique vector $\mathbf{z}$ that has the same l_2 norm as $\mathbf{x}$ and is rotated by an angle of θ is given by multiplying $\mathbf{x}$ by the rotation matrix.

b. Consider the the vector $\mathbf{x} = (1, 1)^t$, which has l_∞ norm 1. If $\theta = \pi/4$, then

$$\begin{bmatrix} \cos\theta & -\sin\theta \\ \sin\theta & \cos\theta \end{bmatrix}\begin{bmatrix} 1 \\ 1 \end{bmatrix} = \begin{bmatrix} \frac{\sqrt{2}}{2} & -\frac{\sqrt{2}}{2} \\ \frac{\sqrt{2}}{2} & \frac{\sqrt{2}}{2} \end{bmatrix}\begin{bmatrix} 1 \\ 1 \end{bmatrix} = \begin{bmatrix} 0 \\ \sqrt{2} \end{bmatrix},$$

which has l_∞ norm $\sqrt{2}$.

8. Let P be a rotation matrix with $p_{ii} = p_{jj} = \cos\theta$ and $p_{ij} = -p_{ji} = \sin\theta$ for $j < i$. Show that for any $n \times n$ matrix A,

$$(AP)_{pq} = \begin{cases} a_{pq}, & \text{if } q \neq i,j \\ a_{pj}\cos\theta + a_{pi}\sin\theta, & \text{if } q = j \\ a_{pi}\cos\theta - a_{pj}\sin\theta, & \text{if } q = i; \end{cases}$$

$$(PA)_{pq} = \begin{cases} a_{pq}, & \text{if } p \neq i,j \\ a_{jq}\cos\theta - a_{iq}\sin\theta, & \text{if } p = j \\ a_{iq}\cos\theta + a_{jq}\sin\theta, & \text{if } p = i. \end{cases}$$

SOLUTION: Let $P = (p_{ij})$ be a rotation matrix with nonzero entries $p_{jj} = p_{ii} = \cos\theta$, $p_{ij} = -p_{ji} = \sin\theta$, and $p_{kk} = 1$, if $k \neq i$ and $k \neq j$. For any $n \times n$ matrix A,

$$(AP)_{rs} = \sum_{k=1}^{n} a_{rk}p_{ks}.$$

If $s \neq i,j$, then $p_{ks} = 0$ unless $k = s$. Thus, $(AP)_{rs} = a_{rs}$.
If $s = j$, then

$$(AP)_{rj} = a_{rj}p_{jj} + a_{ri}p_{ij} = a_{rj}\cos\theta + a_{ri}\sin\theta.$$

If $s = i$, then

$$(AP)_{ri} = a_{rj}p_{ji} + a_{ri}p_{ii} = -a_{rj}\sin\theta + a_{ri}\cos\theta.$$

Similarly, $(PA)_{rs} = \sum_{k=1}^{n} p_{rk}a_{ks}$. If $r \neq i,j$, then $p_{rk} = 0$ unless $r = k$. Thus, $(PA)_{rs} = a_{rs}$.
If $r = i$, then

$$(PA)_{is} = p_{ij}a_{js} + p_{ii}a_{is} = a_{js}\sin\theta + a_{is}\cos\theta.$$

If $r = j$, then

$$(PA)_{js} = p_{jj}a_{js} + p_{ji}a_{is} = a_{js}\cos\theta - a_{is}\sin\theta.$$

11. Jacobi's method for a symmetric matrix is described by

$$A_1 = A, \quad \text{and, for } i \geq 1, \quad A_{i+1} = P_i A_i P_i^t.$$

Develop an algorithm to implement this method.

SOLUTION:
INPUT: dimension n, matrix $A = (a_{ij})$, tolerance *TOL*, maximum number of iterations N.
OUTPUT: eigenvalues $\lambda_1, \ldots, \lambda_n$ of A or a message that the number of iterations was exceeded.
Step 1   Set *FLAG* $= 1$; $k1 = 1$.
          Step 2   While (*FLAG* $= 1$) do Steps 3 – 10
                   Step 3   For $i = 2, \ldots, n$ do Steps 4 – 8.

Step 4   For $j = 1, ..., i-1$ do Steps 5 – 8.
    Step 5   If $a_{ii} = a_{jj}$ then set

$$CO = 0.5\sqrt{2}; \quad (CO \equiv cosine)$$
$$SI = CO \qquad (SI \equiv sine)$$

else set

$$b = |a_{ii} - a_{jj}|;$$
$$c = 2a_{ij}\,\text{sign}(a_{ii} - a_{jj});$$
$$CO = 0.5\left(1 + b/\left(c^2 + b^2\right)^{1/2}\right)^{\frac{1}{2}};$$
$$SI = 0.5c/\left(CO\left(c^2 + b^2\right)^{1/2}\right).$$

    Step 6   For $k = 1, ..., n$
        if $(k \neq i)$ and $(k \neq j)$ then
            set $x = a_{k,j}$;

$$y = a_{k,i};$$
$$a_{k,j} = CO \cdot x + SI \cdot y;$$
$$a_{k,i} = CO \cdot y + SI \cdot x;$$
$$x = a_{j,k};$$
$$y = a_{i,k};$$
$$a_{j,k} = CO \cdot x + SI \cdot y;$$
$$a_{i,k} = CO \cdot y - SI \cdot x.$$

    Step 7   Set $x = a_{j,j}$;

$$y = a_{i,i};$$
$$a_{j,j} = CO \cdot CO \cdot x + 2 \cdot SI \cdot CO \cdot a_{j,i} + SI \cdot SI \cdot y;$$
$$a_{i,i} = SI \cdot SI \cdot x - 2 \cdot SI \cdot CO \cdot a_{i,j} + CO \cdot CO \cdot y.$$

    Step 8   Set $a_{i,j} = 0$; $a_{j,i} = 0$.

Step 9   Set

$$s = \sum_{i=1}^{n}\sum_{\substack{j=1 \\ j \neq i}}^{n} |a_{ij}|.$$

Step 10   If $s < TOL$ then for $i = 1, ..., n$ set
                    $\lambda_i = a_{ii}$;
                    OUTPUT $(\lambda_1, ..., \lambda_n)$;
                    set $FLAG = 0$.
        else set $k1 = k1 + 1$;
        if $k1 > N$ then set $FLAG = 0$.

Step 11   If $k1 > N$ then
    OUTPUT ('Maximum number of iterations exceeded');
    STOP.

Exercise Set 9.6, page 611

1. b. Determine the singular values of

$$A = \begin{bmatrix} 2 & 1 \\ 1 & 1 \\ 0 & 1 \end{bmatrix}$$

SOLUTION: The eigenvalues of

$$A^t A = \begin{bmatrix} 2 & 1 & 0 \\ 1 & 1 & 1 \end{bmatrix}\begin{bmatrix} 2 & 1 \\ 1 & 1 \\ 0 & 1 \end{bmatrix} = \begin{bmatrix} 5 & 3 \\ 3 & 3 \end{bmatrix}$$

are $4 + \sqrt{10}$ and $4 - \sqrt{10}$. So the singular values of A are $s_1 = \sqrt{4 + \sqrt{10}}$ and $s_2 = \sqrt{4 - \sqrt{10}}$.

3. b. Determine a Singular Value Decomposition for the matrix in Exercise 1(b).

SOLUTION: The singular values are $s_1 = \sqrt{4 + \sqrt{10}}$ and $s_2 = \sqrt{4 - \sqrt{10}}$, so

$$S = \begin{bmatrix} \sqrt{4 + \sqrt{10}} & 0 \\ 0 & \sqrt{4 - \sqrt{10}} \\ 0 & 0 \end{bmatrix}.$$

Eigenvectors of $A^t A$ are $(-(1 + \sqrt{(10)})/3, -1)^t$ and $((\sqrt{(10)} - 1)/3, -1)^t$. Normalizing these and interchanging the rows and columns gives

$$V^t \approx \begin{bmatrix} -0.811242 & -0.584710 \\ -0.584710 & -0.8112420 \end{bmatrix}.$$

The eigenvalues of AA^t are

$$0, \quad 4 + \sqrt{10}, \quad \text{and} \quad 4 - \sqrt{10}$$

with eigenvectors, respectively,

$$(1, -2, 1)^t, \quad \left(\frac{1}{3}(5 + 2\sqrt{10}), \frac{1}{3}(4 + \sqrt{10}), 1\right)^t, \quad \text{and} \quad \left(\frac{1}{3}(5 - 2\sqrt{10}), \frac{1}{3}(4 - \sqrt{10}), 1\right)^t$$

Normalizing these and using them for the columns of U gives

$$U = \begin{bmatrix} -0.824736 & 0.391336 & 0.408248 \\ -0.521609 & -0.247502 & -0.816497 \\ -0.218482 & -0.886340 & 0.408248 \end{bmatrix}.$$

A Singular Value Decomposition of A is $U S V^t$.

6. Suppose that A is an $m \times n$ matrix A. Show that Rank(A) is the same as the Rank(A^t).

SOLUTION: The rank of A is the number of linearly independent rows in A, and the rank of A^t is the number of linearly independent rows of A^t, which corresponds to the number of linearly independent columns of A. By Theorem 9.25 the number of linearly independent rows of a matrix is the same as the number of independent columns, so the rank of A is the same as the rank of A^t.

7. Show that Nullity(A) = Nullity(A^t) if and only if A is a square matrix.

SOLUTION: If A is an $n \times m$ matrix, then A^t is $m \times n$. The nullity of A is $n - $ Rank(A) and the nullity of A^t is $m - $ Rank(A^t). By Exercise 6, Rank(A) = Rank(A^t). So the nullity of A and A^t agree precisely when

$$n - \text{Rank}(A) = m - \text{Rank}(A^t) = m - \text{Rank}(A) \quad \text{which holds if and only if } n = m.$$

8. Suppose that A has the Singular Value Decomposition $A = U S V^t$. Determine, with justification, a Singular Value Decomposition of A^t.

SOLUTION: The matrices S and S^t have nonzero values only on their diagonals, and the nonzero eigenvalues of $A^t A$ and AA^t are the same, so the singular values of A^t are on the diagonal of S^t in decreasing order. In addition, the matrices U and V are both orthogonal, so a Singular Value Decomposition of A^t is given by

$$A^t = (U\,S\,V^t)^t = (V^t)^t\,S^t\,U^t = V\,S^t\,U^t$$

11. Suppose that the $n \times n$ matrix A has the Singular Value Decomposition $A = U\,S\,V^t$. Show that A^{-1} exists if and only if S^{-1} exists, and find a Singular Value Decomposition for A^{-1} when it exists.

SOLUTION: The matrices U and V are orthogonal, so they are nonsingular with $U^{-1} = U^t$ and $V^{-1} = V^t$. Since

$$\det A = \det U \cdot \det S \cdot \det V^t,$$

with $\det U$ and $\det V^t$ both nonzero, $\det A = 0$ if and only if $\det S = 0$. Hence A in nonsingular if and only if S is nonsingular.

When A^{-1} exists we have the Singular Value Decomposition of A^{-1} given by

$$A^{-1} = \left(U\,S\,V^t\right)^{-1} = (V^t)^{-1}\,S^{-1}\,U^{-1} = \left(V^{-1}\right)^{-1}\,S^{-1}\,U^t = V\,S^{-1}\,U^t.$$

15. Show that if A is an $n \times n$ nonsingular matrix with singular values $s_1, s_2, \ldots, s_n$, then the l_2 condition number of A is $K_2(A) = s_1/s_n$.

SOLUTION: The condition number is defined on page 470 as

$$K_2(A) = ||A||_2 ||A^{-1}||_2.$$

By Theorem 7.15 on page 447, we have $||A||_2^2 = \rho(A^t A)$, which is the largest eigenvalue of $A^t A$, that is, s_1^2.

In addition, Exercise 15 on page 450 states that the eigenvalues of the inverse of a nonsingular matrix are the reciprocals of the eigenvalues of the matrix, so

$$||A^{-1}||_2^2 = \rho((A^{-1})^t A^{-1}) = \rho((A^t)^{-1} A^{-1}) = \rho((A A^t)^{-1})$$

is the largest eigenvalue of $(A A^t)^{-1}$. This is the reciprocal of the smallest eigenvalue $A A^t$. By Theorem 9.26, the nonzero eigenvalues of $A^t A$ and $A A^t$ are the same, so this value is $1/s_n^2$. As a consequence,

$$K_2(A) = ||A||_2 ||A^{-1}||_2 = \frac{s_1}{s_n}.$$

18. Given the data

x_i	1.0	1.1	1.3	1.5	1.9	2.1
y_i	1.84	1.96	2.21	2.45	2.94	3.18

a. Use the singular value decomposition technique to determine the least squares polynomial of degree 2.

b. Use the singular value decomposition technique to determine the least squares polynomial of degree 3.

SOLUTION: **a.** Use the tabulated values to construct

$$\mathbf{b} = \begin{bmatrix} y_0 \\ y_1 \\ y_2 \\ y_3 \\ y_4 \\ y_5 \end{bmatrix} = \begin{bmatrix} 1.84 \\ 1.96 \\ 2.21 \\ 2.45 \\ 2.94 \\ 3.18 \end{bmatrix}, \quad \text{and} \quad A = \begin{bmatrix} 1 & x_0 & x_0^2 \\ 1 & x_1 & x_1^2 \\ 1 & x_2 & x_2^2 \\ 1 & x_3 & x_3^2 \\ 1 & x_4 & x_4^2 \\ 1 & x_5 & x_5^2 \end{bmatrix} = \begin{bmatrix} 1 & 1.0 & 1.0 \\ 1 & 1.1 & 1.21 \\ 1 & 1.3 & 1.69 \\ 1 & 1.5 & 2.25 \\ 1 & 1.9 & 3.61 \\ 1 & 2.1 & 4.41 \end{bmatrix}.$$

The matrix A has the singular value decomposition $A = U S V^t$, where

$$U = \begin{bmatrix} -0.203339 & -0.550828 & 0.554024 & 0.055615 & -0.177253 & -0.560167 \\ -0.231651 & -0.498430 & 0.185618 & 0.165198 & 0.510822 & 0.612553 \\ -0.294632 & -0.369258 & -0.337742 & -0.711511 & -0.353683 & 0.177288 \\ -0.366088 & -0.20758 & -0.576499 & 0.642950 & -0.264204 & -0.085730 \\ -0.534426 & 0.213281 & -0.200202 & -0.214678 & 0.628127 & -0.433808 \\ -0.631309 & 0.472467 & 0.414851 & 0.062426 & -0.343809 & 0.289864 \end{bmatrix},$$

$$S = \begin{bmatrix} 7.844127 & 0 & 0 \\ 0 & 1.223790 & 0 \\ 0 & 0 & 0.070094 \\ 0 & 0 & 0 \\ 0 & 0 & 0 \end{bmatrix}, \quad \text{and} \quad V^t = \begin{bmatrix} -0.288298 & -0.475702 & -0.831018 \\ -0.768392 & -0.402924 & 0.497218 \\ 0.571365 & -0.781895 & 0.249363 \end{bmatrix}.$$

So

$$\mathbf{c} = U^t \mathbf{b} = \begin{bmatrix} -5.955009 \\ -1.185591 \\ -0.044985 \\ -0.003732 \\ -0.000493 \\ -0.001963 \end{bmatrix},$$

and the components of $\mathbf{z}$ are

$$z_1 = \frac{c_1}{s_1} = \frac{-5.955009}{7.844127} = -0.759168, \quad z_2 = \frac{c_2}{s_2} = \frac{-1.185591}{1.223790} = -0.968786,$$

and

$$z_3 = \frac{c_3}{s_3} = \frac{-0.044985}{0.070094} = -0.641784.$$

This gives the least squares coefficients in $P_2(x) = a_0 + a_1 x + a_2 x^2$ as

$$\begin{bmatrix} a_0 \\ a_1 \\ a_2 \end{bmatrix} = \mathbf{x} = V \mathbf{z} = \begin{bmatrix} 0.596581 \\ 1.253293 \\ -0.010853 \end{bmatrix}.$$

The least squares error using these values uses the last three components of $\mathbf{c}$, and is

$$\|A\mathbf{x} - \mathbf{b}\|_2 = \sqrt{c_4^2 + c_5^2 + c_6^2} = \sqrt{(-0.003732)^2 + (-0.000493)^2 + (-0.001963)^2} = 0.004244.$$

b. Use the tabulated values to construct

$$
\mathbf{b} = \begin{bmatrix} y_0 \\ y_1 \\ y_2 \\ y_3 \\ y_4 \\ y_5 \end{bmatrix} = \begin{bmatrix} 1.84 \\ 1.96 \\ 2.21 \\ 2.45 \\ 2.94 \\ 3.18 \end{bmatrix}, \quad \text{and} \quad A = \begin{bmatrix} 1 & x_0 & x_0^2 & x_0^3 \\ 1 & x_1 & x_1^2 & x_1^3 \\ 1 & x_2 & x_2^2 & x_2^3 \\ 1 & x_3 & x_3^2 & x_3^3 \\ 1 & x_4 & x_4^2 & x_4^3 \\ 1 & x_5 & x_5^2 & x_5^3 \end{bmatrix} = \begin{bmatrix} 1 & 1.0 & 1.0 & 1.0 \\ 1 & 1.1 & 1.21 & 1.331 \\ 1 & 1.3 & 1.69 & 2.197 \\ 1 & 1.5 & 2.25 & 3.375 \\ 1 & 1.9 & 3.61 & 6.859 \\ 1 & 2.1 & 4.41 & 9.261 \end{bmatrix}.
$$

The matrix A has the singular value decomposition $A = U \, S \, V^t$, where

$$
U = \begin{bmatrix}
-0.116086 & -0.514623 & 0.569113 & -0.437866 & -0.381082 & 0.246672 \\
-0.143614 & -0.503586 & 0.266325 & 0.184510 & 0.535306 & 0.578144 \\
-0.212441 & -0.448121 & -0.238475 & 0.48499 & 0.180600 & -0.655247 \\
-0.301963 & -0.339923 & -0.549619 & 0.038581 & -0.573591 & 0.400867 \\
-0.554303 & 0.074101 & -0.306350 & -0.636776 & 0.417792 & -0.115640 \\
-0.722727 & 0.399642 & 0.390359 & 0.363368 & -0.179026 & 0.038548
\end{bmatrix},
$$

$$
S = \begin{bmatrix}
14.506808 & 0 & 0 & 0 \\
0 & 2.084909 & 0 & 0 \\
0 & 0 & 0.198760 & 0 \\
0 & 0 & 0 & 0.868328 \\
0 & 0 & 0 & 0
\end{bmatrix},
$$

and

$$
V^t = \begin{bmatrix}
-0.141391 & -0.246373 & -0.449207 & -0.847067 \\
-0.639122 & -0.566437 & -0.295547 & 0.428163 \\
0.660862 & -0.174510 & -0.667840 & 0.294610 \\
-0.367142 & 0.766807 & -0.514640 & 0.111173
\end{bmatrix}.
$$

So

$$
\mathbf{c} = U^t \mathbf{b} = \begin{bmatrix}
-5.632309 \\
-2.268376 \\
0.036241 \\
0.005717 \\
-0.000845 \\
-0.004086
\end{bmatrix},
$$

and the components of $\mathbf{z}$ are

$$
z_1 = \frac{c_1}{s_1} = \frac{-5.632309}{14.506808} = -0.388253, \quad z_2 = \frac{c_2}{s_2} = \frac{-2.268376}{2.084909} = -1.087998,
$$

$$
z_3 = \frac{c_3}{s_3} = \frac{0.036241}{0.198760} = 0.182336, \quad \text{and} \quad z_4 = \frac{c_4}{s_4} = \frac{0.005717}{0.868328} = 0.65843.
$$

This gives the least squares coefficients in $P_2(x) = a_0 + a_1 x + a_2 x^2 + a_3 x^3$ as

$$
\begin{bmatrix} a_0 \\ a_1 \\ a_2 \\ a_3 \end{bmatrix} = \mathbf{x} = V \, \mathbf{z} = \begin{bmatrix} 0.629019 \\ 1.185010 \\ 0.035333 \\ -0.010047 \end{bmatrix}.
$$

The least squares error using these values uses the last two components of $\mathbf{c}$, and is

$$
\|A\mathbf{x} - \mathbf{b}\|_2 = \sqrt{c_5^2 + c_6^2} = \sqrt{(-0.000845)^2 + (-0.004086)^2} = 0.004172.
$$

Numerical Solutions of Nonlinear Systems of Equations

Exercise Set 10.1, page 636

2. Give an example of a function $\mathbf{F} : \mathbb{R}^2 \to \mathbb{R}^2$ that is continuous at each point of $\mathbb{R}^2$, except at $(1, 0)$.

SOLUTION: There are many examples, you need only ensure that the function is continuous when $(x_1, x_2) \neq (1, 0)$ but undefined at $(1, 0)$. So one example would be

$$\mathbf{F}(x_1, x_2) = \left(1, \frac{1}{(x_1 - 1)^2 + x_2^2}\right)^t.$$

The function in the denominator of the second component, $g(x_1, x_2) = (x_1 - 1)^2 + x_2^2$ is everywhere continuous, so its reciprocal is also continuous, except where it is zero. Since $g(x_1, x_2) = 0$ if and only if $(x_1, x_2) = (1, 0)$, the function $\mathbf{F}$ is continuous if and only if $(x_1, x_2) \neq (1, 0)$.

4. The nonlinear system

$$-x_1(x_1 + 1) + 2x_2 = 18,$$
$$(x_1 - 1)^2 + (x_2 - 6)^2 = 25$$

has two solutions. Approximate the solutions to within 10^{-5} in the l_∞ norm.

SOLUTION: The graphs in the figure below indicate that the solutions are near $(-1.5, 10.5)$ and $(2, 11)$.

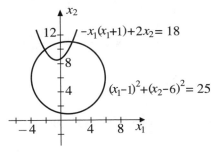

With the initial approximation $\mathbf{x}^{(0)} = (2, 11)^t$, we use

$$\mathbf{G}_1(\mathbf{x}) = \left(-0.5 + \sqrt{2x_2 - 17.75}, 6 + \sqrt{25 - (x_1 - 1)^2}\right)^t.$$

199

This gives $\mathbf{x}^{(9)} = (1.5469466, 10.969994)^t$.

With the initial approximation $\mathbf{x}^{(0)} = (-1.5, 10.5)^t$, we use

$$\mathbf{G}_2(\mathbf{x}) = \left(-0.5 - \sqrt{2x_2 - 17.75}, 6 + \sqrt{25 - (x_1 - 1)^2}\right)^t,$$

which gives $\mathbf{x}^{(34)} = (-2.000003, 9.999996)^t$.

5. a. The nonlinear system

$$x_1^2 - 10x_1 + x_2^2 + 8 = 0,$$
$$x_1 x_2^2 + x_1 - 10x_2 + 8 = 0$$

can be converted into the fixed-point form.

$$x_1 = g_1(x_1, x_2) = \frac{x_1^2 + x_2^2 + 8}{10},$$
$$x_2 = g_2(x_1, x_2) = \frac{x_1 x_2^2 + x_1 + 8}{10}.$$

Show that the problem has a unique fixed point in

$$D = \{(x_1, x_2)^t \mid 0 \le x_1, x_2 \le 1.5\}.$$

b. Apply functional iteration to approximate the solution.

c. Use the Seidel method to see if the approximations improve.

SOLUTION: **a.** Continuity properties are easily shown. Moreover,

$$\frac{8}{10} \le \frac{x_1^2 + x_2^2 + 8}{10} \le 1.25$$

and

$$\frac{8}{10} \le \frac{x_1 x_2^2 + x_1 + 8}{10} \le 1.2875,$$

so $\mathbf{G}(\mathbf{x}) \in D$, whenever $\mathbf{x} \in D$. Further,

$$\frac{\partial g_1}{\partial x_1} = \frac{2x_1}{10}, \quad \text{so} \quad \left|\frac{\partial g_1(\mathbf{x})}{\partial x_1}\right| \le \frac{3}{10}; \quad \frac{\partial g_1}{\partial x_2} = \frac{2x_2}{10}, \quad \text{so} \quad \left|\frac{\partial g_2(\mathbf{x})}{\partial x_2}\right| \le \frac{3}{10};$$

$$\frac{\partial g_2}{\partial x_1} = \frac{x_2^2 + 1}{10}, \quad \text{so} \quad \left|\frac{\partial g_2(\mathbf{x})}{\partial x_1}\right| \le \frac{3.25}{10}; \quad \text{and} \quad \frac{\partial g_2}{\partial x_2} = \frac{2x_1 x_2}{10}, \quad \text{so} \quad \left|\frac{\partial g_2(\mathbf{x})}{\partial x_2}\right| \le \frac{4.5}{10}.$$

Since

$$\left|\frac{\partial g_i(\mathbf{x})}{\partial x_j}\right| \le 0.45 = \frac{0.9}{2},$$

for each $i, j = 1, 2$, all the hypothesis of Theorem 10.6 have been satisfied. Thus, $\mathbf{G}$ has a unique fixed point in D.

b. Functional iteration with $\mathbf{x}^{(0)} = (0, 0)^t$ gives $\mathbf{x}^{(13)} = (0.9999973, 0.9999973)^t$.

c. If we use Seidel iteration with this initial approximation, we obtain
$\mathbf{x}^{(11)} = (0.9999984, 0.9999991)^t$.

12. Show that the function $\mathbf{F}$ is continuous at $\mathbf{x}_0$ precisely when, given any number $\epsilon > 0$, a number $\delta > 0$ exists with

$$\|\mathbf{F}(\mathbf{x}) - \mathbf{F}(\mathbf{x}_0)\| < \epsilon,$$

whenever $\|\mathbf{x} - \mathbf{x}_0\| < \delta$.

SOLUTION: Let $\mathbf{F}(\mathbf{x}) = (f_1(\mathbf{x}), ..., f_n(\mathbf{x}))^t$. Suppose $\mathbf{F}$ is continuous at $\mathbf{x}_0$. By Definition 10.3,

$$\lim_{\mathbf{x} \to \mathbf{x}_0} f_i(\mathbf{x}) = f_i(\mathbf{x}_0), \quad \text{for each } i = 1, ..., n.$$

Given $\epsilon > 0$, there exists $\delta_i > 0$ such that

$$|f_i(\mathbf{x}) - f_i(\mathbf{x}_0)| < \epsilon,$$

whenever $0 < \|\mathbf{x} - \mathbf{x}_0\| < \delta_i$ and $\mathbf{x} \in D$.

Let $\delta = \min_{1 \leq i \leq n} \delta_i$. If $0 < \|\mathbf{x} - \mathbf{x}_0\| < \delta$, then $0 < \|\mathbf{x} - \mathbf{x}_0\| < \delta_i$ and $|f_i(\mathbf{x}) - f_i(\mathbf{x}_0)| < \epsilon$, for each $i = 1, ..., n$, whenever $\mathbf{x} \in D$. This implies that

$$\|\mathbf{F}(\mathbf{x}) - \mathbf{F}(\mathbf{x}_0)\|_\infty < \epsilon,$$

whenever $\|\mathbf{x} - \mathbf{x}_0\| < \delta$ and $\mathbf{x} \in D$. By the equivalence of vector norms, the result holds for all vector norms by suitably adjusting δ.

For the converse, let $\epsilon > 0$ be given. Then there is a $\delta > 0$ such that

$$\|\mathbf{F}(\mathbf{x}) - \mathbf{F}(\mathbf{x}_0)\| < \epsilon,$$

whenever $\mathbf{x} \in D$ and $\|\mathbf{x} - \mathbf{x}_0\| < \delta$. By the equivalence of vector norms, a number $\delta' > 0$ can be found with

$$\|f_i(\mathbf{x}) - f_i(\mathbf{x}_0)\| < \epsilon,$$

whenever $\mathbf{x} \in D$ and $\|\mathbf{x} - \mathbf{x}_0\| < \delta'$.

Thus, $\lim_{\mathbf{x} \to \mathbf{x}_0} f_i(\mathbf{x}) = f_i(\mathbf{x}_0)$, for $i = 1, ..., n$. Since $\mathbf{F}(\mathbf{x}_0)$ is defined, the conditions in Definition 10.3 hold, and $\mathbf{F}$ is continuous at $\mathbf{x}_0$.

Exercise Set 10.2, page 644

2. b. Use Newton's method with $\mathbf{x}^{(0)} = \mathbf{0}$ to compute $\mathbf{x}^{(2)}$ for the nonlinear system

$$x_1^2 + x_2 - 37 = 0,$$
$$x_1 - x_2^2 - 5 = 0,$$
$$x_1 + x_2 + x_3 - 3 = 0.$$

SOLUTION: We first define the component functions

$$f_1(x_1, x_2, x_3) = x_1^2 + x_2 - 37,$$
$$f_2(x_1, x_2, x_3) = x_1 - x_2^2 - 5,$$
$$f_3(x_1, x_2, x_3) = x_1 + x_2 + x_3 - 3.$$

Then we need to construct the Jacobian matrix, which is given by

$$J(\mathbf{x}) = \begin{bmatrix} 2x_1 & 1 & 0 \\ 1 & -2x_2 & 0 \\ 1 & 1 & 1 \end{bmatrix},$$

so

$$J(\mathbf{x}^{(0)}) = \begin{bmatrix} 0 & 1 & 0 \\ 1 & 0 & 0 \\ 1 & 1 & 1 \end{bmatrix} \quad \text{and} \quad J(\mathbf{x}^{(0)})^{-1} = \begin{bmatrix} 0 & 1 & 0 \\ 1 & 0 & 0 \\ -1 & -1 & 1 \end{bmatrix}.$$

Applying Newton's method gives

$$\mathbf{x}^{(1)} = \begin{bmatrix} x_1^{(0)} \\ x_2^{(0)} \\ x_3^{(0)} \end{bmatrix} - J\left(x_1^{(0)}, x_2^{(0)}, x_3^{(0)}\right)^{-1} \begin{bmatrix} \left(x_1^{(0)}\right)^2 + x_2^{(0)} - 37 \\ x_1^{(0)} - \left(x_2^{(0)}\right)^2 - 5 \\ x_1^{(0)} + x_2^{(0)} + x_3^{(0)} - 3 \end{bmatrix},$$

$$= \begin{bmatrix} 0 \\ 0 \\ 0 \end{bmatrix} - \begin{bmatrix} 0 & 1 & 0 \\ 1 & 0 & 0 \\ -1 & -1 & 1 \end{bmatrix} \begin{bmatrix} -37 \\ -5 \\ -3 \end{bmatrix} = \begin{bmatrix} 5 \\ 37 \\ -39 \end{bmatrix},$$

and

$$\mathbf{x}^{(2)} = \begin{bmatrix} 5 \\ 37 \\ -39 \end{bmatrix} - \begin{bmatrix} 10 & 1 & 0 \\ 1 & -74 & 0 \\ 1 & 1 & 1 \end{bmatrix}^{-1} \begin{bmatrix} 25 \\ -1369 \\ 0 \end{bmatrix} = \begin{bmatrix} 4.3508772 \\ 18.491228 \\ -19.842105 \end{bmatrix}.$$

3. c. Plot the graphs of

$$x_1(1 - x_1) + 4x_2 = 12,$$
$$(x_1 - 2)^2 + (2x_2 - 3)^2 = 25$$

using Maple, and approximate the solutions to the nonlinear system.

SOLUTION: Graphs of the two equations are shown below. They indicate that one solution is near $x_1 = -1$ and $x_2 = 3.5$ and the other near $x_1 = 2.5$ and $x_2 = 4$.

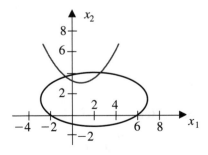

5. c. Use the answers obtained in Exercise 3 as initial approximations to Newton's method. Iterate until $\left\| \mathbf{x}^{(k)} - \mathbf{x}^{(k-1)} \right\|_\infty < 10^{-6}$.

SOLUTION: The first answer from Exercise 3 is $(-1, 3.5)^t$, which is a solution to the problem. Hence, Newton's method converges in one iteration.

The second answer is $(2.5, 4)^t$, which we use as $\mathbf{x}^{(0)}$. We have

$$\mathbf{F}(x_1, x_2) = \begin{bmatrix} x_1(1 - x_1) + 4x_2 - 12 \\ (x_1 - 2)^2 + (2x_2 - 3)^2 - 25 \end{bmatrix},$$

so the Jacobian is

$$J(x_1, x_2) = \begin{bmatrix} 1 - 2x_1 & 4 \\ 2x_1 - 4 & 8x_2 - 12 \end{bmatrix}.$$

The linear system $J\left(\mathbf{x}^{(0)}\right)\mathbf{y}^{(0)} = -\mathbf{F}\left(\mathbf{x}^{(0)}\right)$ is

$$\begin{bmatrix} -4 & 4 \\ 1 & 20 \end{bmatrix} \begin{bmatrix} y_1^{(0)} \\ y_2^{(0)} \end{bmatrix} = -\begin{bmatrix} \frac{1}{4} \\ \frac{1}{4} \end{bmatrix}.$$

The solution is $\mathbf{y}^{(0)} = (0.04761904762, -0.01488095239)^t$, so

$$\mathbf{x}^{(1)} = \mathbf{x}^{(0)} + \mathbf{y}^{(0)} = (2.547619048, 3.985119048)^t.$$

The next linear system is

$$J\left(\mathbf{x}^{(1)}\right)\mathbf{y}^{(1)} = -\begin{bmatrix} -0.00226758 \\ 0.00315335 \end{bmatrix}.$$

The solution is

$$\mathbf{y}^{(1)} = (-0.0006724506978, -0.0001215664286)^t,$$

so

$$\mathbf{x}^{(2)} = \mathbf{x}^{(1)} + \mathbf{y}^{(1)} = (2.546946597, 3.984997482)^t.$$

The next iteration gives

$$\mathbf{y}^{(2)} = \left(-0.1262468243 \times 10^{-6}, -0.1921025377 \times 10^{-7}\right)^t$$

and

$$\mathbf{x}^{(3)} = (2.546946471, 3.984997463)^t.$$

Since

$$\left\| \mathbf{x}^{(3)} - \mathbf{x}^{(2)} \right\|_\infty = 1.26 \times 10^{-7} < 10^{-6},$$

the method stops with $\mathbf{x}^{(3)}$ as the desired approximation.

12. The amount of pressure p needed to sink a circular object of radius r in a homogeneous soil can be approximated by

$$p = k_1 e^{k_2 r} + k_3 r,$$

where k_1, k_2, and k_3 are constants. Determine these constants if $p = 10$ when $r = 1$, $p = 12$ when $r = 2$, and $p = 15$ when $r = 3$. Then approximate the size of the circular plate needed to support a weight of 500 pounds.

SOLUTION: We need to solve the system

$$10 = k_1 e^{k_2} + k_3,$$
$$12 = k_1 e^{2k_2} + 2k_3,$$
$$15 = k_1 e^{3k_2} + 3k_3.$$

Although this is a system of three equations and three unknowns, (and can be solved as such) it can be simplified by solving the first equation for k_3 and substituting this into the second and third equations to give

$$k_3 = 10 - k_1 e^{k_2},$$
$$12 = k_1 e^{k_2} \left(e^{k_2} - 2\right) + 20,$$
$$15 = k_1 e^{k_2} \left(e^{2k_2} - 3\right) + 30.$$

The system can be further simplified by solving the second equation for k_1 and substituting into the third to give

$$k_3 = 10 - k_1 e^{k_2},$$
$$k_1 = \frac{-8}{e^{k_2} \left(e^{k_2} - 2\right)},$$
$$15 = \frac{-8}{\left(e^{k_2} - 2\right)} \left(e^{2k_2} - 3\right) + 30.$$

The problem has now been reduced to solving the third nonlinear equation in a single unknown k_2. Once this has been done, the values of k_1 and k_3 can be found from the second and first equations. The third equation can actually be solved by the quadratic formula, since it can be rewritten as

$$8 \left(e^{k_2}\right)^2 - 15 e^{k_2} + 6 = 0,$$

which has $k_2 = 0.259690$ as its only positive solution. Using this value gives $k_1 = 8.77125$ and $k_3 = -1.37217$.

To find the size of the circular plate needed to support 500 pounds we need to solve for r in the nonlinear equation

$$\frac{500}{\pi r^2} = 8.77125 e^{0.259690 r} - 1.37217 r.$$

Solving this equation for r by one of the methods in Chapter 2 gives $r \approx 3.185$ in.

Exercise Set 10.3, page 652

2. b. Use Broyden's method with $\mathbf{x}^{(0)} = \mathbf{0}$ to compute $\mathbf{x}^{(2)}$ for the nonlinear system

$$x_1^2 + x_2 - 37 = 0,$$
$$x_1 - x_2^2 - 5 = 0,$$
$$x_1 + x_2 + x_3 - 3 = 0.$$

SOLUTION: Since Broyden's method generates $\mathbf{x}^{(1)}$ in the same manner as Newton's method, we have $\mathbf{x}^{(0)} = (0,0,0)^t$ and $\mathbf{x}^{(1)} = (5, 37, -39)^t$. To determine $\mathbf{x}^{(2)}$, we first need the matrix

$$A_1^{-1} = J\left(\mathbf{x}^{(0)}\right)^{-1} + \frac{\left(\mathbf{s}_1 - J\left(\mathbf{x}^{(0)}\right)^{-1}\mathbf{y}_1\right)\mathbf{s}_1^t J\left(\mathbf{x}^{(0)}\right)^{-1}}{\mathbf{s}_1^t J\left(\mathbf{x}^{(0)}\right)^{-1}\mathbf{y}_1},$$

where

$$\mathbf{s}_1 = \mathbf{x}^{(1)} - \mathbf{x}^{(0)} = (5, 37, -39)^t,$$

$$\mathbf{y}_1 = \mathbf{F}(\mathbf{x}^{(1)}) - \mathbf{F}(\mathbf{x}^{(0)}) = \begin{bmatrix} 25 \\ -1369 \\ 0 \end{bmatrix} - \begin{bmatrix} -37 \\ -5 \\ -3 \end{bmatrix} = \begin{bmatrix} 62 \\ -1364 \\ 3 \end{bmatrix},$$

and

$$J(\mathbf{x}^{(0)})^{-1} = \begin{bmatrix} 0 & 1 & 0 \\ 1 & 0 & 0 \\ -1 & -1 & 1 \end{bmatrix}.$$

Thus

$$J\left(\mathbf{x}^{(0)}\right)^{-1}\mathbf{y}_1 = (-1364, 62, 1305)^t,$$

$$\mathbf{s}_1 - J\left(\mathbf{x}^{(0)}\right)^{-1}\mathbf{y}_1 = (1369, -25, -1344)^t,$$

and

$$\mathbf{s}_1^t J\left(\mathbf{x}^{(0)}\right)^{-1} = (76, 44, -39).$$

So

$$\frac{\left(\mathbf{s}_1 - J\left(\mathbf{x}^{(0)}\right)^{-1}\mathbf{y}_1\right)\mathbf{s}_1^t J\left(\mathbf{x}^{(0)}\right)^{-1}}{\mathbf{s}_1^t J\left(\mathbf{x}^{(0)}\right)^{-1}\mathbf{y}_1} = -\frac{1}{55421}\begin{bmatrix} 104044 & 60236 & -53391 \\ -1900 & -1100 & 975 \\ -102144 & -59136 & 52416 \end{bmatrix},$$

$$A_1^{-1} = \frac{1}{55421}\begin{bmatrix} -104044 & -4815 & 53391 \\ 57321 & 1100 & -975 \\ 46723 & 3715 & 3005 \end{bmatrix},$$

and

$$\mathbf{x}^{(2)} = \mathbf{x}^{(1)} - A_1^{-1}\mathbf{F}\left(\mathbf{x}^{(1)}\right)$$

$$= \begin{bmatrix} 5 \\ 37 \\ -39 \end{bmatrix} - \begin{bmatrix} -1.877339 & -0.08688042 & 0.9633713 \\ 1.034283 & 0.01984807 & -0.01759261 \\ 0.8430559 & 0.06703235 & 0.05422132 \end{bmatrix}\begin{bmatrix} 35 \\ -1369 \\ 0 \end{bmatrix} = \begin{bmatrix} -67.00583 \\ 38.31494 \\ 31.69089 \end{bmatrix}.$$

8. Show that if $\mathbf{0} \neq \mathbf{y} \in \mathbb{R}^n$ and $\mathbf{z} \in \mathbb{R}^n$, then $\mathbf{z} = \mathbf{z}_1 + \mathbf{z}_2$, where $\mathbf{z}_1 = \frac{\mathbf{y}^t\mathbf{z}}{\|\mathbf{y}\|_2^2}\mathbf{y}$ is parallel to $\mathbf{y}$ and $\mathbf{z}_2$ is orthogonal to $\mathbf{y}$.

SOLUTION: If $\mathbf{z}^t\mathbf{y} = 0$, then $\mathbf{z} = \mathbf{z}_1 + \mathbf{z}_2$, where $\mathbf{z}_1 = \mathbf{0}$ and $\mathbf{z}_2 = \mathbf{z}$. Otherwise, the nonzero vector

$$\mathbf{z}_1 = \frac{\mathbf{y}^t\mathbf{z}}{\|\mathbf{y}\|_2^2}\mathbf{y}$$

is parallel to $\mathbf{y}$, and we let $\mathbf{z}_2 = \mathbf{z} - \mathbf{z}_1$. Then

$$\mathbf{z}_2^t \mathbf{y} = \mathbf{z}^t \mathbf{y} - \mathbf{z}_1^t \mathbf{y} = \mathbf{z}^t \mathbf{y} - \left[\frac{\mathbf{y}^t \mathbf{z}}{\mathbf{y}^t \mathbf{y}} \mathbf{y}\right]^t \mathbf{y} = \mathbf{z}^t \mathbf{y} - \frac{\mathbf{z}^t \mathbf{y}}{\mathbf{y}^t \mathbf{y}} \mathbf{y}^t \mathbf{y} = 0,$$

and $\mathbf{z}_2$ is orthogonal to $\mathbf{y}$.

9. Show that if $\mathbf{u}, \mathbf{v} \in \mathbb{R}^n$, then $\det (I + \mathbf{u}\mathbf{v}^t) = 1 + \mathbf{v}^t \mathbf{u}$.

SOLUTION: Let λ be an eigenvalue of $M = (I + \mathbf{u}\mathbf{v}^t)$ with eigenvector $\mathbf{x} \neq \mathbf{0}$. Then

$$\lambda \mathbf{x} = M\mathbf{x} = (I + \mathbf{u}\mathbf{v}^t)\mathbf{x} = \mathbf{x} + (\mathbf{v}^t \mathbf{x})\mathbf{u}.$$

Thus

$$(\lambda - 1)\mathbf{x} = (\mathbf{v}^t \mathbf{x})\mathbf{u}.$$

If $\lambda = 1$, then $\mathbf{v}^t \mathbf{x} = 0$. So $\lambda = 1$ is an eigenvalue of M with multiplicity $n-1$ and eigenvectors $\mathbf{x}^{(1)}, \dots, \mathbf{x}^{(n-1)}$, where $\mathbf{v}^t \mathbf{x}^{(j)} = 0$, for $j = 1, \dots, n-1$.

Assuming $\lambda \neq 1$ implies that $\mathbf{x}$ and $\mathbf{u}$ are parallel. Suppose $\mathbf{x} = \alpha \mathbf{u}$. Then

$$(\lambda - 1)\alpha \mathbf{u} = (\mathbf{v}^t (\alpha \mathbf{u}))\mathbf{u}.$$

Thus

$$\alpha(\lambda - 1)\mathbf{u} = \alpha (\mathbf{v}^t \mathbf{u})\mathbf{u},$$

which implies that $\lambda - 1 = \mathbf{v}^t \mathbf{u}$ and $\lambda = 1 + \mathbf{v}^t \mathbf{u}$. Hence, M has eigenvalues λ_i, $1 \leq i \leq n$, where $\lambda_i = 1$, for $i = 1, \dots, n-1$ and $\lambda_n = 1 + \mathbf{v}^t \mathbf{u}$. Since $\det M = \prod_{i=1}^{n} \lambda_i$, we have

$$\det M = 1 + \mathbf{v}^t \mathbf{u}.$$

10. **a.** Use the result in Exercise 9 to show that if A^{-1} exists and $\mathbf{x}, \mathbf{y} \in \mathbb{R}^n$, then $(A + \mathbf{x}\mathbf{y}^t)^{-1}$ exists if and only if $\mathbf{y}^t A^{-1} \mathbf{x} \neq -1$.

b. By multiplying on the right by $A + \mathbf{x}\mathbf{y}^t$, show that when $\mathbf{y}^t A^{-1} \mathbf{x} \neq -1$, we have

$$(A + \mathbf{x}\mathbf{y}^t)^{-1} = A^{-1} - \frac{A^{-1} \mathbf{x}\mathbf{y}^t A^{-1}}{1 + \mathbf{y}^t A^{-1} \mathbf{x}}.$$

SOLUTION: **a.** Since A^{-1} exists, we can write

$$\det (A + \mathbf{x}\mathbf{y}^t) = \det (A + AA^{-1}\mathbf{x}\mathbf{y}^t) = \det A (I + A^{-1}\mathbf{x}\mathbf{y}^t) = \det A \det (I + A^{-1}\mathbf{x}\mathbf{y}^t).$$

But A^{-1} exists, so $\det A \neq 0$. By Exercise 9, $\det I + A^{-1}\mathbf{x}\mathbf{y}^t = 1 + \mathbf{y}^t A^{-1} \mathbf{x}$. So $(A + \mathbf{x}\mathbf{y}^t)^{-1}$ exists if and only if $\mathbf{y}^t A^{-1} \mathbf{x} \neq -1$.

b. Assume $\mathbf{y}^t A^{-1} \mathbf{x} \neq -1$, so that $(A + \mathbf{x}\mathbf{y}^t)^{-1}$ exists. We have

$$\left[A^{-1} - \frac{A^{-1}\mathbf{x}\mathbf{y}^t A^{-1}}{1 + \mathbf{y}^t A^{-1}\mathbf{x}}\right](A + \mathbf{x}\mathbf{y}^t) = A^{-1}A - A^{-1}\mathbf{x}\frac{\mathbf{y}^t A^{-1} A}{1 + \mathbf{y}^t A^{-1}\mathbf{x}} + A^{-1}\mathbf{x}\mathbf{y}^t - \frac{A^{-1}\mathbf{x}\mathbf{y}^t A^{-1}\mathbf{x}\mathbf{y}^t}{1 + \mathbf{y}^t A^{-1}\mathbf{x}}$$

$$= I - \frac{A^{-1}\mathbf{x}\mathbf{y}^t}{1 + \mathbf{y}^t A^{-1}\mathbf{x}} + A^{-1}\mathbf{x}\mathbf{y}^t - \frac{A^{-1}\mathbf{x}\mathbf{y}^t A^{-1}\mathbf{x}\mathbf{y}^t}{1 + \mathbf{y}^t A^{-1}\mathbf{x}}$$

$$= I - \frac{A^{-1}\mathbf{x}\mathbf{y}^t - A^{-1}\mathbf{x}\mathbf{y}^t - \mathbf{y}^t A^{-1}\mathbf{x}A^{-1}\mathbf{x}\mathbf{y}^t + A^{-1}\mathbf{x}\mathbf{y}^t A^{-1}\mathbf{x}\mathbf{y}^t}{1 + \mathbf{y}^t A^{-1}\mathbf{x}}$$

$$= I + \frac{\mathbf{y}^t A^{-1}\mathbf{x}A^{-1}\mathbf{x}\mathbf{y}^t - \mathbf{y}^t A^{-1}\mathbf{x}(A^{-1}\mathbf{x}\mathbf{y}^t)}{1 + \mathbf{y}^t A^{-1}\mathbf{x}} = I.$$

Exercise Set 10.4, page 659

1. a. Use the method of Steepest Descent with $TOL = 0.05$ to approximate the solution to

$$4x_1^2 - 20x_1 + \frac{1}{4}x_2^2 + 8 = 0,$$

$$\frac{1}{2}x_1x_2^2 + 2x_1 - 5x_2 + 8 = 0.$$

SOLUTION: For this problem we have

$$g(x_1, x_2) = \left(4x_1^2 - 20x_1 + \frac{1}{4}x_2^2 + 8\right)^2 + \left(\frac{1}{2}x_1x_2^2 + 2x_1 - 5x_2 + 8\right)^2,$$

so

$$\nabla g(x_1, x_2) = \begin{bmatrix} 2(8x_1 - 20)\left(4x_1^2 - 20x_1 + \frac{1}{4}x_2^2 + 8\right) + 2\left(\frac{1}{2}x^2 + 2\right)\left(\frac{1}{2}x_1x_2^2 + 2x_1 - 5x_2 + 8\right) \\ 2\left(\frac{1}{2}x_2\right)\left(4x_1^2 - 20x_1 + \frac{1}{4}x_2^2 + 8\right) + 2(x_1x_2 - 5)\left(\frac{1}{2}x_1x_2 + 2x_1 - 5x_2 + 8\right) \end{bmatrix}.$$

The initial approximation $\mathbf{x}^{(0)} = (0, 0)^t$ gives $g\left(\mathbf{x}^{(0)}\right) = 128$ and

$$\mathbf{z} = \nabla g\left(\mathbf{x}^{(0)}\right) = \begin{bmatrix} -288 \\ -80 \end{bmatrix}.$$

Thus, we have

$$z_0 = \|\mathbf{z}\|_2 = 298.9047 \quad \text{and} \quad \frac{\mathbf{z}}{z_0} = \begin{bmatrix} -0.9635179 \\ -0.2676439 \end{bmatrix}.$$

To construct the quadratic interpolating polynomial, we can use the following divided-difference table.

$\alpha_1 = 0$	$g_1 = 128$		
$\alpha_2 = 0.25$	$g_2 = 78.05219$	$h_1 = -199.7912$	
$\alpha_3 = 0.5$	$g_3 = 69.35206$	$h_2 = -34.76052$	$h_3 = 330.0614$

This gives the Newton form of the polynomial as

$$P(\alpha) = 128 - 199.7912\alpha + 330.0614\alpha(\alpha - 0.5),$$

which has its only critical point at $\alpha_0 = 0.4276576$ with

$$g_0 = g\left((0, 0)^t - 0.4276576(-0.9635179, -0.2676439)^t\right) = 38.33171.$$

Thus

$$\mathbf{x}^{(1)} = (0, 0)^t - 0.4276576(-0.9635179, -0.2676439)^t = (0.4120558, 0.1144500)^t,$$

with $g\left(\mathbf{x}^{(1)}\right) = 68.33171.$

Subsequent iterations are shown in the following table.

k	$x_1^{(k)}$	$x_2^{(k)}$
2	0.20844885	1.01953237
3	0.42932832	1.06474473
4	0.35551933	1.56337683
5	0.46015625	1.57843277
6	0.43237489	1.79646221
7	0.47963979	1.80244993
8	0.46719397	1.89790884
9	0.48940458	1.90080653
10	0.48322157	1.94653312
11	0.49435415	1.94803972

3. a. Use the results of Exercise 1 and Newton's method to approximate a solution of that system to within 10^{-5}.

SOLUTION: The nonlinear system of equations

$$f_1(x_1, x_2) = 4x_1^2 - 20x_1 + \frac{1}{4}x_2^2 + 8 = 0,$$

$$f_2(x_1, x_2) = \frac{1}{2}x_1 x_2^2 + 2x_1 - 5x_2 + 8 = 0$$

has the Jacobian matrix

$$J(\mathbf{x}) = \begin{bmatrix} 8x_1 - 20 & \frac{1}{2}x_2 \\ \frac{1}{4}x_2^2 + 2 & x_1 x_2 - 5 \end{bmatrix}.$$

With initial approximation $\mathbf{x}^{(0)} = (0.4943542, 1.948040)^t$, the value from the last iteration in Exercise 1, we have

$$\mathbf{x}^{(1)} = \begin{bmatrix} x_1^{(0)} \\ x_2^{(0)} \end{bmatrix} - \begin{bmatrix} 8x_1^{(0)} - 20 & \frac{1}{2}x_2^{(0)} \\ \frac{1}{4}\left(x_2^{(0)}\right)^2 + 2 & x_1^{(0)}x_2^{(0)} - 5 \end{bmatrix}^{-1} \begin{bmatrix} 4\left(x_1^{(0)}\right)^2 - 20x_1^{(0)} + \frac{1}{4}\left(x_2^{(0)}\right)^2 + 8 \\ \frac{1}{2}x_1^{(0)}\left(x_2^{(0)}\right)^2 + 2x_1^{(0)} - 5x_2^{(0)} + 8 \end{bmatrix}$$

$$= \begin{bmatrix} 0.4999947 \\ 1.999972 \end{bmatrix}.$$

Two more iterations produce the exact result $\mathbf{x} = (0.5, 2)^t$.

6. Show that the quadratic polynomial

$$P(\alpha) = g_1 + h_1\alpha + h_3\alpha(\alpha - \alpha_2) \quad \text{interpolates} \quad h(\alpha) = g\left(\mathbf{x}^{(0)}\right) - \alpha\nabla g\left(\mathbf{x}^{(0)}\right)$$

at $\alpha = 0$, α_2, and α_3 and that a critical point of P occurs at

$$\alpha_0 = 0.5\left(\alpha_2 - \frac{h_1}{h_3}\right).$$

SOLUTION: We have $\alpha_1 = 0$,

$$g_1 = g(x_1, ..., x_n) = g\left(\mathbf{x}^{(0)}\right) = h(\alpha_1),$$

$$g_3 = g\left(\mathbf{x}^{(0)} - \alpha_3 \nabla g\left(\mathbf{x}^{(0)}\right)\right) = h(\alpha_3),$$

and

$$g_2 = g\left(\mathbf{x}^{(0)} - \alpha_2 \nabla g\left(\mathbf{x}^{(0)}\right)\right) = h(\alpha_2).$$

So

$$h_1 = \frac{(g_2 - g_1)}{(\alpha_2 - \alpha_1)} = g\left[\mathbf{x}^{(0)} - \alpha_1 \nabla g\left(\mathbf{x}^{(0)}\right), \mathbf{x}^{(0)} - \alpha_2 \nabla g\left(\mathbf{x}^{(0)}\right)\right] = h[\alpha_1, \alpha_2];$$

$$h_2 = \frac{(g_3 - g_2)}{(\alpha_3 - \alpha_2)} = g\left[\mathbf{x}^{(0)} - \alpha_2 \nabla g\left(\mathbf{x}^{(0)}\right), \mathbf{x}^{(0)} - \alpha_3 \nabla g\left(\mathbf{x}^{(0)}\right)\right] = h[\alpha_2, \alpha_3];$$

$$h_3 = \frac{(h_2 - h_1)}{(\alpha_3 - \alpha_1)}$$

$$= g\left[\mathbf{x}^{(0)} - \alpha_1 \nabla g\left(\mathbf{x}^{(0)}\right), \mathbf{x}^{(0)} - \alpha_2 \nabla g\left(\mathbf{x}^{(0)}\right), \mathbf{x}^{(0)} - \alpha_3 \nabla g\left(\mathbf{x}^{(0)}\right)\right]$$

$$= h[\alpha_1, \alpha_2, \alpha_3].$$

The Newton divided-difference form of the second interpolating polynomial is

$$P(\alpha) = h[\alpha_1] + h[\alpha_1, \alpha_2](\alpha - \alpha_1) + h[\alpha_1, \alpha_2, \alpha_3](\alpha - \alpha_1)(\alpha - \alpha_2)$$

$$= g_1 + h_1(\alpha - \alpha_1) + h_3(\alpha - \alpha_1)(\alpha - \alpha_2)$$

$$= g_1 + h_1\alpha + h_3\alpha(\alpha - \alpha_2),$$

so

$$P(\alpha) = g_1 + h_1\alpha + h_3\alpha(\alpha - \alpha_2)$$

interpolates

$$h(\alpha) = g\left(\mathbf{x}^{(0)}\right) - \alpha\nabla g\left(\mathbf{x}^{(0)}\right)$$

at $\alpha = 0, \alpha_2$, and α_3. Since $P'(\alpha) = h_1 - \alpha_2 h_3 + 2h_3\alpha$, we have $P'(\alpha) = 0$ precisely when

$$\alpha_0 = 0.5\left(\alpha_2 - \frac{h_1}{h_3}\right).$$

Exercise Set 10.5, page 666

3. a. Use the continuation method and Euler's method with $N = 2$ on the nonlinear system

$$4x_1^2 - 20x_1 + \frac{1}{4}x_2^2 + 8 = 0,$$

$$\frac{1}{2}x_1 x_2^2 + 2x_1 - 5x_2 + 8 = 0.$$

SOLUTION: We have

$$\mathbf{F}(x_1, x_2) = \begin{bmatrix} 4x_1^2 - 20x_1 + \frac{1}{4}x_2^2 + 8 \\ \frac{1}{2}x_1 x_2^2 + 2x_1 - 5x_2 + 8 \end{bmatrix}$$

and the Jacobian

$$J(x_1, x_2) = \begin{bmatrix} 8x_1 - 20 & \frac{1}{2}x_2 \\ \frac{1}{2}x_2^2 + 2 & x_1x_2 - 5 \end{bmatrix}.$$

Let $\mathbf{x}(0) = (0,0)^t$, so $\mathbf{F}(\mathbf{x}(0)) = (8,8)^t$. Since $N = 2$, we have $h = 1/2$, and so

$$\mathbf{k}_1 = h\left[-J(\mathbf{x}(0))^{-1}\mathbf{F}(\mathbf{x}(0))\right] = \frac{1}{2}\left(-\begin{bmatrix} -20 & 0 \\ 2 & -5 \end{bmatrix}^{-1}\right)\begin{bmatrix} 8 \\ 8 \end{bmatrix} = \begin{bmatrix} 0.2 \\ 0.88 \end{bmatrix}.$$

Thus

$$\mathbf{x}(\lambda_1) = \mathbf{x}(1/2) = \mathbf{x}(0) + \mathbf{k}_1 = (0.2, 0.88)^t.$$

Then

$$\mathbf{k}_1 = h\left[-J\left(\mathbf{x}(1/2)\right)^{-1}\mathbf{F}(x(0))\right] = \frac{1}{2}\left(-\begin{bmatrix} -18.4 & 0.44 \\ 2.3872 & -4.824 \end{bmatrix}^{-1}\right)\begin{bmatrix} 8 \\ 8 \end{bmatrix}$$
$$= (0.2400604748, 0.9479834920)^t$$

and

$$\mathbf{x}(1) = \mathbf{x}(1/2) + \mathbf{k}_1 = (0.4400604748, 1.827983492)^t.$$

5. a. Approximate the solution to

$$x_1(1 - x_1) + 4x_2 = 12,$$
$$(x_1 - 2)^2 + (2x_2 - 3)^2 = 25$$

using the continuation method and the Runge-Kutta Fourth Order method with $N = 1$ and $\mathbf{x}(0) = (2.5, 4)^t$.

SOLUTION: We have

$$\mathbf{F}(x_1, x_2) = \left(x_1(1 - x_1) + 4x_2 - 12, (x_1 - 2)^2 + (2x_2 - 3)^2 - 25\right)^t,$$

so $\mathbf{F}(2.5, 4) = (0.25.0.25)^t$. The Jacobian is

$$J(x_1, x_2) = \begin{bmatrix} 1 - 2x_1 & 4 \\ 2x_1 - 4 & 8x_2 - 12 \end{bmatrix} \quad \text{and} \quad J(2.5, 4) = \begin{bmatrix} -4 & 4 \\ 1 & 20 \end{bmatrix}.$$

Since $N = 1$ and $h = 1$, we have

$$\mathbf{k}_1 = h\left(-J(\mathbf{x}(0))\right)^{-1}\mathbf{F}(\mathbf{x}(0)) = 1\left(-\begin{bmatrix} -4 & 4 \\ 1 & 20 \end{bmatrix}^{-1}\right)\begin{bmatrix} 0.25 \\ 0.25 \end{bmatrix} = (0.04761904762, -0.01488095239)^t;$$

$$\mathbf{k}_2 = h\left(-J\left(\mathbf{x}(0) + \frac{1}{2}\mathbf{k}_1\right)^{-1}\right)\begin{bmatrix} 0.25 \\ 0.25 \end{bmatrix} = -J(2.523809524, 3.992559524)^{-1}\begin{bmatrix} 0.25 \\ 0.25 \end{bmatrix}$$

$$= -\begin{bmatrix} -4.047619048 & 4 \\ 1.047619048 & 19.94047619 \end{bmatrix}^{-1}\begin{bmatrix} 0.25 \\ 0.25 \end{bmatrix} = (0.04693791021, -0.01500330514)^t;$$

$$\mathbf{k}_3 = -J(2.523468955, 3.992498347)^{-1}\begin{bmatrix} 0.25 \\ 0.25 \end{bmatrix} = (0.04694658114, -0.01500252527)^t;$$

$$\mathbf{k}_4 = -J(2.546946581, 3.984997475)^{-1}\begin{bmatrix} 0.25 \\ 0.25 \end{bmatrix} = (0.04629079276, -0.01512261002)^t;$$

and

$$\mathbf{x}(1) = \mathbf{x}(0) + \frac{1}{6}(\mathbf{k}_1 + 2\mathbf{k}_2 + 2\mathbf{k}_3 + \mathbf{k}_4) = (2.546946470, 3.984997462)^t.$$

The result here is nearly equal to the result obtained in Exercise 4(a) of Section 10.2. However, Newton's method in Section 10.2 required 3 iterations or effectively 3 matrix inversions compared to the 4 required in the continuation method.

10. Show that the continuation method and Euler's method with $N = 1$ gives the same result as Newton's method for the first iteration; that is, with $\mathbf{x}(0) = \mathbf{x}^{(0)}$, we always obtain $\mathbf{x}(1) = \mathbf{x}^{(1)}$.

SOLUTION: The system of differential equations to solve by Euler's method is

$$\mathbf{x}'(\lambda) = -[J(\mathbf{x}(\lambda))]^{-1}\mathbf{F}(\mathbf{x}(0)).$$

With $N = 1$, we have $h = 1$ and

$$\begin{aligned}\mathbf{x}(1) &= \mathbf{x}(0) + h[-J(\mathbf{x}(0))]^{-1}\mathbf{F}(\mathbf{x}(0)) \\ &= \mathbf{x}(0) - J(\mathbf{x}(0))^{-1}\mathbf{F}(\mathbf{x}(0)).\end{aligned}$$

However, Newton's method gives

$$\mathbf{x}^{(1)} = \mathbf{x}^{(0)} - J\left(\mathbf{x}^{(0)}\right)^{-1}\mathbf{F}\left(\mathbf{x}^{(0)}\right).$$

Since $\mathbf{x}(0) = \mathbf{x}^{(0)}$, we have $\mathbf{x}(1) = \mathbf{x}^{(1)}$.

Boundary-Value Problems for Ordinary Differential Equations

Exercise Set 11.1, page 677

1. a. Use the Linear Shooting method with $h = 1/2$ to approximate the solution to

$$y'' = 4(y - x), \qquad 0 \le x \le 1, \quad y(0) = 0, \quad y(1) = 2.$$

SOLUTION: The two initial-value problems we need to solve are

$$y_1'' = 4y_1 - 4x, \qquad 0 \le x \le 1, \quad y_1(0) = 0, \quad y_1'(0) = 0;$$
$$y_2'' = 4y_2, \qquad 0 \le x \le 1, \quad y_2(0) = 0, \quad y_2'(0) = 1.$$

These are written as systems of first order equations

$$u_1' = u_2, \qquad 0 \le x \le 1, \quad u_1(0) = 0, \quad u_2(0) = 0;$$
$$u_2' = 4u_1 - 4x;$$
$$v_1' = v_2, \qquad 0 \le x \le 1, \quad v_1(0) = 0, \quad v_2(0) = 1;$$
$$v_2' = 4v_1.$$

Let $h = 1/2$. This gives $x_0 = 0$, $x_1 = \frac{1}{2}$, and $x_2 = 1$. Then the first iteration of the Runge-Kutta method for systems applied simultaneously to the two systems gives

$$
\begin{aligned}
u_1(x_0) &= 0, & u_2(x_0) &= 0, \\
k_{11} &= 0, & k_{12} &= 0, \\
k_{21} &= 0, & k_{22} &= -0.5, \\
k_{31} &= -0.125, & k_{32} &= -0.5, \\
k_{41} &= -0.25, & k_{42} &= -1.25, \\
u_1(x_1) &\approx -0.08333333, & u_2(x_1) &\approx -0.5416667,
\end{aligned}
$$

213

$$v_1(x_0) = 0, \qquad v_2(x_0) = 1,$$
$$k'_{11} = 0.5, \qquad k'_{12} = 0,$$
$$k'_{21} = 0.5, \qquad k'_{22} = 0.5,$$
$$k'_{31} = 0.625, \qquad k'_{32} = 0.5,$$
$$k'_{41} = 0.75, \qquad k'_{42} = 1.25,$$
$$v_1(x_1) \approx 0.5833333, \, v_2(x_1) \approx 1.5416667.$$

Subsequent approximations are given in the following table.

i	x_i	$u_1(x_i)$	$u_2(x_i)$	$v_1(x_i)$	$v_2(x_i)$
2	1	-0.79861111	-2.3778472	1.7986111	3.7378472

Thus, $y_1(1) = -0.79861111$ and $y_2(1) = 1.7986111$, so

$$y(x) = y_1(x) + \frac{\beta - y_1(1)}{y_2(1)} y_2(x)$$

gives the approximations

$$y(x_i) \approx u_1(x_i) + \frac{2.7986111}{1.7986111} v_1(x_i) \quad \text{and} \quad y'(x_i) \approx u_2(x_i) + \frac{2.7986111}{1.7986111} v_2(x_i).$$

The results are shown in the following table.

i	x_i	w_{1i}	$y(x_i)$
1	0.5	0.82432432	0.82402713

5. Use the Linear Shooting Algorithm to approximate the solution $y = e^{-10x}$ to the boundary-value problem

$$y'' = 100y, \quad 0 \le x \le 1, \quad y(0) = 1, \quad y(1) = e^{-10}.$$

Use $h = 0.1$ and 0.05.

SOLUTION: We have two initial value problems to solve. They are

$$y'' = 100y, \quad 0 \le x \le 1, \quad y(0) = 1, y'(0) = 0,$$

whose solution is denoted as $y_1(x)$, and

$$y'' = 100y, \quad 0 \le x \le 1, \quad y(0) = 0, y'(0) = 1,$$

whose solution is denoted as $y_2(x)$.

Each second-order equation is written as a system of first-order differential equations. The first differential equation with $u_1(x) = y_1(x)$ and $u_2(x) = y_1'(x)$ gives

$$u_1'(x) = u_2(x), \quad u_1(0) = 1, \quad \text{and} \quad u_2'(x) = 100u_1(x), \quad u_2(0) = 0.$$

The second differential equation with $v_1(x) = y_2(x)$ and $v_2(x) = y_2'(x)$ gives

$$v_1'(x) = v_2(x), \quad v_1(0) = 0, \quad \text{and} \quad v_2'(x) = 100v_1(x), \quad v_2(0) = 1.$$

The solution to the original problems is

$$y(x) = y_1(x) + \frac{y(1) - y_1(1)}{y_2(1)} \cdot y_2(x) = y_1(x) + \frac{e^{-10} - y_1(1)}{y_2(1)} \cdot y_2(x).$$

Using the Runge-Kutta method of order 4 with $h = 0.1$ gives the results in the following table.

i	x_i	$u_1(x_i)$	$u_2(x_i)$	$v_1(x_i)$	$v_2(x_i)$	$y(x_i)$	Actual $y(x_i)$
1	0.1	1.541667	11.66667	0.1166667	1.541667	0.3750000	0.3678794
2	0.2	3.737847	35.97222	0.3597222	3.737847	0.1406250	0.1353353
3	0.3	9.959274	99.06539	0.9906539	9.959274	0.0527344	0.0497870
4	0.4	26.91151	268.9173	2.689173	26.91151	0.0197754	0.0183156
5	0.5	72.86227	728.5485	7.285485	72.86227	0.0074156	0.0067379
6	0.6	197.3267	1973.239	19.73239	197.3267	0.0027807	0.0024788
7	0.7	534.4231	5344.221	53.44221	534.4231	0.0010425	0.0009119
8	0.8	1447.395	14473.94	144.7394	1447.395	0.0003900	0.0003355
9	0.9	3920.027	39200.27	392.0027	3920.027	0.0001430	0.0001234
10	1.0	10616.74	106167.4	1061.674	10616.74	0.0000500	0.0000454

Using the Runge-Kutta method of order 4 with $h = 0.05$ gives the comparable results in the following table.

i	x_i	$u_1(x_i)$	$u_2(x_i)$	$v_1(x_i)$	$v_2(x_i)$	$y(x_i)$	Actual $y(x_i)$
2	0.1	1.542759	11.74588	0.1174588	1.542759	0.3681708	0.3678794
4	0.2	3.759760	36.24210	0.3624210	3.759760	0.1355498	0.1353353
6	0.3	10.05735	100.0745	1.000745	10.05735	0.0499055	0.0497870
8	0.4	27.27070	272.5232	2.725232	7.270696	0.0183737	0.0183156
10	0.5	74.08234	740.7557	7.407558	122.1166	0.0067647	0.0067379
12	0.6	201.2994	2012.969	20.12970	201.2994	0.0024905	0.0024788
14	0.7	546.9973	5469.964	54.69964	46.99728	0.0009168	0.0009119
16	0.8	1486.380	14863.80	148.6380	1486.380	0.0003380	0.0003355
18	0.9	4039.008	40390.08	403.9008	4039.008	0.0001260	0.0001234
20	1.0	10975.38	109753.8	1097.540	10975.38	0.0000500	0.0000454

6. Write the second-order initial-value problems (11.3) and (11.4) as first order systems, and derive the equations needed to solve the systems, using the fourth-order Runge-Kutta method for systems.

SOLUTION: For Eq. (11.3), let $u_1(x) = y$ and $u_2(x) = y'$. Then

$$u_1'(x) = u_2(x), \quad a \leq x \leq b, \quad u_1(a) = \alpha$$

and

$$u_2'(x) = p(x)u_2(x) + q(x)u_1(x) + r(x), \quad a \leq x \leq b, \quad u_2(a) = 0.$$

For Eq. (11.4), let $v_1(x) = y$ and $v_2(x) = y'$. Then

$$v_1'(x) = v_2(x), \quad a \leq x \leq b, \quad v_1(a) = 0$$

and

$$v_2'(x) = p(x)v_2(x) + q(x)v_1(x), \quad a \leq x \leq b, \quad v_2(a) = 1.$$

Using the notation $u_{1,i} = u_1(x_i), u_{2,i} = u_2(x_i), v_{1,i} = v_1(x_i)$ and $v_{2,i} = v_2(x_i)$ leads to the equations in Step 4 of Algorithm 11.1.

8. Show that if y_2 is a solution to $y'' = p(x)y' + q(x)y$ and $y_2(a) = y_2(b) = 0$, then $y_2 \equiv 0$.

SOLUTION: Since $y_2(a) = 0$ and $y_2(b) = 0$, the boundary value problem

$$y'' = p(x)y' + q(x)y, \quad a \leq x \leq b \quad, \quad y(a) = 0, \quad y(b) = 0$$

has $y = 0$ as a unique solution. This implies that $y_2 \equiv 0$.

10. Attempt to apply Exercise 9 to the boundary-value problem

$$y'' - y = 0, \quad 0 \leq x \leq b, \quad y(0) = 0, \quad y(b) = B.$$

SOLUTION: The unique solution to this problem is

$$y(x) = B\frac{e^x - e^{-x}}{e^b - e^{-b}}.$$

Corollary 11.2 applies to this problem, but not to Exercise 9, since in that case we have $q(x) < 0$.

Exercise Set 11.2, page 684

1. Use the Nonlinear Shooting method with $h = 0.5$ to approximate the solution to

$$y'' = -(y')^2 - y + \ln x, \quad 1 \leq x \leq 2, \quad y(1) = 0, \quad y(2) = \ln 2.$$

SOLUTION: The two initial-value problems we need to solve are

$$y'' = -(y')^2 - y + \ln x, \quad 1 \leq x \leq 2, \quad y(1) = 0, \quad y'(1) = t_k;$$
$$z'' = (-2y')z' - z, \quad 1 \leq x \leq 2, \quad z(1) = 0, \quad z'(1) = 1.$$

These are written as systems of first-order equations

$$u_1' = u_2, \qquad\qquad 1 \le x \le 2, \quad u_1(1) = 0, \quad u_2(1) = t_k;$$
$$u_2' = -u_2^2 - u_1 + \ln x;$$
$$v_1' = v_2, \qquad\qquad 1 \le x \le 2, \quad v_1(1) = 0, \quad v_2(1) = 1;$$
$$v_2' = -2u_2v_2 - v_1.$$

Now set

$$t_0 = \frac{\ln 2 - 0}{2 - 1} = \ln 2.$$

The Runge-Kutta method for systems gives

$$u_1(1.5) \approx 0.3020796, \quad v_1(1.5) \approx 0.3269534,$$

$$u_1(2.0) \approx 0.5535444, \quad v_1(2.0) \approx 0.4115828.$$

Thus

$$t_1 = \ln 2 - \frac{0.5535444 - \ln 2}{0.4115825} = 1.032332.$$

Since $|0.553544 - \ln 2| > 10^{-4}$, we continue iterating the Runge-Kutta approximations. Convergence occurs in 8 iterations with

$$u_1(1.5) \approx 0.4055354 \quad \text{and} \quad u_1(2.0) \approx 0.6931944.$$

4. a. Use the Nonlinear Shooting method with $TOL = 10^{-4}$ and $h = 0.1$ to approximate the solution to

$$y'' = y^3 - yy', \quad 1 \le x \le 2, \quad y(1) = \frac{1}{2}, \quad y(2) = \frac{1}{3},$$

whose actual solution is $y(x) = (x+1)^{-1}$.

SOLUTION: For this problem we have

$$f(x, y, y') = y^3 - y \cdot y', \quad \text{so} \quad \frac{\partial f}{\partial y}(x, y, y') = 3y^2 - y' \text{ and } \frac{\partial f}{\partial y'}(x, y, y') = -y.$$

The two initial value problems to be solved are

$$y''(x) = y(x)^3 - y(x)y'(x), \quad 1 \le x \le 2, \quad y(1) = \frac{1}{3}, \, y'(1) = t_k,$$

and

$$z''(x) = \left(3y(x)^2 - y'(x)\right)z(x) - y(x)z'(x), \quad 1 \le x \le 2, \quad z(1) = 0, \, z'(1) = 1.$$

Applying the Runge-Kutta method of order 4 with $h = 0.1$ to both equations gives the following results and the comparison of the approximation to the actual solution.

i	x_i	w_{1i}	$y(x_i)$	$|y(x_i) - w_{1i}|$
1	1.1	0.47619079	0.47619048	0.00000031
2	1.2	0.45454606	0.45454545	0.00000060
3	1.3	0.43478349	0.43478261	0.00000088
4	1.4	0.41666781	0.41666667	0.00000114
5	1.5	0.40000140	0.40000000	0.00000140
6	1.6	0.38461703	0.38461538	0.00000165
7	1.7	0.37037226	0.37037037	0.00000189
8	1.8	0.35714500	0.35714286	0.00000214
9	1.9	0.34482998	0.34482759	0.00000239
10	2.0	0.33333597	0.33333333	0.00000264

Convergence was in 3 iterations with $t_3 = -0.2.4999738$.

5. **a.** Change Algorithm 11.2 to incorporate the Secant method with

$$t_0 = \frac{\beta - \alpha}{b - a} \quad \text{and} \quad t_1 = \frac{t_0 + (\beta - y(b, t_0))}{b - a}.$$

SOLUTION: Modify Algorithm 11.2 as follows:

Step 1   Set $h = (b - a)/N$;
$\quad\quad\quad\quad\ k = 2$;
$\quad\quad\quad\quad\ TK1 = (\beta - \alpha)/(b - a)$.

Step 2   Set $w_{1,0} = \alpha$;
$\quad\quad\quad\quad\ w_{2,0} = TK1$.

Step 3   For $i = 1, \ldots, N$ do Steps 4 and 5.

$\quad$ *Step 4*   Set $x = a + (i - 1)h$.

$\quad$ *Step 5*   Set

$\quad\quad\quad\quad k_{1,1} = hw_{2,i-1}$;
$\quad\quad\quad\quad k_{1,2} = hf(x, w_{1,i-1}, w_{2,i-1})$;
$\quad\quad\quad\quad k_{2,1} = h(w_{2,i-1} + k_{1,2}/2)$;
$\quad\quad\quad\quad k_{2,2} = hf(x + h/2, w_{1,i-1} + k_{1,1}/2, w_{2,i-1} + k_{1,2}/2)$;
$\quad\quad\quad\quad k_{3,1} = h(w_{2,i-1} + k_{2,2}/2)$;
$\quad\quad\quad\quad k_{3,2} = hf(x + h/2, w_{1,i-1} + k_{2,1}/2, w_{2,i-1} + k_{2,2}/2)$;
$\quad\quad\quad\quad k_{4,1} = h(w_{2,i-1} + k_{3,2}/2)$;
$\quad\quad\quad\quad k_{4,2} = hf(x + h/2, w_{1,i-1} + k_{3,1}, w_{2,i-1} + k_{3,2})$;
$\quad\quad\quad\quad w_{1,i} = w_{1,i-1} + (k_{1,1} + 2k_{2,1} + 2k_{3,1} + k_{4,1})/6$;
$\quad\quad\quad\quad w_{2,i} = w_{2,i-1} + (k_{1,2} + 2k_{2,2} + 2k_{3,2} + k_{4,2})/6$.

Step 6   Set $TK2 = TK1 + (\beta - w_{1,N})/(b - a)$.

Step 7   While $(k \leq M)$ do Steps 8–15.

Step 8   Set $w_{2,0} = TK2$;
$HOLD = w_{1,N}$.

Step 9   For $i = 1, \ldots, N$ do Steps 10 and 11.

    Step 10   (Same as Step 4)

    Step 11   (Same as Step 5)

Step 12   If $|w_{1,N} - \beta| \leq TOL$ then do Steps 13 and 14.

    Step 13   For $i = 0, \ldots, N$ set $x = a + ih$;
        OUTPUT$(x, w_{1,i}, w_{2,i})$.

    Step 14   STOP.

Step 15   Set
$TK = TK2 - (w_{1,N} - \beta)(TK2 - TK1)/(w_{1.N} - HOLD)$;
$TK1 = TK2$;
$TK2 = TK$;
$k = k + 1$.

Step 16   OUTPUT('Maximum number of iterations exceeded.');
STOP.

Exercise Set 11.3, page 689

1. Use the Linear Finite-Difference method to approximate the solution to

$$y'' = 4(y - x), \quad 0 \leq x \leq 1, \quad y(0) = 0, \quad y(1) = 2;$$

a. with $h = 1/2$,

b. with $h = 1/4$,

c. with extrapolation to approximate $y(1/2)$.

SOLUTION: Since $p(x) = 0$, $q(x) = 4$, and $r(x) = -4x$, the difference equations for this boundary-value problem are

$$-w_{i-1} + \left(2 + 4h^2\right) w_i - w_{i+1} = 4h^2 x_i,$$

for each $i = 1, 2, \ldots, N$.

a. With $h = 1/2$, we have $N = 2$, $x_1 = 1/2$, $w_0 = 0$, and $w_2 = 2$. The system reduces to the single equation for $i = 1$ to

$$-w_0 + \left(2 + 4(1/2)^2\right) w_1 - w_2 = 4(1/2)^2(x_1).$$

Substituting $w_0 = 0$, $w_2 = 2$, and $x_1 = 1/2$ yields

$$w_1 = 5/6 = 0.83333333.$$

b. With $h = 1/4$, we have $N = 3$, $x_1 = 1/4$, $x_2 = 1/2$, $x_3 = 3/4$, $w_0 = 0$, and $w_4 = 2$. The resulting system is

$$-w_0 + \left(2 + 4h^2\right) w_1 - w_2 = 4h^2 x_1,$$
$$-w_1 + \left(2 + 4h^2\right) w_2 - w_3 = 4h^2 x_2,$$
$$-w_2 + \left(2 + 4h^2\right) w_3 - w_4 = 4h^2 x_3.$$

Making the appropriate substitutions gives

$$\frac{9}{4}w_1 \quad -w_2 \qquad\qquad = \frac{1}{16},$$
$$-w_1 + \frac{9}{4}w_2 \quad -w_3 = \frac{1}{18},$$
$$-w_2 + \frac{9}{4}w_3 = \frac{35}{16}.$$

Solving the system gives $w_1 = 0.39512472$, $w_2 = 0.82653061$, and $w_3 = 1.33956916$.

We construct the following table.

i	x_i	w_{1i}	$y(x_i)$
1	0.25	0.39512472	0.39367669
2	0.50	0.82653061	0.82402714
3	0.75	1.33956916	1.33708613

c. Using extrapolation gives

$$y(1/2) \approx \frac{4(0.82653061) - 0.83333333}{3} \approx 0.82426304.$$

5. Use the Linear Finite-Difference Algorithm to approximate the solution $y = e^{-10x}$ to the boundary-value problem

$$y'' = 100y, \quad 0 \le x \le 1, \quad y(0) = 1, \quad y(1) = e^{-10}.$$

Use $h = 0.1$ and 0.05.

SOLUTION: For this problem we have

$p(x) = 0$, $q(x) = 100$, $r(x) = 0$, $a = 0$, $b = 1$, $\alpha = 1$, $\beta = e^{-10} \approx 4.54 \times 10^{-5}$, $N = 9$, and $h = 0.1$.

In the linear system $A\mathbf{w} = \mathbf{b}$ that needs to be solved we have A the tridiagonal matrix with diagonal entries are all 3, and whose nonzero off-diagonal entries are all -1, The vector $\mathbf{b}$ has all zero entries except the first, with $b_1 = 1$, and the last, with $b_{10} = \beta$.

i	x_i	w_i	$y(x_i)$	$|y(x_i) - w_i|$
1	0.1	0.38196601	0.36787944	0.01408657
2	0.2	0.14589802	0.13533528	0.01056274
3	0.3	0.05572807	0.04978707	0.00594100
4	0.4	0.02128617	0.01831564	0.00297053
5	0.5	0.00813045	0.00673795	0.00139250
6	0.6	0.00310518	0.00247875	0.00062643
7	0.7	0.00118509	0.00091188	0.00027321
8	0.8	0.00045008	0.00033546	0.00011462
9	0.9	0.00016516	0.00012341	0.00004175

Applying Algorithm to this problem with $h = 0.05$ gives the comparable results in the following table.

i	x_i	w_i	$y(x_i)$	$|y(x_i) - w_i|$
1	0.2	0.37162654	0.36787944	0.00374710
2	0.4	0.13810629	0.13533528	0.00277100
3	0.6	0.05132396	0.04978707	0.00153689
4	0.8	0.01907333	0.01831564	0.00075769
5	1.0	0.00708813	0.00673795	0.00035018
6	1.2	0.00263406	0.00247875	0.00015530
7	1.4	0.00097867	0.00091188	0.00006679
8	1.6	0.00036312	0.00033546	0.00002766
9	1.8	0.00013340	0.00012341	0.00000999

9. Prove that if p, q, and r are continuous and $q(x) \geq 0$, then the tridiagonal system in (11.19) has a unique solution provided that $h < 2/L$, where $L = \max_{a \leq x \leq b}$.

SOLUTION: We will prove this result be using Theorem 6.31. First note that

$$\left| \frac{h}{2} p(x_i) \right| \leq \frac{hL}{2} < 1,$$

so

$$\left| -1 - \frac{h}{2} p(x_i) \right| = 1 + \frac{h}{2} p(x_i) \quad \text{and} \quad \left| -1 + \frac{h}{2} p(x_i) \right| = 1 - \frac{h}{2} p(x_i).$$

Therefore,

$$\left| -1 - \frac{h}{2} p(x_i) \right| + \left| -1 + \frac{h}{2} p(x_i) \right| = 2 \leq 2 + h^2 q(x_i),$$

for $2 \leq i \leq N - 1$. Since

$$\left| -1 + \frac{h}{2} p(x_1) \right| < 2 \leq 2 + h^2 q(x_1) \quad \text{and} \quad \left| -1 - \frac{h}{2} p(x_N) \right| < 2 \leq 2 + h^2 q(x_N),$$

Theorem 6.31 on page 424 implies that the linear system (11.19) has a unique solution.

10. Show that if $y \in C^6[a, b]$, $q(x) > 0$, and $w_0, w_1, \ldots, w_{N+1}$ satisfy (11.18), then

$$w_i - y(x_i) = Ah^2 + O\left(h^4\right),$$

where A is independent of h, provided $q(x) \geq w > 0$ for some w.

SOLUTION: Let $q(x) \geq w > 0$ on $[a, b]$. Using the sixth Taylor polynomial gives

$$\frac{y(x_{i+1}) - y(x_{i-1})}{2h} = y'(x_i) + \frac{h^2}{6} y'''(x_i) + \frac{h^4}{120} y^{(5)}(x_i) + O\left(h^5\right)$$

and

$$\frac{y(x_{i+1}) - 2y(x_i) + y(x_{i-1})}{h^2} = y''(x_i) + \frac{h^2}{12} y^{(4)}(x_i) + O\left(h^4\right).$$

Thus,

$$\left(2 + h^2 q(x_i)\right) y(x_i) - \left(1 - \frac{h}{2} p(x_i)\right) y(x_{i+1}) - \left(1 + \frac{h}{2} p(x_i)\right) y(x_{i-1}) + h^2 r(x_i)$$

$$= p(x_i) \frac{h^4}{6} y'''(x_i) - \frac{h^4}{12} y^{(4)}(x_i) + O\left(h^6\right).$$

Subtracting h^2 times Equation (11.18) gives

$$\left(2 + h^2 q(x_i)\right) (y(x_i) - w_i) = \left(1 - \frac{h}{2} p(x_i)\right) (y(x_{i+1}) - w_{i+1}) + \left(1 + \frac{h}{2} p(x_i)\right) (y(x_{i-1}) - w_{i-1})$$

$$+ \left[\frac{p(x_i)}{6} y'''(x_i) - \frac{1}{12} y^{(4)}(x_i)\right] h^4 + O\left(h^6\right).$$

Let E denote the maximum approximation error; that is,

$$E = \max_{0 \le i \le N+1} |y(x_i) - w_i|.$$

Then since $\left|\frac{h}{2} p(x_i)\right| < 1$, we have

$$\left(2 + h^2 q(x_i)\right) (y(x_i) - w_i) \le 2E + h^4 \left|\frac{p(x_i)}{6} y'''(x_i) - \frac{1}{12} y^{(4)}(x_i)\right| + O\left(h^6\right).$$

Now let $K_1 = \max_{a \le x \le b} |y'''(x)|$ and $K_2 = \max_{a \le x \le b} |y^{(4)}(x)|$. If $q(x_i) \ge w$, then

$$\left(2 + h^2 w\right) E \le 2E + h^4 \left[\frac{L K_1}{6} + \frac{K_2}{12}\right] + O\left(h^6\right)$$

and

$$E = \max_{0 \le i \le N+1} |y(x_i) - w_i| \le h^2 \left[\frac{2 L K_1 + K_2}{12w}\right] + O\left(h^4\right).$$

Exercise Set 11.4, page 696

1. Use the Nonlinear Finite-Difference Algorithm with $h = 0.5$ to approximate the solution to

$$y'' = -(y')^2 - y + \ln x, \quad 1 \le x \le 2, \quad y(1) = 0, \quad y(2) = \ln 2.$$

SOLUTION: Since $h = 0.5$ and $h = (b - a)/(N + 1)$, we have $N = 1$. The difference equation becomes

$$w_0 = \alpha = 0, \quad w_2 = \beta = \ln 2,$$

and

$$-\frac{w_2 - 2w_1 + w_0}{h^2} + f\left(x_1, w_1, \frac{w_2 - w_0}{2h}\right) = 0.$$

Thus

$$-w_2 + 2w_1 - w_0 + \frac{1}{4} f(1.5, w_1, \ln 2) = 0$$

and

$$-\ln 2 + 2w_1 + \frac{1}{4}\left[-(\ln 2)^2 - w_1 + \ln(1.5)\right] = 0,$$

which gives $w_1 = 0.406797$.

4. a. Use the Nonlinear Finite-Difference Algorithm with $TOL = 10^{-4}$ to approximate the solution to

$$y'' = y^3 - yy', \quad 1 \le x \le 2, \quad y(1) = \frac{1}{2}, \quad y(2) = \frac{1}{3},$$

whose actual solution is $y(x) = (x+1)^{-1}$.

SOLUTION: For this problem we have

$$f(x, y, y') = y^3 - y \cdot y', \quad a = 1, \quad b = 2, \quad \alpha = 0.5, \quad \beta = \frac{1}{3}, \quad N = 4, \quad \text{and} \quad h = \frac{b-a}{N+1} = 0.2.$$

The finite-differences method requires that the following nonlinear system be solved:

$$0 = 2w_1 - w_2 + h^2 f\left(x_1, w_1, \frac{1}{2}\frac{w_2 - \alpha}{h}\right) - \alpha = 2w_1 - w_2 + 0.04w_1^3 - 0.04w_1(2.5w_2 - 1.25) - 0.5,$$

$$0 = -w_1 + 2w_2 - w_3 + h^2 f\left(x_2, w_2, \frac{1}{2}\frac{w_3 - w_2}{h}\right) = -w_1 + 2w_2 - w_3 + 0.04w_2^3 - 0.04w_2(2.5w_3 - 2.5w_1),$$

$$0 = -w_2 + 2w_3 - w_4 + h^2 f\left(x_3, w_3, \frac{1}{2}\frac{w_4 - w_3}{h}\right) = -w_2 + 2w_3 - w_4 + 0.04w_3^3 - 0.04w_3(2.5w_4 - 2.5w_2),$$

$$0 = -w_3 + 2w_4 + h^2 f\left(x_4, w_4, \frac{1}{2}\frac{\beta - w_3}{h}\right) - \beta = -w_3 + 2w_4 + 0.04w_4^3 - 0.04w_4\left(\frac{5}{6} - 2.5w_3\right) - \frac{1}{3}.$$

Using the Nonlinear Finite Difference Algorithm 11.4, which uses Newton's method for nonlinear systems, gives the approximations in the table.

i	x_i	w_i	$y(x_i)$	$|y(x_i) - w_i|$
0	1.0	0.50000000		
1	1.2	0.45458862	0.45454545	0.00004317
2	1.4	0.41672067	0.41666667	0.00005400
3	1.6	0.38466137	0.38461538	0.00004599
4	1.8	0.35716943	0.35714286	0.00002657
5	2.0	0.33333333		

5. a. Repeat Exercise 4(a) using extrapolation.

SOLUTION: Repeating the method in Exercise 4(a) with $h = 0.1$ and $h = 0.05$ gives the following results.

x_i	$w_i(h = 0.2)$	$w_i(h = 0.1)$	$w_i(h = 0.05)$
1.00	0.5000000000	0.5000000000	0.5000000000
1.05			0.4878058215
1.10		0.4761972439	0.4761921720
1.15			0.4651185619
1.20	0.4545886201	0.4545563382	0.4545481813
1.25			0.4444474908
1.30		0.4347956122	0.4347858663
1.35			0.4255352892
1.40	0.4167206681	0.4166802725	0.4166700749
1.45			0.4081666349
1.50		0.4000130439	0.4000032672
1.55			0.3921599714
1.60	0.3846613728	0.3846269650	0.3846182851
1.65			0.3773611389
1.70		0.3703797826	0.3703727277
1.75			0.3636383953
1.80	0.3571694307	0.3571495457	0.3571445322
1.85			0.3508784839
1.90		0.3448311085	0.3448284683
1.95			0.3389835019
2.00	0.3333333333	0.3333333333	0.3333333333

The first extrapolations

$$\text{Ext}_{i1}(0.2, 0.1) = \frac{1}{3}(4w_i(0.1) - w_i(0.2)), \quad \text{and} \quad \text{Ext}_{i2}(0.1, 0.05) = \frac{1}{3}(4w_i(0.1) - w_i(0.2)),$$

and the final extrapolation

$$\text{Ext}_{i3}(0.2, 0.1, 0.05) = \frac{1}{15}(16\text{Ext}_i 2(0.1, 0.05) - \text{Ext}_i 1(0.2, 0.1))$$

are shown in the table

x_i	Ext_{i1}	Ext_{i2}	Ext_{i3}	$y(x_i)$
1.20	0.4545455776	0.4545454623	0.4545454546	0.4545454545
1.30		0.4347826177		
1.40	0.4166668073	0.4166666757	0.4166666670	0.4166666667
1.50		0.4000000083		
1.60	0.3846154957	0.3846153918	0.3846153848	0.3846153846
1.70		0.3703703761		
1.80	0.3571429174	0.3571428610	0.3571428573	0.3571428571
1.90		0.3448275883		

Extrapolation made a significant accuracy difference even over the approximations with $h = 0.05$.

7. Show that the hypotheses listed at the beginning of the section ensure the nonsingularity of the Jacobian matrix for $h < 2/L$.

SOLUTION: We will apply Theorem 6.31 on page 424 to prove the result. The Jacobian matrix $J = (a_{i,j})$ is tridiagonal with entries given in (11.21). So

$$a_{1,1} = 2 + h^2 f_y \left(x_1, w_1, \frac{1}{2h}(w_2 - \alpha) \right),$$

$$a_{1,2} = -1 + \frac{h}{2} f_{y'} \left(x_1, w_1, \frac{1}{2h}(w_2 - \alpha) \right),$$

$$a_{i,i-1} = -1 - \frac{h}{2} f_{y'} \left(x_i, w_i, \frac{1}{2h}(w_{i+1} - w_{i-1}) \right), \quad \text{for } 2 \le i \le N - 1;$$

$$a_{i,i} = 2 + h^2 f_y \left(x_i, w_i, \frac{1}{2h}(w_{i+1} - w_{i-1}) \right), \quad \text{for } 2 \le i \le N - 1;$$

$$a_{i,i+1} = -1 + \frac{h}{2} f_{y'} \left(x_i, w_i, \frac{1}{2h}(w_{i+1} - w_{i-1}) \right), \quad \text{for } 2 \le i \le N - 1;$$

$$a_{N,N-1} = -1 - \frac{h}{2} f_{y'} \left(x_N, w_N, \frac{1}{2h}(\beta - w_{N-1}) \right),$$

$$a_{N,N} = 2 + h^2 f_y \left(x_N, w_N, \frac{1}{2h}(\beta - w_{N-1}) \right).$$

Thus, $|a_{i,i}| \ge 2 + h^2 \delta$, for $i = 1, \ldots, N$. Since $|f_{y'}(x, y, y')| \le L$ and $h < 2/L$, we have

$$\left| \frac{h}{2} f_{y'}(x, y, y') \right| \le \frac{hL}{2} < 1.$$

So

$$|a_{1,2}| = \left| -1 + \frac{h}{2} f_{y'} \left(x_1, w_1, \frac{1}{2h}(w_2 - \alpha) \right) \right| < 2 < |a_{1,1}|,$$

$$|a_{i,i-1}| + |a_{i,i+1}| = -a_{i,i-1} - a_{i,i+1}$$

$$= 1 + \frac{h}{2} f_{y'} \left(x_i, w_i, \frac{1}{2h}(w_{i+1} - w_{i-1}) \right) + 1 - \frac{h}{2} f_{y'} \left(x_i, w_i, \frac{1}{2h}(w_{i+1} - w_{i-1}) \right)$$

$$= 2 \le |a_{i,i}|,$$

and

$$|a_{N,N-1}| = -a_{N,N-1} = 1 + \frac{h}{2} f_{y'} \left(x_N, w_N, \frac{1}{2h}(\beta - w_{N-1}) \right) < 2 < |a_{N,N}|.$$

By Theorem 6.31, J is nonsingular.

Exercise Set 11.5, page 710

1. Use the Piecewise Linear Shooting method with $x_0 = 0$, $x_1 = 0.3$, $x_2 = 0.7$, and $x_3 = 1$ to approximate the solution to

$$y'' + \frac{\pi^2}{4} y = \frac{\pi^2}{16} \cos \frac{\pi}{4} x, \quad 0 \le x \le 1, \quad y(0) = y(1) = 0.$$

0.4761904814

SOLUTION: We have $x_0 = 0$, $x_1 = 0.3$, $x_2 = 0.7$, and $x_3 = 1$, so $n = 2$, $h_0 = 0.3$, $h_1 = 0.4$, and $h_2 = 0.3$. The basis functions are

$$\phi_1 = \begin{cases} \dfrac{x}{0.3}, & \text{if } 0 < x \le 0.3 \\[2mm] \dfrac{0.7 - x}{0.4}, & \text{if } 0.3 < x \le 0.7 \\[2mm] 0, & \text{elsewhere;} \end{cases} \quad \text{and} \quad \phi_2 = \begin{cases} \dfrac{x - 0.3}{0.4}, & \text{if } 0.3 < x \le 0.7 \\[2mm] \dfrac{1.0 - x}{0.3}, & \text{if } 0.7 < x \le 1.0 \\[2mm] 0, & \text{elsewhere.} \end{cases}$$

Since

$$p(x) = 1, \quad q(x) = -\frac{1}{4}\pi^2, \quad \text{and} \quad f(x) = -\frac{\pi^2}{16}\cos\frac{\pi}{4}x,$$

we have

$$Q_{11} = \frac{1}{h_1^2}\int_{x_1}^{x_2}(x_2 - x)(x - x_1)\left(\frac{-\pi^2}{4}\right)dx = -0.1644934,$$

$$Q_{21} = \frac{1}{h_0^2}\int_{x_0}^{x_1}(x - x_0)^2\left(\frac{-\pi^2}{4}\right)dx = -0.2467401,$$

$$Q_{22} = \frac{1}{h_1^2}\int_{x_1}^{x_2}(x - x_1)^2\left(\frac{-\pi^2}{4}\right)dx = -0.3289868,$$

$$Q_{31} = \frac{1}{h_1^2}\int_{x_1}^{x_2}(x_2 - x)^2\left(\frac{-\pi^2}{4}\right)dx = -0.3289868,$$

$$Q_{32} = \frac{1}{h_2^2}\int_{x_2}^{x_3}(x_3 - x)^2\left(\frac{-\pi^2}{4}\right)dx = -0.2467401,$$

$$Q_{41} = \frac{1}{h_0^2}\int_{x_0}^{x_1}dx = 3.\overline{3},$$

$$Q_{42} = \frac{1}{h_1^2}\int_{x_1}^{x_2}dx = 2.5,$$

$$Q_{43} = \frac{1}{h_2^2}\int_{x_2}^{x_3}dx = 3.\overline{3},$$

$$Q_{51} = \frac{1}{h_0}\int_{x_0}^{x_1}(x - x_0)\left(\frac{-\pi^2}{16}\cos\frac{\pi}{4}x\right)dx = -0.09124729,$$

$$Q_{52} = \frac{1}{h_1}\int_{x_1}^{x_2}(x - x_1)\left(\frac{-\pi^2}{16}\cos\frac{\pi}{4}x\right)dx = -0.1110450,$$

$$Q_{61} = \frac{1}{h_1}\int_{x_1}^{x_2}(x_2 - x)\left(\frac{-\pi^2}{16}\cos\frac{\pi}{4}x\right)dx = -0.1159768,$$

$$Q_{62} = \frac{1}{h_2}\int_{x_2}^{x_3}(x_3 - x)\left(\frac{-\pi^2}{16}\cos\frac{\pi}{4}x\right)dx = -0.07474186.$$

So

$$a_{11} = Q_{41} + Q_{42} + Q_{21} + Q_{31} = 5.257606,$$
$$a_{22} = Q_{42} + Q_{43} + Q_{22} + Q_{32} = 5.257606,$$
$$a_{12} = -Q_{42} + Q_{11} = -2.664493,$$
$$a_{21} = -Q_{42} + Q_{11} = -2.664493,$$
$$b_1 = Q_{51} + Q_{61} = -0.2072241,$$
$$b_2 = Q_{52} + Q_{62} = -0.1857869.$$

The solution to the linear system $A\mathbf{c} = \mathbf{b}$ is $c_1 = -0.07713274$ and $c_2 = -0.07442678$. The following table illustrates the accuracy of the approximations.

x	$\phi(x)$	$y(x)$
0.3	-0.07713274	-0.07988545
0.7	-0.07442678	-0.07712903

4. a. Use the Cubic Spline method with $n = 3$ to approximate the solution to the boundary-value problem in Exercise 1.

SOLUTION: Since $n = 3$, we have $h = 1/(n + 1) = 0.25$. Thus, $x_0 = 0$, $x_1 = 0.25$, $x_2 = 0.5$, $x_3 = 0.75$, and $x_4 = 1$. The basis functions are

$$\phi_0(x) = S(4x) - 4S(4x + 1),$$
$$\phi_1(x) = S(4x - 1) - S(4x + 1),$$
$$\phi_2(x) = S(4x - 2),$$
$$\phi_3(x) = S(4x - 3) - S(4x - 5),$$
$$\phi_4(x) = S(4x - 4) - 4S(4x - 5).$$

The entries in the matrix A are given by the following, where $p(x) = 1$ and $q(x) = -\pi^2/4$.

$$a_{00} = \int_0^{x_2} \left[p(x)(\phi_0'(x))^2 + q(x)(\phi_0(x))^2 \right] \, dx = 0.1628725,$$

$$a_{01} = \int_0^{x_2} \left[p(x)\phi_0'(x)\phi_1'(x) + q(x)\phi_0(x)\phi_1(x) \right] \, dx = a_{10} = 0.08367032,$$

$$a_{02} = \int_0^{x_2} \left[p(x)\phi_0'(x)\phi_2'(x) + q(x)\phi_0(x)\phi_2(x) \right] \, dx = a_{20} = -0.04255400,$$

$$a_{03} = \int_{x_1}^{x_2} \left[p(x)\phi_0'(x)\phi_3'(x) + q(x)\phi_0(x)\phi_3(x) \right] \, dx = a_{30} = -0.002090983,$$

$$a_{11} = \int_0^{x_3} \left[p(x)(\phi_1'(x))^2 + q(x)(\phi_1(x))^2 \right] \, dx = 0.1991036,$$

$$a_{12} = \int_0^{x_3} \left[p(x)\phi_1'(x)\phi_2'(x) + q(x)\phi_1(x)\phi_2(x) \right] \, dx = a_{21} = -0.03826949,$$

$$a_{13} = \int_{x_1}^{x_3} \left[p(x)\phi_1'(x)\phi_3'(x) + q(x)\phi_1(x)\phi_3(x) \right] \, dx = a_{31} = -0.05091793,$$

$$a_{14} = \int_{x_2}^{x_3} \left[p(x)\phi_1'(x)\phi_4'(x) + q(x)\phi_1(x)\phi_4(x) \right] \, dx = a_{41} = -0.002090983,$$

$$a_{22} = \int_0^1 \left[p(x)(\phi_2'(x))^2 + q(x)(\phi_2(x))^2 \right] \, dx = 0.1481856,$$

$$a_{23} = \int_{x_1}^1 \left[p(x)\phi_2'(x)\phi_3'(x) + q(x)\phi_2(x)\phi_3(x) \right] \, dx = a_{32} = -0.03826949,$$

$$a_{24} = \int_{x_2}^1 \left[p(x)\phi_2'(x)\phi_4'(x) + q(x)\phi_2(x)\phi_4(x) \right] \, dx = a_{42} = -0.04255400,$$

$$a_{33} = \int_{x_1}^1 \left[p(x)(\phi_3'(x))^2 + q(x)(\phi_3(x))^2 \right] \, dx = 0.1991036,$$

$$a_{34} = \int_{x_2}^1 \left[p(x)\phi_3'(x)\phi_4'(x) + q(x)\phi_3(x)\phi_4(x) \right] \, dx = a_{43} = 0.08367032,$$

$$a_{44} = \int_{x_2}^1 \left[p(x)(\phi_4'(x))^2 + q(x)(\phi_4(x))^2 \right] \, dx = 0.1628725.$$

With $f(x) = -\dfrac{\pi^2}{16} \cos \dfrac{\pi}{4} x$, we have

$$b_0 = \int_0^{x_2} f(x)\phi_0(x) \, dx = -0.01273576,$$

$$b_1 = \int_0^{x_3} f(x)\phi_1(x) \, dx = -0.03436138,$$

$$b_2 = \int_0^1 f(x)\phi_2(x) \, dx = -0.03539018,$$

$$b_3 = \int_{x_1}^1 f(x)\phi_3(x) \, dx = -0.02958148,$$

$$b_4 = \int_{x_2}^1 f(x)\phi_4(x) \, dx = -0.01007119.$$

The 5×5 linear system $Ac = b$ has the solution $c_0 = -0.02628760$, $c_1 = -0.3179557$, $c_2 = -0.4116851$, $c_3 = -0.3014723$, and $c_4 = -0.01860715$.

The following table shows the accuracy of the approximations $\phi(x) = \sum_{i=0}^{4} c_i \phi_i(x)$.

j	x_j	$\phi(x_j)$	$y(x_j)$
1	0.25	-0.07124149	-0.07123077
2	0.5	-0.09442369	-0.09440908
3	0.75	-0.06817423	-0.06816510

6. Show that

$$-\frac{d}{dx}(p(x)y') + q(x)y = f(x), \quad 0 \le x \le 1, \quad y(0) = \alpha, \quad y(1) = \beta$$

can be transformed into

$$-\frac{d}{dx}(p(x)z') + q(x)z = F(x), \quad 0 \le x \le 1, \quad z(0) = 0, \quad z(1) = 0.$$

SOLUTION: Define

$$z = y + (\alpha - \beta)x - \alpha.$$

Then

$$z(0) = y(0) - \alpha = \alpha - \alpha = 0$$

and

$$z(1) = y(1) + (\alpha - \beta) - \alpha = y(1) - \beta = \beta - \beta = 0.$$

Since

$$z' = y' + (\alpha - \beta),$$

we have

$$-\frac{d}{dx}(p(x)z') + q(x)z = -\frac{d}{dx}\left(p(x)(y' + (\alpha - \beta))\right) + q(x)(y + (\alpha - \beta)x - \alpha)$$

$$= -\frac{d}{dx}(p(x)y') + q(x)y + (\beta - \alpha)p'(x) + ((\alpha - \beta)x - \alpha)q(x)$$

$$= F(x) + (\beta - \alpha)p'(x) + ((\alpha - \beta)x - \alpha)q(x).$$

Hence, if we define

$$z = y + (\alpha - \beta)x - \alpha$$

and

$$F(x) = f(x) + (\beta - \alpha)p'(x) + ((\alpha - \beta)x - \alpha)q(x),$$

we have

$$-\frac{d}{dx}(p(x)z') + q(x)z = F(x), \quad 0 \le x \le 1, \quad z(0) = 0, \quad z(1) = 0.$$

9. Show that

$$-\frac{d}{dx}(p(x)y') + q(x)y = f(x), \quad a \le x \le b, \quad y(a) = \alpha, \quad y(b) = \beta$$

can be transformed into

$$-\frac{d}{dw}(p(w)z') + q(w)z = F(w), \quad 0 \le w \le 1, \quad z(0) = 0, \quad z(1) = 0.$$

SOLUTION: With $w = (x - a)/(b - a)$, the chain rule gives

$$\frac{dy}{dx} = \frac{dy}{dw}\frac{dw}{dx} = \frac{1}{b-a}\frac{dy}{dw}$$

and

$$\begin{aligned}
\frac{d}{dx}\left[p(x)\frac{dy}{dx}\right] &= \frac{d}{dw}\left[p(x)\frac{dy}{dx}\right]\frac{dw}{dx} \\
&= \frac{d}{dw}\left[p((b-a)w+a)\left(\frac{1}{b-a}\right)\frac{dy}{dw}\right]\left(\frac{1}{b-a}\right) \\
&= \left(\frac{1}{b-a}\right)^2\frac{d}{dw}\left[p((b-a)w+a)\frac{dy}{dw}\right].
\end{aligned}$$

Substituting into the differential equation gives

$$-\left(\frac{1}{b-a}\right)^2\frac{d}{dw}\left[p((b-a)w+a)\frac{dy}{dw}\right] + q((b-a)w+a)y = f((b-a)w+a),$$

so

$$-\frac{d}{dw}\left[p((b-a)w+a)\frac{dy}{dw}\right] + (b-a)^2q((b-a)w+a)y = (b-a)^2f((b-a)w+a).$$

Letting

$$\tilde{p}(w) = p((b-a)w+a),$$
$$\tilde{q}(w) = (b-a)^2q((b-a)w+a),$$

and

$$\tilde{f}(w) = (b-a)^2f((b-a)w+a)$$

results in the boundary value problem

$$-\frac{d}{dw}\left[\tilde{p}(w)\frac{dy}{dw}\right] + \tilde{q}(w)y = \tilde{f}(w),$$

with $0 < w < 1$, $y(w = 0) = y(x = a) = \alpha$ and $y(w = 1) = y(x = b) = \beta$. Exercise 6 can then be used to complete the solution.

10. Show that $\{\phi_i\}_{i=1}^n$ is a linear independent set for the functions defined in Eq. (11.29).

SOLUTION: If $\sum_{i=1}^n c_i \phi_i(x) = 0$ for all values of x in $[0, 1]$, then, in particular, for each j we have

$$\sum_{i=1}^n c_i \phi_i(x_j) = 0.$$

But, for each i and j we have

$$\phi_i(x_j) = \begin{cases} 0, & \text{if } i \neq j, \\ 1, & \text{if } i = j. \end{cases}$$

So for each $j = 0, 1, \ldots, n$ we have $c_j \phi_j(x_j) = c_j = 0$. Hence the set of functions $\{\phi_i\}_{i=1}^n$ is linearly independent.

12. Show that the matrix given by the piecewise linear basis functions is positive definite.

SOLUTION: Let $\mathbf{c} = (c_1, \ldots, c_n)^t$ be any vector and let $\phi(x) = \sum_{j=1}^n c_j \phi_j(x)$. Then

$$
\begin{aligned}
\mathbf{c}^t A \mathbf{c} &= \sum_{i=1}^n \sum_{j=1}^n a_{ij} c_i c_j = \sum_{i=1}^n \sum_{j=i-1}^{i+1} a_{ij} c_i c_j \\
&= \sum_{i=1}^n \left[\int_0^1 \{ p(x) c_i \phi_i'(x) c_{i-1} \phi_{i-1}'(x) + q(x) c_i \phi_i(x) c_{i-1} \phi_{i-1}(x) \} \, dx \right. \\
&\quad + \int_0^1 \{ p(x) c_i^2 [\phi_i'(x)]^2 + q(x) c_i^2 [\phi_i'(x)]^2 \} \, dx \\
&\quad + \left. \int_0^1 \{ p(x) c_i \phi_i'(x) c_{i+1} \phi_{i+1}'(x) + q(x) c_i \phi_i(x) c_{i+1} \phi_{i+1}(x) \} \, dx \right] \\
&= \int_0^1 \{ p(x) [\phi'(x)]^2 + q(x) [\phi(x)]^2 \} \, dx.
\end{aligned}
$$

So $\mathbf{c}^t A \mathbf{c} \geq 0$ with equality only if $\mathbf{c} = \mathbf{0}$. Since A is also symmetric, the matrix is positive definite.

13. Show that the matrix in (11.32) on page 706 given by the cubic spline basis functions is positive definite.

SOLUTION: For $\mathbf{c} = (c_0, c_1, \ldots, c_{n+1})^t$ and $\phi(x) = \sum_{i=0}^{n+1} c_i \phi_i(x)$, we have

$$\mathbf{c}^t A \mathbf{c} = \int_0^1 p(x) [\phi'(x)]^2 + q(x) [\phi(x)]^2 \, dx.$$

But $p(x) > 0$ and $q(x) [\phi(x)]^2 \geq 0$, so $\mathbf{c}^t A \mathbf{c} \geq 0$. It can be zero only if $\phi'(x) = 0$ on $[0, 1]$. However, $\{\phi_0', \phi_1', \ldots, \phi_{n+1}'\}$ is linearly independent, so $\mathbf{c}^t A \mathbf{c} = 0$ if and only if $\mathbf{c} = \mathbf{0}$. This implies that the matrix is positive definite since A is also symmetric.

Numerical Solutions to Partial Differential Equations

Exercise Set 12.1, page 723

1. Use the Finite Difference method with $h = 0.5$ and $k = 0.5$ to approximate the solution to

$$\frac{\partial^2 u}{\partial x^2} + \frac{\partial^2 u}{\partial y^2} = 4, \quad 0 < x < 1, \quad 0 < y < 2;$$

$$u(x, 0) = x^2, \quad u(x, 2) = (x - 2)^2, \quad 0 \leq x \leq 1;$$
$$u(0, y) = y^2, \quad u(1, y) = (y - 1)^2, \quad 0 \leq y \leq 2.$$

SOLUTION: The difference equations for this problem are

$$4w_{1,1} - w_{1,2} = -h^2 f(x_1, y_1) + w_{1,0} + w_{2,1} + w_{0,1},$$
$$4w_{1,2} - w_{1,3} - w_{1,1} = -h^2 f(x_1, y_2) + w_{2,2} + w_{0,2},$$
$$4w_{1,3} - w_{1,2} = -h^2 f(x_1, y_3) + w_{1,4} + w_{2,3} + w_{0,3}.$$

Thus

$$4w_{1,1} - w_{1,2} = -1 + (0.5)^2 + (-0.5)^2 + (0.5)^2 = -0.25,$$
$$4w_{1,2} - w_{1,3} - w_{1,1} = -1 + 0^2 + 1^2 = 0,$$
$$4w_{1,3} - w_{1,2} = -1 + (-1.5)^2 + (0.5)^2 + (1.5)^2 = 3.75.$$

This linear system has the solution $w_{1,1} = 0$, $w_{1,2} = 0.25$, and $w_{1,3} = 1$.

5. Construct an algorithm similar to Algorithm 12.1 that uses the SOR method with optimal ω to solve the linear system.

SOLUTION: Make the following changes in Algorithm 12.1.

Step 1 Set
$$h = (b - a)/n;$$
$$k = (d - c)/m;$$

$$\omega = 4\left(2 + \sqrt{4 - [\cos(\pi/n)]^2 - [\cos(\pi/m)]^2}\right)^{-1};$$
$$\omega_0 = 1 - \omega.$$

In each of Steps 7, 8, 9, 11, 12, 13, 14, 15, and 16 after

$$\text{Set } z = \ldots$$

insert

$\quad$ Set $E = \omega_{\alpha,\beta} - z$;

$\qquad$ if $(\ |E| > NORM)$ then set $NORM = |E|$;

$\quad$ Set $\omega_{\alpha,\beta} = \omega_0 E + z$.

where α and β depend on which Step is being changed.

Exercise Set 12.2, page 736

2. Use the Backward-Difference method with $m = 3$, $T = 0.1$ and $N = 2$ to approximate the solution to

$$\frac{\partial u}{\partial t} - \frac{1}{16}\frac{\partial^2 u}{\partial x^2} = 0, \quad 0 < x < 1, \quad 0 < t;$$

$$u(0, t) = u(1, t) = 0, \quad 0 < t, \qquad u(x, 0) = 2\sin 2\pi x, \quad 0 \le x \le 1.$$

SOLUTION: For this problem, we have $\alpha = 1/4$, $m = 3$, $T = 0.1$, $N = 2$, and $l = 1$. Thus, $h = l/m = 1/3$ and $k = T/N = 0.05$.

The difference equations are given by

$$\frac{w_{i,j} - w_{i,j-1}}{0.05} - \frac{1}{16}\left[\frac{w_{i+1,j} - 2w_{i,j} + w_{i-1,j}}{\frac{1}{9}}\right] = 0,$$

for each $i = 1, 2$ and $j = 1, 2$.

For $j = 1$, we have the equations

$$\frac{169}{160}w_{1,1} - \frac{9}{320}w_{2,1} - \frac{9}{320}w_{0,1} = w_{1,0} = 2\sin\frac{2\pi}{3} = \sqrt{3},$$

$$\frac{169}{160}w_{2,1} - \frac{9}{320}w_{3,1} - \frac{9}{320}w_{1,1} = w_{2,0} = 2\sin\frac{4\pi}{3} = -\sqrt{3}.$$

Since $w_{0,1} = w_{3,1} = 0$, we have the linear system

$$\begin{bmatrix} \frac{169}{160} & -\frac{9}{320} \\ -\frac{9}{320} & \frac{169}{160} \end{bmatrix}\begin{bmatrix} w_{1,1} \\ w_{2,1} \end{bmatrix} = \begin{bmatrix} \sqrt{3} \\ -\sqrt{3} \end{bmatrix},$$

which has the solution $w_{1,1} = 1.59728$ and $w_{1,1} = -1.59728$. For $j = 2$, we have the equations

$$\frac{169}{160}w_{1,2} - \frac{9}{320}w_{2,2} - \frac{9}{320}w_{0,2} = w_{1,1} = 1.59728,$$

$$\frac{169}{160}w_{2,2} - \frac{9}{320}w_{3,2} - \frac{9}{320}w_{1,2} = w_{2,1} = -1.59728.$$

Since $w_{0,2} = w_{3,2} = 0$, we have the linear system

$$\begin{bmatrix} \frac{169}{160} & -\frac{9}{320} \\ -\frac{9}{320} & \frac{169}{160} \end{bmatrix} \begin{bmatrix} w_{1,2} \\ w_{2,2} \end{bmatrix} = \begin{bmatrix} 1.59728 \\ -1.59728 \end{bmatrix},$$

which has the solution $w_{1,2} = 1.47300$ and $w_{2,2} = -1.47300$.

The following table summarizes the results.

i	j	x_i	t_j	w_{ij}	$u(x_i, t_j)$
1	1	1/3	0.05	1.59728	1.53102
2	1	2/3	0.05	-1.59728	- 1.53102
1	2	1/3	0.1	1.47300	1.35333
2	2	2/3	0.1	-1.47300	- 1.35333

4. Repeat Exercise 2 using the Crank-Nicolson method.

SOLUTION: For this problem, we have $\alpha = 1/4$, $m = 3$, $T = 0.1$, $N = 2$, $l = 1$, $h = 1/3$, and $k = 0.05$.

The difference equations are given by

$$\frac{w_{i,j+1} - w_{i,j}}{0.05} - \frac{1}{32} \left[\frac{w_{i+1,j} - 2w_{i,j} + w_{i-1,j}}{\frac{1}{9}} + \frac{w_{i+1,j+1} - 2w_{i,j+1} + w_{i-1,j+1}}{\frac{1}{9}} \right] = 0.$$

For $j = 0$, we have, for $i = 1$ and 2,

$$-\frac{9}{640} w_{i-1,1} + \frac{329}{320} w_{i,1} - \frac{9}{640} w_{i+1,1} = \frac{9}{640} w_{i-1,0} + \frac{311}{320} w_{i,0} + \frac{9}{640} w_{i+1,0}.$$

Thus,

$$-\frac{9}{640} w_{0,1} + \frac{329}{320} w_{1,1} - \frac{9}{640} w_{2,1} = \frac{9}{640} w_{0,0} + \frac{311}{320} w_{1,0} + \frac{9}{640} w_{2,0}$$

$$-\frac{9}{640} w_{1,1} + \frac{329}{320} w_{2,1} - \frac{9}{640} w_{3,1} = \frac{9}{640} w_{1,0} + \frac{311}{320} w_{2,0} + \frac{9}{640} w_{3,0}.$$

Since $w_{0,0} = w_{3,0} = w_{0,1} = w_{3,1} = 0$, $w_{1,0} = \sqrt{3}$, and $w_{2,0} = -\sqrt{3}$, we have the linear system

$$\begin{bmatrix} \frac{329}{320} & -\frac{9}{640} \\ -\frac{9}{640} & \frac{329}{320} \end{bmatrix} \begin{bmatrix} w_{1,1} \\ w_{2,1} \end{bmatrix} = \begin{bmatrix} \frac{311}{320}\sqrt{3} - \frac{9}{640}\sqrt{3} \\ \frac{9}{640}\sqrt{3} - \frac{311}{320}\sqrt{3} \end{bmatrix},$$

which has the solution $w_{1,1} = 1.591825$ and $w_{2,1} = -1.591825$.

For $j = 1$, we have, for $i = 1$ and 2,

$$-\frac{9}{640} w_{0,2} + \frac{329}{320} w_{1,2} - \frac{9}{640} w_{2,2} = \frac{9}{640} w_{0,1} + \frac{311}{320} w_{1,1} + \frac{9}{640} w_{2,1}$$

$$-\frac{9}{640} w_{1,2} + \frac{329}{320} w_{2,2} - \frac{9}{640} w_{3,2} = \frac{9}{640} w_{1,1} + \frac{311}{320} w_{2,1} + \frac{9}{640} w_{3,1}.$$

Since $w_{0,1} = w_{3,1} = w_{0,2} = w_{3,2} = 0$, we have the linear system

$$\begin{bmatrix} \frac{329}{320} & -\frac{9}{640} \\ -\frac{9}{640} & \frac{329}{320} \end{bmatrix} \begin{bmatrix} w_{1,2} \\ w_{2,2} \end{bmatrix} = \begin{bmatrix} \frac{311}{320}(1.591825) - \frac{9}{640}(1.591825) \\ \frac{9}{640}(1.591825) - \frac{311}{320}(1.591825) \end{bmatrix},$$

which has the solution $w_{1,2} = 1.462951$ and $w_{2,2} = -1.462951$.

The following table summarizes the results.

i	j	x_i	t_j	w_{ij}	$u(x_i, t_j)$
1	1	$\frac{1}{3}$	0.05	1.591825	1.53102
2	1	$\frac{2}{3}$	0.05	-1.591825	-1.53102
1	2	$\frac{1}{3}$	0.1	1.462951	1.35333
2	2	$\frac{2}{3}$	0.1	-1.462951	-1.35333

13. Determine the eigenvalues and associated eigenvectors for the tridiagonal matrix A given by

$$a_{ij} = \begin{cases} \lambda, & \text{if } j = i-1 \text{ or } j = i+1, \\ 1 - 2\lambda, & \text{if } j = i, \\ 0, & \text{otherwise.} \end{cases}$$

SOLUTION: We have

$$a_{11}v_1^{(i)} + a_{12}v_2^{(i)} = (1-2\lambda)\sin\frac{i\pi}{m} + \lambda\sin\frac{2\pi i}{m}$$

and

$$\begin{aligned} \mu_i v_1^{(i)} &= \left[1 - 4\lambda\left(\sin\frac{i\pi}{2m}\right)^2\right]\sin\frac{i\pi}{m} \\ &= \left[1 - 4\lambda\left(\sin\frac{i\pi}{2m}\right)^2\right]\left(2\sin\frac{i\pi}{2m}\cos\frac{i\pi}{2m}\right) \\ &= 2\sin\frac{i\pi}{2m}\cos\frac{i\pi}{2m} - 8\lambda\left(\sin\frac{i\pi}{2m}\right)^3\cos\frac{i\pi}{2m}. \end{aligned}$$

However,

$$\begin{aligned} (1-2\lambda)\sin\frac{i\pi}{m} + \lambda\sin\frac{2\pi i}{m} &= 2(1-2\lambda)\sin\frac{i\pi}{2m}\cos\frac{i\pi}{2m} + 2\lambda\sin\frac{i\pi}{m}\cos\frac{i\pi}{m} \\ &= 2(1-2\lambda)\sin\frac{i\pi}{2m}\cos\frac{i\pi}{2m} + 2\lambda\left[2\sin\frac{i\pi}{2m}\cos\frac{i\pi}{2m}\right]\left[1 - 2\left(\sin\frac{i\pi}{2m}\right)^2\right], \end{aligned}$$

so

$$(1-2\lambda)\sin\frac{i\pi}{m} + \lambda\sin\frac{2\pi i}{m} = 2\sin\frac{i\pi}{2m}\cos\frac{i\pi}{2m} - 8\lambda\cos\frac{i\pi}{2m}\left[\sin\frac{i\pi}{2m}\right]^3.$$

Thus

$$a_{11}v_1^{(i)} + a_{12}v_2^{(i)} = \mu_i v_1^{(i)}.$$

Further,

$$a_{j,j-1}v_{j-1}^{(i)} + a_{j,j}v_j^{(i)} + a_{j,j+1}v_{j+1}^{(i)} = \lambda \sin \frac{i(j-1)\pi}{m} + (1-2\lambda) \sin \frac{ij\pi}{m} + \lambda \sin \frac{i(j+1)\pi}{m}$$

$$= \lambda \left(\sin \frac{ij\pi}{m} \cos \frac{i\pi}{m} - \sin \frac{i\pi}{m} \cos \frac{ij\pi}{m} \right) + (1-2\lambda) \sin \frac{ij\pi}{m}$$

$$+ \lambda \left(\sin \frac{ij\pi}{m} \cos \frac{i\pi}{m} + \sin \frac{i\pi}{m} \cos \frac{ij\pi}{m} \right)$$

$$= \sin \frac{ij\pi}{m} - 2\lambda \sin \frac{ij\pi}{m} + 2\lambda \sin \frac{ij\pi}{m} \cos \frac{i\pi}{m}$$

$$= \sin \frac{ij\pi}{m} + 2\lambda \sin \frac{ij\pi}{m} \left(\cos \frac{i\pi}{m} - 1 \right)$$

and

$$\mu_i v_j^{(i)} = \left[1 - 4\lambda \sin^2 \frac{i\pi}{2m} \right] \sin \frac{ij\pi}{m} = \left[1 - 4\lambda \left(\frac{1}{2} - \frac{1}{2} \cos \frac{i\pi}{m} \right) \right] \sin \frac{ij\pi}{m}$$

$$= \left[1 + 2\lambda \left(\cos \frac{i\pi}{m} - 1 \right) \right] \sin \frac{ij\pi}{m},$$

so

$$a_{j,j-1}v_{j-1}^{(i)} + a_{j,j}v_j^{(i)} + a_{j,j+1}v_j^{(i)} = \mu_i v_j^{(i)}.$$

Similarly,

$$a_{m-2,m-1}v_{m-2}^{(i)} + a_{m-1,m-1}v_{m-1}^{(i)} = \mu_i v_{m-1}^{(i)},$$

so $A\mathbf{v}^{(i)} = \mu_i \mathbf{v}^{(i)}$.

14. Determine the eigenvalues and associated eigenvectors for the tridiagonal matrix A given by

$$a_{ij} = \begin{cases} -\lambda, & \text{if } j = i-1 \text{ or } j = i+1, \\ 1+2\lambda, & \text{if } j = i, \\ 0, & \text{otherwise}, \end{cases}$$

where $\lambda > 0$, and show that A is positive definite and diagonally dominant.

SOLUTION: We have

$$a_{11}v_1^{(i)} + a_{12}v_2^{(i)} = (1+2\lambda) \sin \frac{i\pi}{m} - \lambda \sin \frac{2i\pi}{m} = (1+2\lambda) \sin \frac{i\pi}{m} - 2\lambda \sin \frac{i\pi}{m} \cos \frac{i\pi}{m}$$

$$= \sin \frac{i\pi}{m} \left[1 + 2\lambda \left(1 - \cos \frac{i\pi}{m} \right) \right]$$

and

$$\mu_i v_1^{(i)} = \left(1 + 4\lambda \sin^2 \frac{i\pi}{2m} \right) \sin \frac{i\pi}{m} = \left[1 + 2\lambda \left(1 - \cos \frac{i\pi}{m} \right) \right] \sin \frac{i\pi}{m} = a_{11}v_1^{(i)} + a_{12}v_2^{(i)}.$$

In general,

$$a_{j,j-1}v_{j-1}^{(i)} + a_{j,j}v_j^{(i)} + a_{j,j+1}v_{j+1}^{(i)} = -\lambda \sin \frac{i(j-1)\pi}{m} + (1+2\lambda)\sin\frac{ij\pi}{m} - \lambda \sin\frac{i(j+1)\pi}{m}$$

$$= -\lambda \left(\sin\frac{ij\pi}{m}\cos\frac{i\pi}{m} - \sin\frac{i\pi}{m}\cos\frac{ij\pi}{m} \right) + (1+2\lambda)\sin\frac{ij\pi}{m}$$

$$-\lambda \left(\sin\frac{ij\pi}{m}\cos\frac{i\pi}{m} + \sin\frac{i\pi}{m}\cos\frac{ij\pi}{m} \right)$$

$$= -2\lambda \sin\frac{ij\pi}{m}\cos\frac{i\pi}{m} + (1+2\lambda)\sin\frac{ij\pi}{m}$$

$$= \left[1 + 2\lambda\left(1 - \cos\frac{i\pi}{m}\right)\right]\sin\frac{ij\pi}{m} = \mu_i v_j^{(i)}.$$

Similarly,

$$a_{m-2,m-1}v_{m-2}^{(i)} + a_{m-1,m-1}v_{m-1}^{(i)} = \mu_i v_{m-1}^{(i)},$$

so $A\mathbf{v}^{(i)} = \mu_i \mathbf{v}^{(i)}$. Since A is symmetric with positive eigenvalues, A is positive definite. Further,

$$\sum_{j=1,j\neq i}^{n}|a_{ij}| = 2\lambda < 1 + 2\lambda = |a_{ii}|, \quad \text{for } 1 \le i \le n,$$

so that A is diagonally dominant.

15. Change Algorithms 12.2 and 12.3 to solve

$$\frac{\partial u}{\partial t} - \alpha^2 \frac{\partial^2 u}{\partial x^2} = 0, \quad 0 < x < l, \quad 0 < t;$$

$$u(0,t) = \phi(t), \quad u(l,t) = \psi(t), \quad 0 < t;$$

$$u(x,0) = f(x), \quad 0 \le x \le l,$$

where $f(0) = \phi(0)$ and $f(l) = \psi(0)$.

SOLUTION: To modify Algorithm 12.2, change the following:

Step 7   Set
$$t = jk;$$
$$w_0 = \phi(t);$$
$$z_1 = (w_1 + \lambda w_0)/l_1.$$
$$w_m = \psi(t).$$

Step 8   For $i = 2, \ldots, m-2$ set
$$z_i = (w_i + \lambda z_{i-1})/l_i;$$
Set
$$z_{m-1} = (w_{m-1} + \lambda w_m + \lambda z_{m-2})/l_{m-1}.$$

Step 11 OUTPUT (t);
For $i = 0, \ldots, m$ set $x = ih$;
OUTPUT (x, w_i).

To modify Algorithm 12.3, change the following:

Step 1 Set
$$h = l/m;$$
$$k = T/N;$$
$$\lambda = \alpha^2 k/h^2;$$
$$w_m = \psi(0);$$
$$w_0 = \phi(0).$$

Step 7 Set
$$t = jk;$$
$$z_1 = \left[(1-\lambda)w_1 + \tfrac{\lambda}{2}w_2 + \tfrac{\lambda}{2}0 + \tfrac{\lambda}{2}\phi(t)\right]/l_1;$$
$$w_0 = \phi(t).$$

Step 8 For $i = 2, \ldots, m-2$ set
$$z_i = \left[(1-\lambda)w_i + \tfrac{\lambda}{2}(w_{i+1} + w_{i-1} + z_{i-1})\right]/l_i;$$
Set
$$z_{m-1} = \left[(1-\lambda)w_{m-1} + \tfrac{\lambda}{2}(w_m + w_{m-2} + z_{m-2} + \psi(t))\right]/l_{m-1};$$
$$w_m = \psi(t).$$

Step 11 OUTPUT (t);
For $i = 0, \ldots, m$ set $x = ih$;
OUTPUT (x, w_i).

Exercise Set 12.3, page 744

1. Use the Finite-Difference method with $m = 4$, $N = 4$, and $T = 1.0$ to approximate the solution to

$$\frac{\partial^2 u}{\partial t^2} - \frac{\partial^2 u}{\partial x^2} = 0, \quad 0 < x < 1, \quad 0 < t;$$

$$u(0, t) = u(1, t) = 0, \quad 0 < t;$$

$$u(x, 0) = \sin \pi x, \quad \text{and} \quad u_t(x, 0) = 0, \quad 0 \le x \le 1.$$

SOLUTION: For this problem $\alpha = 1$, $m = 4$, $l = 1$, $T = 1$, and $N = 4$, so $h = m/l = 0.25$ and $k = T/N = 0.25$. Since $\lambda = (\alpha k)/h = 1$, the difference equations become

$$w_{i,j+1} - 2w_{i,j} + w_{i,j-1} - w_{i+1,j} + 2w_{i,j} - w_{i-1,j} = 0.$$

This gives the matrix equation

$$\begin{bmatrix} w_{1,j+1} \\ w_{2,j+1} \\ w_{3,j+1} \end{bmatrix} = \begin{bmatrix} 0 & 1 & 0 \\ 1 & 0 & 1 \\ 0 & 1 & 0 \end{bmatrix} \begin{bmatrix} w_{1,j} \\ w_{2,j} \\ w_{3,j} \end{bmatrix} - \begin{bmatrix} w_{1,j-1} \\ w_{2,j-1} \\ w_{3,j-1} \end{bmatrix}.$$

Since $f(x) = \sin \pi x$, we have

$$w_{1,0} = \sin \frac{\pi}{4} = \frac{\sqrt{2}}{2}, \quad w_{2,0} = \sin \frac{\pi}{2} = 1, \quad \text{and} \quad w_{3,0} = \sin \frac{3\pi}{4} = \frac{\sqrt{2}}{2}.$$

Using the approximations

$$w_{i,1} = (1 - \lambda^2)f(x_i) + \frac{\lambda^2}{2}f(x_{i+1}) + \frac{\lambda^2}{2}f(x_{i-1}) + kg(x_i)$$

and the fact that $g(x) = 0$ gives

$$w_{1,1} = \frac{1}{2}f(x_2) + \frac{1}{2}f(x_0) = \frac{1}{2}\sin\frac{\pi}{2} = \frac{1}{2}$$

$$w_{2,1} = \frac{1}{2}f(x_3) + \frac{1}{2}f(x_1) = \frac{1}{2}\sin\frac{3\pi}{4} + \frac{1}{2}\sin\frac{\pi}{4} = \frac{\sqrt{2}}{2}$$

$$w_{3,1} = \frac{1}{2}f(x_2) + \frac{1}{2}f(x_4) = \frac{1}{2}\sin\frac{\pi}{2} = \frac{1}{2}.$$

Thus

$$\begin{bmatrix} w_{1,2} \\ w_{2,2} \\ w_{3,2} \end{bmatrix} = \begin{bmatrix} 0 & 1 & 0 \\ 1 & 0 & 1 \\ 0 & 1 & 0 \end{bmatrix} \begin{bmatrix} \frac{1}{2} \\ \frac{\sqrt{2}}{2} \\ \frac{1}{2} \end{bmatrix} - \begin{bmatrix} \frac{\sqrt{2}}{2} \\ 1 \\ \frac{\sqrt{2}}{2} \end{bmatrix} = \begin{bmatrix} 0 \\ 0 \\ 0 \end{bmatrix},$$

$$\begin{bmatrix} w_{1,3} \\ w_{2,3} \\ w_{3,3} \end{bmatrix} = \begin{bmatrix} 0 & 1 & 0 \\ 1 & 0 & 1 \\ 0 & 1 & 0 \end{bmatrix} \begin{bmatrix} 0 \\ 0 \\ 0 \end{bmatrix} - \begin{bmatrix} \frac{1}{2} \\ \frac{\sqrt{2}}{2} \\ \frac{1}{2} \end{bmatrix} = \begin{bmatrix} -\frac{1}{2} \\ -\frac{\sqrt{2}}{2} \\ -\frac{1}{2} \end{bmatrix},$$

and

$$\begin{bmatrix} w_{1,4} \\ w_{2,4} \\ w_{3,4} \end{bmatrix} = \begin{bmatrix} 0 & 1 & 0 \\ 1 & 0 & 1 \\ 0 & 1 & 0 \end{bmatrix} \begin{bmatrix} -\frac{1}{2} \\ -\frac{\sqrt{2}}{2} \\ -\frac{1}{2} \end{bmatrix} - \begin{bmatrix} 0 \\ 0 \\ 0 \end{bmatrix} = \begin{bmatrix} -\frac{\sqrt{2}}{2} \\ 1 \\ -\frac{\sqrt{2}}{2} \end{bmatrix}.$$

This gives the results in the following table.

i	j	x_i	t_j	w_{ij}
2	4	0.25	1.0	-0.7071068
3	4	0.50	1.0	-1.0000000
4	4	0.75	1.0	-0.7071068

8. In an electric transmission line of length l that carries alternating current of high frequency (called a "lossless" line), the voltage V and current i are described by

$$\frac{\partial^2 V}{\partial x^2} = LC\frac{\partial^2 V}{\partial t^2}, \quad 0 < x < l,\ 0 < t;$$

$$\frac{\partial^2 i}{\partial x^2} = LC\frac{\partial^2 i}{\partial t^2}, \quad 0 < x < l,\ 0 < t;$$

where L is the inductance per unit length, and C is the capacitance per unit length. Suppose the line is 200 ft long and the constants C and L are given by $C = 0.1$ farads/ft and $L = 0.3$ henries/ft.

Suppose the voltage and current also satisfy

$$V(0,t) = V(200,t) = 0, \quad 0 < t;$$

$$V(x,0) = 110 \sin \frac{\pi x}{200}, \quad 0 \le x \le 200;$$

$$\frac{\partial V}{\partial t}(x,0) = 0, \quad 0 \le x \le 200;$$

$$i(0,t) = i(200,t) = 0, \quad 0 < t;$$

$$i(x,0) = 5.5 \cos \frac{\pi x}{200}, \quad 0 \le x \le 200;$$

and

$$\frac{\partial i}{\partial t}(x,0) = 0, \quad 0 \le x \le 200.$$

Approximate the voltage and current at $t = 0.2$ and $t = 0.5$ using Algorithm 12.4 with $h = 10$ and $k = 0.1$.

SOLUTION: For the voltage equations we have

$$\alpha = \sqrt{LC} = \sqrt{0.03}, \; l = 200, \; t_1 = 0.2, \; t_2 = 0.5, \; f(x) = 110 \sin\left(\frac{\pi x}{200}\right), \; g(x) = 0,$$

and for $h = 10$, $k = 0.1$ we need $m = 20$, $N_1 = 2$, and $N_2 = 5$ for the two runs. We also have

$$\lambda = \frac{0.1\alpha}{10} = 0.001732050808.$$

The initial conditions are

$\mathbf{w}_0 = (17.20779116, 33.99186938, 49.93895498, 64.65637776, 77.78174594, 88.99186938, 98.01071767,$
$\quad 104.6162168, 108.6457175, 110.0, 108.6457175, 104.6162168, 98.01071765, 88.99186935,$
$\quad 77.78174594, 64.65637774, 49.93895492, 33.99186930)^t$

and

$\mathbf{w}_1 = (1 - \lambda^2)f(x - i) + \lambda^2(f(x_{i+1} + f(x_{i-1}))$
$\quad = (34.41553007, 67.98363554, 99.87775829, 129.3125592, 155.5632557, 177.9834685, 196.0211377,$
$\quad 209.2321159, 217.2911050, 219.9996660, 217.2911049, 209.2321159, 196.0211376, 177.9834685,$
$\quad 155.5632557, 129.3125591, 99.87775818, 67.98363536, 34.41553004)^t$

Let A be the matrix the 19 by 19 tridiagonal matrix with diagonal entries 1.999994 and nonzero off-diagonal entries 0.000003. Then the iterations are given by

$$\mathbf{w}_{i+1} = A\mathbf{w}_i - \mathbf{w}_{i-1},$$

for $i = 1, 2, \ldots, N - 1$. The results are shown in the table, as are the corresponding results for the current values.

i	x_i	$w_{i,2}$(Voltage)	$w_{i,5}$(Voltage)	$w_{i,2}$(Current)	$w_{i,5}$(Current)
0	0	0.00000000	0.00000000	0.00000000	0.00000000
1	10	17.20778834	17.20777511	5.43226857	5.43211584
2	20	33.99186378	33.99183732	5.23081008	5.23080605
3	30	49.93894679	49.93890794	4.90053518	4.90053137
4	40	64.65636730	64.65631761	4.44959284	4.44958939
5	50	77.78173331	77.78167241	3.88908678	3.88908376
6	60	88.99185516	88.99178615	3.23281848	3.23281598
7	70	98.01070227	98.01062621	2.49694747	2.49694553
8	80	104.61620090	104.61612110	1.69959333	1.69959201
9	90	108.64570100	108.64561700	0.86038956	0.86038890
10	100	109.99998370	109.99989850	0.00000015	0.00000015
11	110	108.64570200	108.64561800	−0.86038927	−0.86038860
12	120	104.61620270	104.61612290	−1.69959305	−1.69959173
13	130	98.01070490	98.01062858	−2.49694721	−2.49694527
14	140	88.99185870	88.99178971	−3.23281824	−3.23281574
15	150	77.78173754	77.78167678	−3.88908657	−3.88908356
16	160	64.65637203	64.65632208	−4.44959267	−4.44958922
17	170	49.93895203	49.93891318	−4.90053504	−4.90053123
18	180	33.99186938	33.99184292	−5.23080999	−5.23080598
19	190	17.20779416	17.20778093	−5.43226852	−5.43211578
20	200	0.00000000	0.00000000	0.00000000	0.00000000

Exercise Set 12.4, page 758

1. Use Algorithm 12.5 with $M = 2$ to approximate the solution to

$$\frac{\partial}{\partial x}\left(y^2\frac{\partial u}{\partial x}(x,y)\right) + \frac{\partial}{\partial y}\left(y^2\frac{\partial u}{\partial y}(x,y)\right) - yu(x,y) = -x, \quad (x,y) \in D,$$

$$u(x,0.5) = 2x, \quad 0 \le x \le 0.5, \quad u(0,y) = 0, \quad 0.5 \le y \le 1,$$

$$y^2\frac{\partial u}{\partial x}(x,y)\cos\theta_1 + y^2\frac{\partial u}{\partial y}(x,y)\cos\theta_2 = \frac{\sqrt{2}}{2}(y - x), \quad \text{for } (x,y) \in \mathcal{S}_\in.$$

SOLUTION: With $E_1 = (0.25, 0.75)$, $E_2 = (0, 1)$, $E_3 = (0.5, 0.5)$, and $E_4 = (0, 0.5)$, the basis functions are

$$\phi_1(x,y) = \begin{cases} 4x & \text{on } T_1 \\ -2 + 4y & \text{on } T_2, \end{cases} \qquad \phi_2(x,y) = \begin{cases} -1 - 2x + 2y & \text{on } T_1 \\ 0 & \text{on } T_2, \end{cases}$$

$$\phi_3(x,y) = \begin{cases} 0 & \text{on } T_1 \\ 1 + 2x - 2y & \text{on } T_2, \end{cases} \qquad \phi_4(x,y) = \begin{cases} 2 - 2x - 2y & \text{on } T_1 \\ 2 - 2x - 2y & \text{on } T_2, \end{cases}$$

and $\gamma_1 = 0.323825$, $\gamma_2 = 0$, $\gamma_3 = 1.0000$, and $\gamma_4 = 0$.

The approximation is

$$
\begin{aligned}
\phi(x,y) &= \gamma_1\phi_1(x,y) + \gamma_2\phi_2(x,y) + \gamma_3\phi_3(x,y) + \gamma_4\phi_4(x,y) \\
&= 0.323825\phi_1(x,y) + \phi_3(x,y) \\
&= \begin{cases} 1.2953x, & \text{on } T_1 \\ 0.35235 + 2x - 0.7047y, & \text{on } T_2 . \end{cases}
\end{aligned}
$$

Results are shown in the following table.

x	y	$\phi(x,y)$	$u(x,y) = x/y$
0.25	0.75	0.323825	0.333333
0	1	0	0
0.5	0.5	1	1
0	0.5	0	0

5. A silver plate in the shape of a trapezoid (see the accompanying figure) has heat being uniformly generated at each point at the rate $q = 1.5$ cal/cm$^3 \cdot$ s. The steady-state temperature $u(x,y)$ of the plate satisfies the Poisson equation

$$
\frac{\partial^2 u}{\partial x^2}(x,y) + \frac{\partial^2 u}{\partial y^2}(x,y) = \frac{-q}{k},
$$

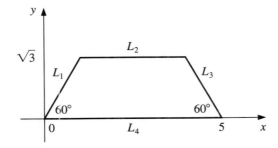

where k, the thermal conductivity, is 1.04 cal/cm·deg·s. Assume that the temperature is held at 15°C on L_2, that heat is lost on the slanted edges L_1 and L_3 according to the boundary condition $\partial u/\partial n = 4$, and that no heat is lost on L_4; that is, $\partial u/\partial n = 0$. Approximate the temperature of the plate at $(1,0)$, $(4,0)$, and $\left(\frac{5}{2}, \sqrt{3}/2\right)$ by using Algorithm 12.5.

SOLUTION: The Finite-Element Algorithm with $K = 0$, $N = 12$, $M = 32$, $n = 20$, $m = 27$, and $NL = 14$ gives the following results, where the labeling is as shown in the diagram.

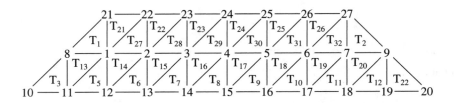

$$\gamma_1 = 21.40335, \quad \gamma_8 = 24.19855, \quad \gamma_{15} = 20.23334, \quad \gamma_{22} = 15,$$
$$\gamma_2 = 19.87372, \quad \gamma_9 = 24.16799, \quad \gamma_{16} = 20.50056, \quad \gamma_{23} = 15,$$
$$\gamma_3 = 19.10019, \quad \gamma_{10} = 27.55237, \quad \gamma_{17} = 21.35070, \quad \gamma_{24} = 15,$$
$$\gamma_4 = 18.85895, \quad \gamma_{11} = 25.11508, \quad \gamma_{18} = 22.84663, \quad \gamma_{25} = 15,$$
$$\gamma_5 = 19.08533, \quad \gamma_{12} = 22.92824, \quad \gamma_{19} = 24.98178, \quad \gamma_{26} = 15,$$
$$\gamma_6 = 19.84115, \quad \gamma_{13} = 21.39741, \quad \gamma_{20} = 27.41907, \quad \gamma_{27} = 15,$$
$$\gamma_7 = 21.34694, \quad \gamma_{14} = 20.52179, \quad \gamma_{21} = 15$$

$$u(1,0) \approx 22.92824, \, u(4,0) \approx 22.84663, \text{ and } u\left(\tfrac{5}{2}, \tfrac{\sqrt{3}}{2}\right) \approx 18.85895.$$